Behavioral Neuroendocrinology

Behavioral Neuroendocrinology

Edited by

Barry R. Komisaruk
Gabriela González-Mariscal

CRC Press is an imprint of the
Taylor & Francis Group, an **informa** business

CRC Press
Taylor & Francis Group
6000 Broken Sound Parkway NW, Suite 300
Boca Raton, FL 33487-2742

First issued in paperback 2020

ISBN-13: 978-1-4987-3191-1 (hbk)
ISBN-13: 978-0-367-73637-8 (pbk)

Library of Congress Cataloging-in-Publication Data

Names: Komisaruk, Barry R., editor. | González-Mariscal, Gabriela, editor.
Title: Behavioral neuroendocrinology / edited by Barry R. Komisaruk and Gabriela González-Mariscal.
Description: Boca Raton : CRC Press, 2016. | Includes bibliographical references and index.
Identifiers: LCCN 2016026418| ISBN 9781498731911 (hbk : alk. paper) | ISBN 9781315388069 (ebook)
Subjects: LCSH: Psychoneuroendocrinology.
Classification: LCC QP356.45 .B45 2016 | DDC 612.4/05--dc23
LC record available at https://lccn.loc.gov/2016026418

Visit the Taylor & Francis Web site at
http://www.taylorandfrancis.com

and the CRC Press Web site at
http://www.crcpress.com

Contents

Preface ix
Editors xiii
Contributors xv

SECTION I Neuroendocrinology of Sexual Behavior

Chapter 1 Pioneering Studies on the Effects of Steroid Hormone Metabolism on the Brain and Behavior 3

María Luisa Cruz, Miguel Cervantes, and Gabriela Moralí

Chapter 2 Neuroendocrine and Behavioral Role of the Medial Preoptic Area in Rabbits: Recollections of Collaboration with Carlos Beyer 33

Anders Ågmo and Knut Larsson

Chapter 3 Hormonal Regulation of the Copulatory Motor Pattern in Mammals 45

Gabriela Moralí, Knut Larsson, José Luis Contreras, and Miguel Cervantes

Chapter 4 Neuronal and Neurochemical Correlates of Copulatory Motor Patterns in Male Rats 65

Marisela Hernández González and Miguel Ángel Guevara

Chapter 5 Male Sexual Satiety and the Coolidge Effect in Rats: Relation between Behavioral and Seminal Parameters 83

Rosa Angélica Lucio, Alonso Fernández-Guasti, and Knut Larsson

Chapter 6 The Sexual Cerebellum 103

Jorge Manzo, Porfirio Carrillo, Genaro A. Coria-Avila, and Luis I. Garcia

Chapter 7 Intracellular Signaling Involved in Progestin Regulation of Female Sexual Behavior in Rodents 113

Oscar González-Flores, Ignacio Camacho-Arroyo, and Anne M. Etgen

Chapter 8 The Delicate Line between "Wanting" (Desire) and "Liking" (Reward) Sexual Behavior 131

Elisa Ventura-Aquino, Jorge Baños-Araujo, Alonso Fernández-Guasti, and Raúl G. Paredes

SECTION II Neuroendocrinology of Maternal Behavior and Brain Development

Chapter 9 The Rabbit Doe as a Model of Neuroendocrine Synchronization 149

Mario Caba, Teresa Morales, and Enrique Meza

Chapter 10 The Domestic Rabbit (*Oryctolagus cuniculus*) as a Model Animal: From Reproductive Neurobiology to Developmental Psychobiology 163

Kurt Leroy Hoffman, Robyn Hudson, Margarita Martínez-Gómez, and Amando Bautista

Chapter 11 A View of Rabbit Maternal Behavior from the Perspectives of Complex Systems and Chronostasis 187

Gabriela González-Mariscal and Raúl Aguilar-Roblero

Chapter 12 Multisignaling Approach to the Study of Sexual Differentiation of Brain and Behavior in Mammals 201

María Cruz Rodríguez Del Cerro, Carmen Pérez-Laso, Francisco Gómez, Antonio Guillamón, and Santiago Segovia

Chapter 13 Ubiquitous Modulators of Brain Activity: GABA and Carlos Beyer-Flores, PhD 231

Margaret M. McCarthy and Rae Silver

SECTION III Neuroendocrine Insights toward Development of Therapeutic Agents

Chapter 14 From Reproductive Neuroendocrinology and Lactation to Vasoinhibins and Angiogenesis 251

Carmen Clapp and Gonzalo Martínez de la Escalera

Chapter 15 Neuroendocrine and Molecular Aspects of the Physiology and Pathology of the Prostate........ 263

Maria Elena Hernández, Gonzalo E. Aranda-Abreu, and Fausto Rojas-Durán

Chapter 16 From Sexual Behavior to Analgesia to an Antinociceptive Agent: Glycinamide........ 279

Porfirio Gómora-Arrati, Galicia-Aguas Y., Oscar González-Flores, and Barry R. Komisaruk

SECTION IV Epilogue

Chapter 17 How Carlos Beyer Influenced Our Lives 289

Barry R. Komisaruk and Beverly Whipple

Chapter 18 Photo Gallery 313

Complete Bibliography of Carlos Beyer (-Flores) 321

Index 333

Preface

We dedicate this book *In Memoriam* Carlos Beyer, our dear colleague, friend, and mentor, for his immense scientific contributions throughout his distinguished career. The contributors of the chapters include his former students who are now independent researchers, his earliest colleagues, and later colleagues with whom he developed collaborations and lasting friendships in Mexico and in many international universities. From the earliest beginning of his scientific career, Carlos cultivated the area of behavioral neuroendocrinology, from different perspectives, in multiple mammalian species, and using innovative methodologies. This book, therefore, reflects both the scientific development of Carlos' profound understanding of neural, hormonal, and biochemical mechanisms underlying behavior and his ability to teach and inspire those with whom he came in contact to extend his insights, thereby expanding and transforming the field of behavioral neuroendocrinology over a span of more than 50 years. The published works cited here are referenced in the "Complete Bibliography of Carlos Beyer" found on pages 321–332.

Carlos began his scientific career at the Institute for Biomedical Studies (National Autonomous University of Mexico—UNAM) while he was still an undergraduate student of biology. His first publication (on the physiological mechanism by which scorpion venom provokes hyperglycemia) appeared, 3 years before receiving his baccalaureate, in the Institute's Bulletin (1). Carlos' first international publication (6) was on the neuroendocrine control of lactation. Initially with Flavio Mena (11, 13, 15), this line of research would be extended by Flavio for four decades through his students, as exemplified in Chapter 14 by Carmen Clapp and Gonzalo Martínez de la Escalera. Carlos' finding that electrical stimulation of the cingulate gyrus induced the release of oxytocin was originally presented at a conference in Mexico, catching the attention of a colleague of Professor Charles Sawyer. Shortly afterward, Sawyer contacted Carlos, encouraging him to publish that research in the *American Journal of Physiology* (6) and inviting Carlos to his laboratory in the Brain Research Institute at the University of California, Los Angeles.

In the course of several stays in Sawyer's laboratory, Carlos developed the method of multiunit activity recording in nonanesthetized mammals. Through an early use of time-series analysis, he and his colleagues showed that estradiol and progesterone modulated, with a very short latency, evoked potentials in the cerebral cortex of cats and rabbits (23). Shortly afterward, they found similar effects provoked by corticosteroids and progestins (46, 47), and later they reported changes in neuronal activity provoked by genital stimulation and modified by steroids in freely moving cats (41). These works (see Chapter 1 by Cruz, Cervantes, and Moralí) paved the way for exploring the relevance of steroid modulation of brain excitability in the regulation of reproductive function in mammals. Two models were used. In the first—the inhibition of gonadotropin release—Carlos and colleagues showed for the first time that the 5-alpha reduction of testosterone played a major role in the secretion of LH and FSH in female rats (52, 63). This provided evidence that, as in peripheral target tissues, a specific metabolic transformation of testosterone had a major role in the functioning of the hypothalamic–pituitary axis. In the second model—stimulation of sexual behavior—in a series of studies performed between 1971 and 1973, Carlos and his colleagues showed that estradiol per se, rather than its metabolites, was the effective hormone for stimulating female sexual behavior and that the capacity of some androgens to provoke both male and female sexual behavior depended on whether they could be aromatized (i.e., transformed to estrogens) (42, 43, 45, 50, 55, 57). Additional support for this proposal came from the finding that, in both rats and rabbits, male sexual behavior could be induced by administering estradiol plus dihydrotestosterone (56, 58, 69). The latter findings included research performed with Knut Larsson during his many visits to the laboratories that Carlos had established in Mexico City (Mexican Institute of Social Security and Autonomous Metropolitan University; see Chapter 2 by Agmo and Larsson).

A series of publications in the early 1980s by Carlos and his students provided evidence that the stimulation of female sexual behavior in rats involved a nongenomic mechanism that could be triggered by a variety of nonsteroidal agents (82, 84, 85, 87). These findings led to the proposal that the stimulation of estrous behavior by progesterone required an increase in hypothalamic cyclic AMP and the consequent activation of protein phosphorylation (105). Later work provided evidence for the involvement of an additional second messenger system: cyclic GMP (90). This insightful proposal, at a time when the prevalent model of steroid hormone action was the classic genomic mechanism, was developed much more elaborately over the years by Carlos and his collaborators until his death (see Chapter 7 by González-Flores, Camacho, and Etgen).

Carlos re-engaged in exploring lactation (in rabbits) many years after his initial studies, in his new laboratory in the state of Tlaxcala, through a program he launched together with Jay Rosenblatt (director of the Institute of Animal Behavior, Rutgers University, New Jersey). This collaboration added a new dimension to the study of lactation, as it considered not only the endocrine and physiological control of milk synthesis and output, but also the crucial role of the rabbit's maternal behavior. As this species nurses the young only once a day and only for about 3 to 5 minutes, this model offered the possibility of studying the role of mother–young contact for the maintenance of maternal responsiveness and lactation, and the chronobiological regulation of a complex mammalian behavior pattern (see Chapters 9 and 11 by Caba, Morales and Meza; González-Mariscal and Aguilar-Roblero). The large colony of rabbits in Tlaxcala also allowed Carlos to re-explore the neuroendocrine regulation of female reproductive behavior in that species. Together with Robyn Hudson, who came from Münich at the invitation of Jay Rosenblatt, Carlos and his colleagues explored the endocrine control of scent marking and the emission of the so-called nipple pheromone, an olfactory signal that guides the young to locate the mother's nipples and suckle. The richness of the mother–young relationship in rabbits opened this area of research, incorporating new young students and colleagues from Mexico and the United States and adding new topics of study such as the interactions among siblings and nest building as a stereotypic motor pattern (see Chapter 10 by Hoffman, Hudson, Martínez-Gómez and Bautista).

Carlos and Barry first met when they happened to converge as postdoctoral fellows independently in Sawyer's laboratory in 1965. The collaboration that they developed there became permanent, through continuing joint research and book projects and an exchange program. That program brought many of Carlos' students and faculty colleagues to the Institute of Animal Behavior at Rutgers University, and Barry's students and faculty colleagues to Universidad Autónoma de Tlaxcala (UAT), Mexico, CINVESTAV, Mexico City, Mexico, and the University of Veracruz (UV) at Xalapa, Mexico, for joint research projects. The collaborations fostered by the program resulted in more than 30 joint publications. The collaborative research projects, based on their joint interests, led to a genuine cross-fertilization of ideas on a wide range of reproductive, physiological, and behavioral processes (see, for instance, Chapter 13 by McCarthy and Silver). These included testosterone effects on female rat sexual behavior and uterus (42, 48, 50, 51), mapping neuronal olfactory input into the brain (49), identification of neural pathways in sexual behavior and analgesia in rats (111, 124, 134, 136, 137), behavioral and pharmacological analysis, blockage, and augmentation of the sex behavioral and analgesic effect of vaginal stimulation (106, 107, 110, 113, 122, 129, 145), mechanisms of allodynia and analgesia (96, 97, 121, and Chapter 16 by Gómora, Galicia-Aguas, González-Flores, and Komisaruk), interactions between estrogen, progesterone, GABA, and second messengers in sexual behavior (112, 117, 119, 170, 176), and more recently, reviews of hormonal, pharmacological, neural, and cultural aspects of sexual response and orgasm in humans in collaboration with Beverly Whipple (168, 177, 178, and Chapter 17 by Komisaruk and Whipple). Carlos' invitation to Barry to collaborate in Mexico, and Barry's invitation to Carlos to collaborate as a visiting professor at the Institute of Animal Behavior, led to collaboration with Jay Rosenblatt, which developed into a highly active and successful separate line of investigation with Gaby and her colleagues in Tlaxcala. Moreover, the interaction between Rutgers, CINVESTAV, UAT, and UV has led to the exploration of new aspects of the physiology and psychobiology of sexual behavior (see Chapters 6, 8 and 15 by Manzo, Carrillo, Coria-Avila and García; Hernández, Aranda-Abreu and Rojas-Durán; Ventura-Aquino, Baños-Araujo, Fenández-Guasti and Paredes) that enrich the studies on the neurohormonal regulation of copulatory motor patterns (see Chapters 3 and 4 by Moralí, Larsson, Contreras and Cervantes; Hernández González and Guevara). Two further chance convergences of Carlos at Rutgers led to the development of extensive research collaborations: with Knut Larsson on the role of aromatization in sexual behavior (see Chapters 2 and 5 by Agmo and Larsson, and by Lucio, Fernández-Guasti and Larsson) and in 1990 with María Cruz Rodríguez del Cerro, who was a visiting fellow from Madrid at Jay's invitation. The latter developed into an extensive fruitful line of research on the pharmacological, hormonal, and neural mechanisms underlying the pre- and perinatal sexual differentiation of maternal behavior with the research group at UNED (National University of Distance Education) in Madrid (see Chapter 12 by Del Cerro, Pérez-Laso, Segovia, and Guillamón).

It was Carlos' magnetic personality, erudition, scholarliness, compassion, intellect, sparkling sense of humor, and scientific brilliance that galvanized and energized such a remarkably creative, wide-ranging, and eminent group of young and seasoned scientists. Carlos, we miss you! Your legacy lives on in the myriad and wonderful ways that you touched the lives of us, your students, and your colleagues. We have tried to capture your spirit and impact on our thinking, our work, and indeed our lives, in the pages of this book, which we dedicate to you.

It is with deep sadness that we wish that Jay and Knut, both of whom passed away during the preparation of this book, could join us in this dedication.

Barry R. Komisaruk
Rutgers University-Newark

Gabriela González-Mariscal
CINVESTAV-Autonomous University of Tlaxcala

Editors

Barry R. Komisaruk is a distinguished professor of psychology at Rutgers University-Newark, New Jersey, having graduated from the City University of New York with a BS in biology in 1961, from Rutgers University with a PhD in psychobiology in 1965, and a postdoctoral National Institutes of Health (NIH) fellowship in neuroendocrinology at UCLA in 1966.

He has served as a program director at the NIH, on the Psychobiology Review Panel at the National Science Foundation (NSF), and as associate editor of the *Journal of Sexual Medicine* and *Sexual Medicine Reviews*. His research firsts include identification of brain regions activated during orgasm in women, the role of the vagus nerves in conveying genital sensation in women with severed spinal cord, and the phenomenon and mechanism of the pain-blocking action of vaginal stimulation. The research has been funded by grants from the NIH, NSF, and state and private foundations and resulted in more than 160 research and 3 books, including *The Science of Orgasm* and *The Orgasm Answer Guide* coauthored with Carlos Beyer, Beverly Whipple, and Sara Nasserzadeh and published in seven languages. He has shared the Hugo F. Beigel Research Award in Sexuality and the Bullough Award of the Foundation for the Scientific Study of Sexuality and supervised the doctoral dissertations of 26 PhDs and 20 postdoctoral scholars.

Gabriela González-Mariscal graduated in 1978 from the Metropolitan Autonomous University, Mexico with a BS in biology. She earned a master's degree in 1982 from the National Autonomous University of Mexico and a PhD in physiology in 1990 from the Center for Research and Advanced Studies of the National Polytechnic Institute, CINVESTAV. As a full professor and the director of the Center for Research in Animal Reproduction (CINVESTAV-Autonomous University of Tlaxcala), her main area of research is the neuroendocrine regulation of sexual and maternal behavior in rabbits, which has been funded by NIH, National Council of Science and Technology (Mexico), and CINVESTAV. She is particularly interested in connecting behavioral neuroendocrinology with animal science, especially in rabbits. She has been the section editor of *Animal: International Journal of the Biosciences* and *World Rabbit Science,* and she also serves on the editorial boards of *Developmental Psychobiology* and *Hormones and Behavior.*

Contributors

Anders Ågmo
Department of Psychology
University of Tromsø
Tromsø, Norway

Raúl Aguilar-Roblero
Instituto de Fisiología Celular
Universidad Nacional Autónoma de México
Mexico City, Mexico

Gonzalo E. Aranda-Abreu
Centro de Investigaciones Cerebrales
Universidad Veracruzana
Xalapa, Veracruz, Mexico

Jorge Baños-Araujo
Escuela de Medicina
Universidad Anáhuac México Norte
Mexico City, Mexico

Amando Bautista
Centro Tlaxcala de la Biología de la Conducta
Universidad Autónoma de Tlaxcala
Tlaxcala, Mexico

Mario Caba
Centro de Investigaciones Biomédicas
Universidad Veracruzana
Xalapa, Veracruz, Mexico

Ignacio Camacho-Arroyo
Unidad de Investigación en Reproducción Humana
Instituto Nacional de Perinatología-Facultad de Química
Universidad Nacional Autónoma de México
Mexico City, Mexico

Porfirio Carrillo
Cuerpo Academico de Neurociencias
and
Instituto de Neuroetologia
Universidad Veracruzana
Xalapa, Veracruz, Mexico

Miguel Cervantes
Laboratorio de Neurociencias
División de Estudios de Posgrado
Facultad de Ciencias Médicas y Biológicas "Dr. Ignacio Chávez"
UMSNH
Morelia, Michoacán, Mexico

Carmen Clapp
Instituto de Neurobiología
Universidad Nacional Autónoma de México
Campus UNAM-Juriquilla
Querétaro, México

José Luis Contreras
Departamento de Biología de la Reproducción
Universidad Autónoma Metropolitana-Iztapalapa
Mexico City, Mexico

Genaro A. Coria-Avila
Centro de Investigaciones Cerebrales
and
Cuerpo Academico de Neurociencias
Universidad Veracruzana
Xalapa, Veracruz, Mexico

María Luisa Cruz
Unidad de Investigación Médica en Farmacología
UMAE Hospital de Especialidades
Centro Médico Nacional Siglo XXI
Mexico City, Mexico

Gonzalo Martínez de la Escalera
Instituto de Neurobiología
Universidad Nacional Autónoma de México
Campus UNAM-Juriquilla
Querétaro, Mexico

María Cruz Rodríguez Del Cerro
Departamento de Psicobiología
Universidad Nacional de Educación a Distancia
Madrid, Spain

Anne M. Etgen
Dominick P. Purpura Department of Neuroscience
Albert Einstein College of Medicine
Bronx, New York

Alonso Fernández-Guasti
Departamento de Farmacobiología
CINVESTAV-Sede Sur
Mexico City, Mexico

Galicia-Aguas Y.
Centro de Investigación en Reproducción Animal
Universidad Autónoma de Tlaxcala-CINVESTAV
Tlaxcala, Mexico

Luis I. Garcia
Centro de Investigaciones Cerebrales
and
Cuerpo Academico de Neurociencias
Universidad Veracruzana
Xalapa, Veracruz, Mexico

Francisco Gómez
Centro de Salud Joaquín Rodrigo
Servicio Madrileño de Salud
Madrid, Spain

Porfirio Gómora-Arrati
Centro de Investigación en Reproducción Animal
Universidad Autónoma de Tlaxcala-CINVESTAV
Tlaxcala, Mexico

Oscar González-Flores
Centro de Investigación en Reproducción Animal
Universidad Autónoma de Tlaxcala-CINVESTAV
Tlaxcala, Mexico

Gabriela González-Mariscal
Centro de Investigación en Reproducción Animal
CINVESTAV-Universidad Autónoma de Tlaxcala
Tlaxcala, Mexico

Miguel Ángel Guevara
Instituto de Neurociencias
CUCBA
Universidad de Guadalajara
Guadalajara, Jalisco, Mexico

Antonio Guillamón
Departamento de Psicobiología
Universidad Nacional de Educación a Distancia
Madrid, Spain

Maria Elena Hernandez
Centro de Investigaciones Cerebrales
Universidad Veracruzana
Xalapa, Veracruz, Mexico

Marisela Hernández González
Instituto de Neurociencias
CUCBA
Universidad de Guadalajara
Guadalajara, Jalisco, Mexico

Kurt Leroy Hoffman
Centro de Investigación en Reproducción Animal
CINVESTAV-Universidad Autónoma de Tlaxcala
Tlaxcala, Mexico

Robyn Hudson
Instituto de Investigaciones Biomédicas
Universidad Nacional Autónoma de México
Mexico City, Mexico

Barry R. Komisaruk
Department of Psychology
Rutgers University-Newark
Newark, New Jersey

Knut Larsson
Department of Psychology
University of Gothenburg, Gothenburg, Sweden

Rosa Angélica Lucio
Centro Tlaxcala de Biología de la Conducta
Universidad Autónoma de Tlaxcala
Tlaxcala, Mexico

Jorge Manzo
Centro de Investigaciones Cerebrales
and
Cuerpo Academico de Neurociencias
Universidad Veracruzana
Xalapa, Veracruz, Mexico

Margarita Martínez-Gómez
Instituto de Investigaciones Biomédicas
Universidad Nacional Autónoma de México
Mexico City, Mexico

and

Centro Tlaxcala de la Biología de la Conducta
Universidad Autónoma de Tlaxcala
Tlaxcala, Mexico

Margaret M. McCarthy
Department of Pharmacology
University of Maryland School of Medicine
Baltimore, Maryland

Enrique Meza
Centro de Investigaciones Biomédicas
Universidad Veracruzana
Xalapa, Veracruz, Mexico

Teresa Morales
Instituto de Neurobiología
Universidad Nacional Autónoma de México
Querétaro, Mexico

Gabriela Moralí
Unidad de Investigación Médica en Farmacología
UMAE Hospital de Especialidades
Centro Médico Nacional Siglo XXI
Mexico City, Mexico

Raúl G. Paredes
Instituto de Neurobiología
Universidad Nacional Autónoma de México
Mexico City, Mexico

Carmen Pérez-Laso
Departamento de Psicobiología
Universidad Nacional de Educación a Distancia
Madrid, Spain

Fausto Rojas-Duran
Centro de Investigaciones Cerebrales
Universidad Veracruzana
Xalapa, Veracruz, Mexico

Santiago Segovia
Departamento de Psicobiología
Universidad Nacional de Educación a Distancia
Madrid, Spain

Rae Silver
Department of Psychology
Barnard College
and
Department of Psychology
Columbia University
and
Department of Pathology and Cell Biology
Columbia School of Medicine
New York City, New York

Elisa Ventura-Aquino
Departamento de Farmacobiología
CINVESTAV-Sede Sur
and
Instituto de Neurobiología
Universidad Nacional Autónoma de México
Mexico City, Mexico

Beverly Whipple
College of Nursing
Rutgers University-Newark
Newark, New Jersey

Section I

Neuroendocrinology of Sexual Behavior

1 Pioneering Studies on the Effects of Steroid Hormone Metabolism on the Brain and Behavior

María Luisa Cruz, Miguel Cervantes, and Gabriela Moralí

CONTENTS

1.1 Introduction 4
1.2 Neural Regulation of Lactation and Maternal Behavior in the Female Cat and Rabbit 4
1.2.1 Background 4
1.2.2 Studies on Neural and Hormonal Factors Regulating Lactation and Maternal Behavior in Cats and Rabbits 5
1.2.2.1 Neural Factors Involved in the Secretion of Oxytocin and Prolactin 5
1.2.2.2 Role of Ovarian Steroid Hormones in Maintaining Lactation 6
1.2.2.3 Effect of Steroid Hormones on the Mammary Glands of Adult Rabbits 6
1.2.2.4 Effect of Septal Lesions on Maternal Behavior and Lactation in Rabbits 6
1.3 Hormonal Regulation of Female Sexual Behavior of Rabbits and Rats 7
1.3.1 Lordosis and Pseudomale Behavior in Rabbits 7
1.3.1.1 Display of Pseudomale Behavior in Rabbits 7
1.3.1.2 Estrous Behavior during Various Reproductive Conditions 8
1.3.1.3 Effect of Removing Endocrine Glands on Estrous Behavior 8
1.3.1.4 Estrogen–Progesterone Interactions in the Expression of Estrous Behavior 9
1.3.1.5 Effect of Androgen Replacement on Estrous Behavior of Rabbits and Rats; Role of Aromatization 9
1.3.2 Role of 5-Alpha-Reduction 13
1.3.2.1 Effect of Testosterone (T) and 5-Alpha-Dihydrotestosterone (DHT) on Gonadotropin Secretion and Lordosis Behavior of Rats 13

1.3.2.2 Effect of 5-Alpha-Dihydrotestosterone and Some of Its Metabolites on Sexual Differentiation of the Brain in the Female Rat 15
1.4 Hormonal Regulation of Male Sexual Behavior of Rabbits and Rats 17
1.4.1 Effect of Androgen Replacement on Male Sexual Behavior of Rabbits and Rats; Role of 5-Alpha Reduction 17
1.4.2 Role of Aromatization in the Effect of Testosterone on Male Sexual Behavior of Rabbits and Rats 18
1.5 Electrophysiological Correlates of Neural Processes Involved in Reproduction-Related Behavioral Phenomena: Actions of Some Neuroactive Steroids on the Brain 21
1.5.1 Effect of Estrogen on Responsiveness to Sensory or to Sexually Relevant Stimuli 21
1.5.2 Electrophysiological Expressions of "Relaxation Behavior" 23
1.5.2.1 Characterization of "Relaxation Behavior" in Response to Sexually Relevant and Other Pleasant Stimuli 23
1.5.2.2 Facilitatory Effect of Progesterone and Some Neuroactive Drugs on EEG Synchronization during Milk Drinking 24
1.5.2.3 Electrographic Signs of Relaxation Behavior in Lactating Mothers 25
1.5.3 Effect of Progesterone on Brain Electrical Activity; Functional Role of Some Metabolic Pathways 26
References 28

1.1 INTRODUCTION

With all our affection and respect for Dr. Carlos Beyer Flores, as some of his first coworkers in his scientific research, we present our recollections. We begin by referring to four pioneering, internationally recognized research lines developed by Dr. Beyer starting around 1961, in various fields of behavioral neuroendocrinology: (1) the neural regulation of lactation and maternal behavior in the female cat and rabbit, (2) the hormonal regulation of female sexual behavior of rabbits and rats, (3) the hormonal regulation of male sexual behavior of rabbits and rats, and (4) the electrophysiological approach to the effects of neuroactive steroids on the central nervous system, characterizing some electrophysiological correlates of behavioral patterns of reproduction-related phenomena.

1.2 NEURAL REGULATION OF LACTATION AND MATERNAL BEHAVIOR IN THE FEMALE CAT AND RABBIT

1.2.1 Background

Together with his first student, Flavio Mena, Dr. Beyer became interested in the effects of stimulation of the cerebral cortex on the secretory activity of the neural pituitary. At that time, it had just been suggested that hormones of the neural lobe of the pituitary gland were produced by the hypothalamic supraoptic and paraventricular nuclei (Shealey and Peele 1957). The effect on oxytocin release of electrical stimulation of the cingulate cortex, which is anatomically connected with these nuclei, was then studied by recording uterine contractility in cats. Electrical stimulation of the cingulate cortex induced clear contractile uterine responses. The possible role of a neural efferent pathway was ruled out since contractile uterine responses persisted in spite of vagotomy or spinal cord transection. Furthermore, uterine responses to cortical stimulation were very similar in their slope and shape to the very characteristic contractile responses induced by intravenous administration of oxytocin. Moreover, in two lactating cats, cerebral cortex stimulation elicited an evident milk ejection response (Beyer et al. 1961; Mena et al. 1961). These results indicated that electrical stimulation of the anterior portion of the cingulate cortex resulted in oxytocin release from the neural pituitary. This evidence added information to that already existent about the limbic circuit, the various brain structures of this circuit being involved in the secretory mechanism of neural pituitary hormones. This was, for Dr. Beyer and Dr. Mena, the starting point of an integrative and detailed research line on the neural regulation of lactation.

Dr. Charles Sawyer, an internationally renowned researcher in the fields of neurophysiology and neuroendocrinology, learned of these results and invited Dr. Beyer to collaborate in research with him at the Department of Anatomy of the School of Medicine and the Brain Research Institute of the University of California at Los Angeles. Significant collaborative research derived from that academic interlude. By 1963, having returned to México, Dr. Beyer continued a fruitful research career at the Instituto de Estudios Médicos y Biológicos (National Autonomous University of Mexico) with Dr. Mena as his main coworker.

1.2.2 Studies on Neural and Hormonal Factors Regulating Lactation and Maternal Behavior in Cats and Rabbits

1.2.2.1 Neural Factors Involved in the Secretion of Oxytocin and Prolactin

By the early 1960s, it was recognized that both milk ejection and milk secretion were dependent on the activation of the neuroendocrine reflex initiated by suckling. This stimulus was able to initiate or prolong established lactation (Uyldert 1946; Bruce 1958), and Beyer and Mena investigated the neural pathway—initiated in sensory receptors of the mammary glands—involved in the synthesis and release of prolactin and oxytocin. They performed electrolytic lesions in several regions of the CNS in cats and found that lesioning the periventricular region of the caudal hypothalamus interrupted established lactation. This effect was due to interference with milk removal, since oxytocin administration allowed lactation to continue, though at a reduced level (Beyer and Mena 1965a). On the other hand, total lesion of the telencephalon in rabbits induced alveolar growth and milk secretion (Beyer and Mena 1965c), results interpreted as due to the secretion of prolactin and possibly also ACTH, provoked by the removal of an inhibitory control of brain structures on hypothalamic areas regulating the secretion of these hormones.

Beyer and Mena also performed spinal cord lesions in rabbit dams during the daily nursing period. Under these conditions, the milk ejection reflex in response to suckling was suppressed but, provided the mammary glands were emptied by oxytocin administration, milk production continued (Mena and Beyer 1963; Tindall et al. 1963; Beyer and Mena 1970). These results suggested that, in the rabbit, afferent information provided by the suckling stimulus to the hypothalamic–pituitary system was not essential for the release of prolactin and other hormones of the anterior pituitary involved in maintaining milk secretion. Rather, the suckling stimulus seemed to play an indirect role in this response by inducing the release of oxytocin from the neurohypophysis, thus emptying of the mammary gland, and thereby enabling subsequent milk secretion.

1.2.2.2 Role of Ovarian Steroid Hormones in Maintaining Lactation

Based on the above-mentioned results, and knowing that estradiol and progesterone may stimulate the release of prolactin from the pituitary, Beyer and Mena investigated the effect of ovariectomy upon maintenance of lactation in rabbits and cats to assess whether, in anesthetized animals thus deprived of suckling stimulation, milk secretion could be maintained during lactation by ovarian steroids (Beyer and Mena 1965b). Anesthesia suppressed the milk ejection reflex and interfered with milk removal from the mammary glands, but this was restored by subsequent administration of oxytocin. On successive days, milk secretion continued, provided it was evacuated every day by oxytocin administration. As exemplified in Figure 1.1, milk production was diminished gradually to about 50% of control values, but persisted even after ovariectomy within the first postpartum week. These findings indicated that ovarian steroids are not indispensable for maintaining lactation in rabbits and cats.

1.2.2.3 Effect of Steroid Hormones on the Mammary Glands of Adult Rabbits

A study of the effects of hormones on lactation in the rabbit included the assessment, with Martínez Manautou, of the effects of chlormadinone acetate, a synthetic progestin. Administration to ovariectomized rabbits did not induce tubular development in the mammary glands, denoting its lack of estrogenic activity, but confirming its high progestational potency by inducing an intense tubuloalveolar development in the mammary glands, as well as milk secretion when administered in association with estrone (Beyer et al. 1970a).

1.2.2.4 Effect of Septal Lesions on Maternal Behavior and Lactation in Rabbits

Besides studying the neural factors involved in the secretion of oxytocin and prolactin, the effects of septal lesions on several reproductive parameters and in particular on maternal behavior were assessed in rabbits. These lesions did not interfere with receptive behavior, ovulation, gestation, or parturition. However, they resulted in clear deficiencies in nest building, care of the young, and willingness of the dams to feed them, in addition to inducing a high level of cannibalism (Cruz and Beyer 1972). Some of these expressions of maternal behavior improved after a further gestation

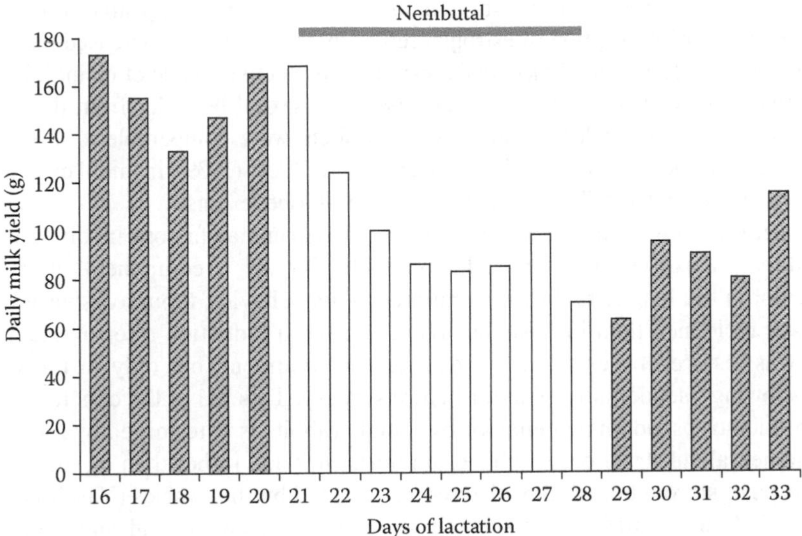

FIGURE 1.1 Daily milk yield of an ovariectomized rabbit that was suckled under pentobarbital (Nembutal) anesthesia from the 21st to 28th day of lactation. Solid bars indicate the quantity of milk obtained without oxytocin administration and without anesthesia. Empty bars indicate the quantity of milk obtained following oxytocin administration in anesthetized rabbits. Note that milk secretion persists, though at a subnormal rate, in spite of the pharmacological deafferentation of the hypothalamic–pituitary system. (Modified from Beyer, C. and F. Mena, *Bol Inst Estud Med Biol Univ Nac Auton Mex*, 23, 89–99, 1965b.)

in some does. Alterations of maternal behavior were specific, not associated with overt motor or emotional perturbation. These behavioral deficits of lesioned rabbits resulted in a higher mortality of their young during the lactation period compared with control rabbits. Mortality of pups was not associated with neuroendocrine disruptions, since lactogenesis, galactopoiesis, and milk ejection were not altered by the septal lesions.

1.3 HORMONAL REGULATION OF FEMALE SEXUAL BEHAVIOR OF RABBITS AND RATS

1.3.1 Lordosis and Pseudomale Behavior in Rabbits

1.3.1.1 Display of Pseudomale Behavior in Rabbits

Though pseudomale (mounting) behavior is commonly displayed by females of several mammalian species including rabbits (Brooks 1937; Beach 1968), Beyer and Mena observed that temporal lobe lesions in female rabbits, mainly involving the entorhinal cortex and some amygdaloid nuclei, increased the incidence of pseudomale behavior. Lesioned rabbits directed this behavior in an indiscriminate manner not only to other female rabbits, but also to inanimate objects and even to male cats, eliciting aggressive behavior in the latter (Beyer et al. 1964). These behavioral alterations were dependent on estrogen, as they gradually disappeared after ovariectomy or after interruption of estrogen replacement treatment. Increased mounting behavior in the lesioned rabbits was interpreted as a consequence of disinhibition of hypothalamic centers related to the expression of sexual behavior from the inhibitory influence of the telencephalon. These effects were comparable to alterations described by the classical study of Klüver and Bucy (1939) in monkeys, which showed "hypersexuality" resulting from temporal lobe lesions.

From the above study, the investigation of hormonal factors involved in the spontaneous expression of pseudomale behavior, as a component of estrous behavior in the female rabbit, was initiated. This behavior, though displayed with a lower incidence than lordosis behavior, showed a reduction after ovariectomy, and was restored by estradiol or testosterone treatment, but only in those rabbits showing pseudomale behavior before surgery. This led to the conclusion that expression of pseudomale behavior by female rabbits is determined by the nature of the neural substrate on which steroid hormones act, rather than by the amount and type of steroid being administered or secreted by the ovaries (Yaschine et al. 1967). Individual differences seemed to exist in the anatomical and functional organization of the neural substrate responsible for the display of pseudomale behavior in the female rabbit. An additional possibility exists to account for such variability—that is, that undetermined levels of circulating steroids may exert a virilizing effect on the developing female brain, increasing the propensity to pseudomale behavior in adulthood.

1.3.1.2 Estrous Behavior during Various Reproductive Conditions

Later studies explored the incidence of lordosis and mounting behavior in pregnant and lactating rabbits (Beyer and Rivaud 1969). While these responses diminish abruptly 12 or 24 h after mating, they gradually reappear in some rabbits with a low incidence in the final third of gestation. Most females showed a postpartum estrus lasting 24 to 48 h, followed by a reduction of both lordosis and mounting responses during the lactation period; approximately 50% of the rabbits showed these behaviors sporadically. Results from this study have been confirmed and expanded recently by the finding that the degree of sexual behavior inhibition is directly related to the number of suckling pups (García-Dalmán and González-Mariscal 2012).

1.3.1.3 Effect of Removing Endocrine Glands on Estrous Behavior

The effect of ovariectomy and subsequent adrenalectomy on female and pseudomale behavior of rabbits was also analyzed. Ovariectomy reduced the expression of lordosis and mounting behavior; only 25% of rabbits showing these responses, which were also low in number (Beyer et al. 1969a). When ovariectomized rabbits were subsequently adrenalectomized, incidence of sexual responses further decreased, being displayed only by those rabbits retaining some lordosis or mounting activity after ovariectomy. These findings were interpreted as evidence that adrenal steroids contributed to the persistence of sexual responses shown by some of the ovariectomized rabbits, but also that some other factors, possibly extraglandular sources of sexual steroids (skin, adipose tissue), might have played a role in the expression of sexual activity after ovariectomy/adrenalectomy. No information existed at that time on the possibility of neurosteroidogenesis, and, to our knowledge, this issue has still not been investigated in rabbits.

The effects of ovariectomy performed at 30, 60, or 90–100 days of age were also studied in rabbits to evaluate the ability of the neural substrate to initiate the expression of sexual behavior in the absence of ovarian hormones. The onset and incidence of sexual activity was compared in relation to nonovariectomized rabbits (Beyer et al. 1970c). The expression of lordosis and mounting behavior was assessed in these groups of rabbits starting at 90 to 107 days of age. From 33% to 45% of rabbits that were ovariectomized in the infantile, early juvenile, or late juvenile stages showed sporadic lordosis responses. On the other hand, 0%, 9%, and 28%, respectively, displayed mounts. The mean age at which rabbits showed lordosis for the first time (144, 138, and 123 days, respectively) and the frequency of lordosis displayed (the lordosis quotient [LQ], number of lordosis/number of mounts received = 0.038, 0.043, 0.015) were similar in the three groups. However, the incidence of lordosis and mounts was significantly lower than the responses shown by intact nonovariectomized rabbits. Taken together, these findings showed that lordosis and mounting responses are initiated at the typical age for this species, even in the absence of ovarian steroids, suggesting that hormones from other sources, such as the adrenal glands, skin, adipose tissue, and liver, can increase the responsiveness of the neural substrate of sexual behavior in rabbits. Based on these findings, previous sexual experience is evidently not required for the expression of sexual behavior after ovariectomy (Beyer et al. 1970c).

1.3.1.4 Estrogen–Progesterone Interactions in the Expression of Estrous Behavior

The inhibitory effect of progesterone on estrogen-induced lordosis and mounting behavior in female rabbits was investigated in collaboration with Peter McDonald (Beyer et al. 1969b). Maximal inhibition of lordosis behavior was observed at 24 h of progesterone administration, irrespective of the dose, and this effect persisted in spite of the continuous administration of estrogen. The duration of inhibition, however, was correlated with the dose of progesterone injected. Behavioral inhibition by progesterone was consistent with the inhibition of lordosis and mounting behavior previously described to occur in pregnant rabbits (Beyer and Rivaud 1969). Administration of 20-alpha-hydroxy-pregn-4-en-3-one, known to be secreted in large amounts by the rabbit ovary after coitus (Hilliard et al. 1963), did not inhibit receptive or mounting behavior when given as a single dose. This may indicate that this progestin does not play a role in behavioral inhibition after coitus. However, a possible inhibitory effect of a sustained secretion of this hormone by the ovary, as occurs after coitus, cannot be ruled out.

1.3.1.5 Effect of Androgen Replacement on Estrous Behavior of Rabbits and Rats; Role of Aromatization

The effect of androgens on receptive and mounting behavior was also assessed in the female rabbit. The threshold dose, latency, and duration of the period in which sexual behavior (lordosis and mounting) was displayed were determined in ovariectomized rabbits, after being treated with different doses of estradiol benzoate (EB) or testosterone propionate (TP) (McDonald et al. 1970b). As illustrated in Figure 1.2, the percentage of female rabbits showing lordosis, as well as the duration of both lordosis and mounting behavior, was dependent on the dose of these hormones. However, the latency to show this behavior was similar for the various doses, and also for TP compared to EB, suggesting that both hormones activate the neural structures involved in mating behavior via a similar mechanism. On a body weight basis, the amount of EB (1 μg/day for 10 days) required for eliciting and maintaining a high level of sexual receptivity in the rabbit was lower than that required for eliciting a similar response in other species (e.g., cat, rat, and mouse). TP was not more effective than EB to elicit mounting, and the expression of this behavior was less likely to occur than lordosis, either after EB or TP treatment. This finding is consistent with previous suggestions by Palka and Sawyer (1966) and Yaschine et al. (1967) that the neural mechanism for the expression of the heterotypical (mounting) behavior pattern in the female is less sensitive to gonadal hormones than that of the homotypical (lordosis) behavior.

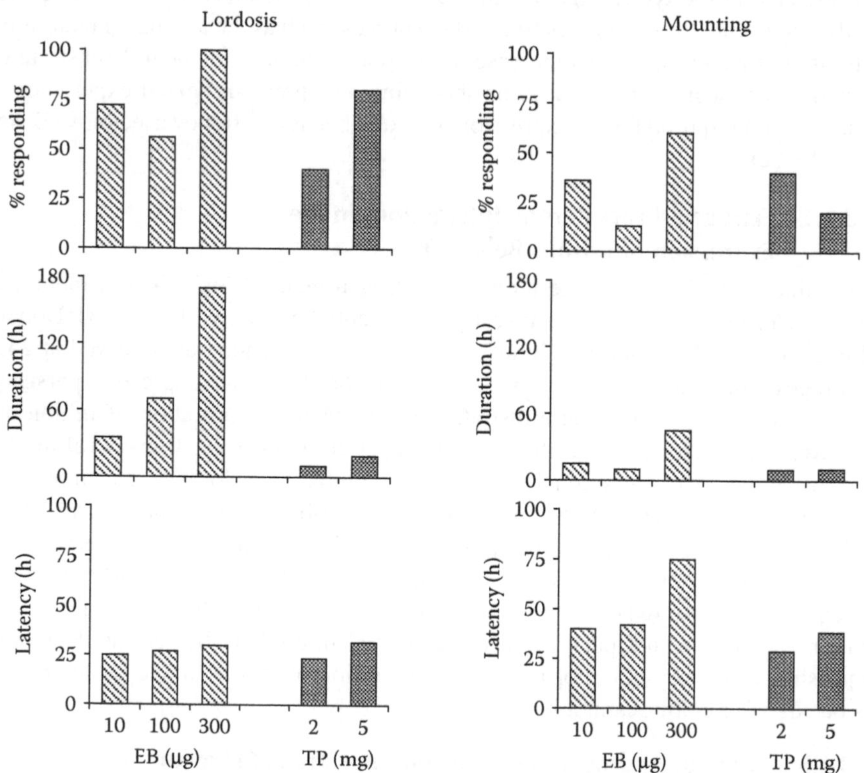

FIGURE 1.2 Percentages of ovariectomized rabbits showing lordosis or mounting behavior, latency to respond, and duration of the period with response, after a single administration of 10 (n = 11), 100 (n = 16), or 300 (n = 5) μg of estradiol benzoate (EB), or 2 (n = 5) or 5 (n = 5) mg testosterone propionate (TP). (Adapted from McDonald, P. G. et al., *Horm Behav*, 1, 161–72, 1970b.)

A functional role of testosterone (T) metabolites for their effect on the brain was suspected, as T was known to affect gonadotropin secretion and to stimulate estrous behavior in females of several mammalian species after ovariectomy (Greep 1961; Young 1961). The effect of this androgen on estrous behavior had been proposed to be due to its conversion to estrogenic metabolites (Young 1961). In accordance with this possibility a productive line of research was developed by Beyer, in collaboration with Komisaruk and McDonald, determining the involvement of the two main bioconversion pathways of T, that is, 5-alpha-reduction and aromatization, for the expression of sexual behavior. Thus, the effects of 5α-dihydrotestosterone (DHT), with potent androgenic actions on peripheral effector tissues but not being susceptible to aromatization, were compared to those of T in ovariectomized rabbits. As hypothesized, T (0.5 mg/day for 5 days) effectively stimulated lordosis and mounting behavior, while DHT, even at higher doses (1, 2, 6.5, or 10 mg/day), failed to stimulate these behaviors (Beyer et al. 1970b). Moreover, as shown in Figure 1.3,

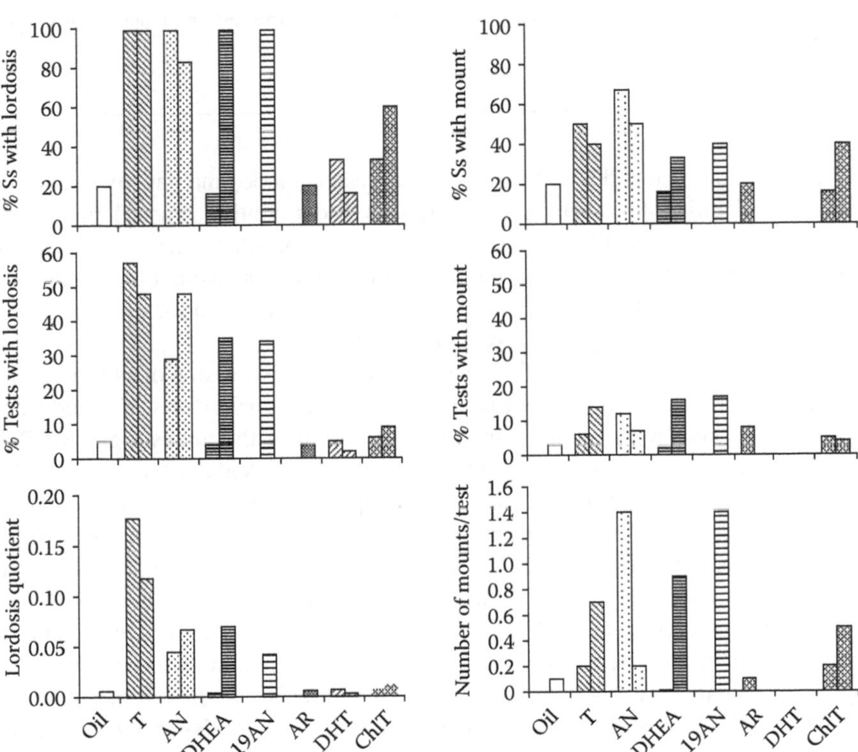

FIGURE 1.3 Effect of diverse androgens on lordosis and mounting behavior in ovariectomized rabbits. T, testosterone, 1 and 2 mg; AN, androstenedione, 2 and 4 mg; DHEA, dehydroepiandrosterone, 10 and 40 mg; 19AN, 19-hydroxyandrostenedione, 4 mg; AR, androsterone, 2 and 20 mg; DHT, 5α-dihydrotestosterone, 2 and 20 mg; ChlT, chlorotestosterone, 10 and 20 mg. Note that aromatizable androgens (T, AN, 19AN, and DHEA) stimulated high levels of lordosis and, to a lesser degree, mounting behavior, while AR, DHT, and ChlT were ineffective. (Adapted from Beyer, C. et al., *Endocrinology*, 87, 1386–9, 1970d.)

TABLE 1.1
Effect of MER-25 on Estrous Behavior Induced by Gonadal Hormones in Ovariectomized Rabbits

Treatment	No. of Rabbits	Females Mating (%)	Receptivity Quotient[a]	Females Mounting (%)	Average Mounts per Test
EB	8	100	0.150	50	0.7
TP	8	100	0.133	25	0.3
EB + 5 mg MER-25	5	80	0.055*	60	1.2
TP + 5 mg MER-25	5	60	0.022*	40	0.8
EB + 50 mg MER-25	6	0	0.000*	0	0
TP + 50 mg MER-25	6	16	0.004*	16	0.3

Source: Beyer, C. and N. Vidal, *J Endocrinol*, 51, 401–2, 1971.
[a] Lordosis/male mounts: calculated from the ten tests during treatment.
**P* < 0.01 compared with group not treated with MER-25 (chi square test on proportion of mounted animals showing lordosis).
EB, estradiol benzoate; TP, testosterone propionate.

testosterone, androstenedione, 19-hydroxy-androstenedione, and dehydroepiandrosterone stimulated lordosis behavior in all rabbits, and mounting in a high proportion, while androsterone, 5α-dihydrotestosterone, and chlorotestosterone were ineffective in stimulating these behaviors. As the three latter androgens cannot be metabolized to estrogen, the results supported a critical role of aromatization for the induction of estrous behavior by androgens (Beyer et al. 1970d).

Regarding a possible role of T aromatization for the expression of estrous behavior, as shown in Table 1.1, it was found that the stimulatory effects of T and estradiol were blocked by the concomitant administration of the antiestrogen MER-25 (ethamoxytriphetol) to ovariectomized rabbits (Beyer and Vidal 1971).

Using a different experimental model (i.e., applying cervical stimulation with a glass rod), the capacity of diverse natural aromatizable and nonaromatizable androgens to induce lordosis behavior in ovariectomized rats was explored and compared to that of EB (Beyer and Komisaruk 1971). After being treated for 10 days with TP, DHT propionate, androsterone, androstenedione, or EB, the number of responding rats, and the magnitude of dorsiflexion in response to cervical probing were evaluated. As shown in Figure 1.4, the intensity of the lordosis response to this stimulus was significantly increased only in TP- or EB-treated rats. Some rats treated with androstenedione also showed some lordosis responses but of low intensity. Androsterone or DHT failed to stimulate the lordosis response. The efficacy of T to facilitate the lordosis response was interpreted as due to its conversion to estradiol (E_2) (aromatization), in contrast to androsterone and DHT (nonaromatizable androgens) that failed to stimulate this response. The lower effect of androstenedione compared to T was accounted for by the fact that its aromatization leads to estrone, which is a weaker estrogen than E_2. These data provided further support for an important role of T aromatization for its effects on sexual behavior.

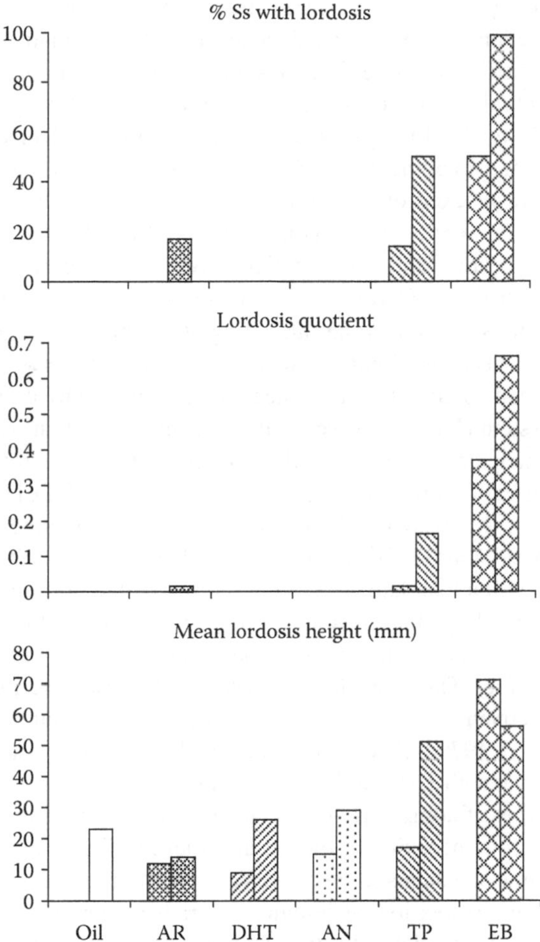

FIGURE 1.4 Effect of sex steroids on lordosis behavior (percent of subjects with lordosis, and lordosis quotients), and lordosis reflex in response to cervical stimulation (mean lordosis height), of ovariectomized rats. Steroids were daily administered for 10 days. AR, androsterone, 5 and 15 mg; DHT, 5α-dihydrotestosterone, 2 and 6 mg; AN, androstenedione, 5 and 15 mg; TP, testosterone propionate, 2 and 6 mg; EB, estradiol benzoate, 2 and 6 μg. Only EB and TP treatments stimulated the display of lordosis behavior and increased the lordosis response to cervical stimulation. (Adapted from Beyer, C. and B. Komisaruk, *Horm Behav,* 2, 217–25, 1971.)

1.3.2 Role of 5-Alpha-Reduction

1.3.2.1 Effect of Testosterone (T) and 5-Alpha-Dihydrotestosterone (DHT) on Gonadotropin Secretion and Lordosis Behavior of Rats

To further explore other effects of androgens on the brain, the efficacy of DHT in regulating gonadotropin secretion was explored in the model of inhibition of ovarian compensatory hypertrophy in hemi-ovariectomized rats (Beyer et al. 1971c).

Ovarian hypertrophy occurs after hemi-ovariectomy, reflecting the increase of gonadotropin secretion in response to a decreased level of circulating ovarian steroids (i.e., a reduction in the negative feedback to the hypothalamo–pituitary–ovarian axis). After unilateral ovariectomy, rats were either treated with DHT propionate (50, 200, or 1,000 μg) for 15 days, or uni- or bi-laterally implanted with free DHT into the ventromedial hypothalamus. Administration of DHT propionate in doses of 200 or 1,000 μg blocked ovarian compensatory hypertrophy. Bilateral implantation of DHT in the ventromedial hypothalamus also inhibited ovarian compensatory hypertrophy, while unilateral implantation of the hormone did not prevent this response (Beyer et al. 1971c). The results demonstrated the ability of DHT to inhibit gonadotropin secretion in the rat, acting at a central level rather than on the ovary itself. As T exerts its effects in several of its target tissues after being converted to 5α-reduced metabolites, and since the mammalian hypothalamus contains a 5α-reductase system (Jaffe 1969; Pérez-Palacios et al. 1970), it was suggested that 5α-reduction participates in the inhibition of gonadotropin secretion by T. In the same study, the effect of a prolonged (20 day) treatment with DHT (1 or 5 mg/day) on lordosis behavior was evaluated in ovariectomized rats. Treatment with DHT propionate induced weak lordosis behavior (LQ = 0.02 to 0.17) in 60% and 40% of rats, respectively, and with a long latency (15 to 20 days). Bilateral implantation of DHT into the ventromedial hypothalamus induced lordosis in several rats starting on day 10 after implantation, and the maximal response (LQ = 0.21) was observed on day 20 (Beyer et al. 1971c). These results contrasted with those described above, which showed the ineffectiveness of DHT in stimulating female sexual behavior in rabbits and rats (Beyer et al. 1970d; Beyer and Komisaruk 1971). They indicated that more prolonged treatment with DHT induces clear but weak lordosis behavior in some rats. In any case, these findings provided evidence that aromatization is not essential for stimulating estrous behavior in the rat and that DHT can alter the functioning of the hypothalamic mechanism involved both in the regulation of gonadotropin secretion and in the expression of female sexual behavior. A possible alternative explanation for the effect of DHT to stimulate estrous behavior in the rat is a mechanism in which some metabolites of this hormone, in particular 3β,5α-androstanediol (3β-diol), can bind to the intracellular estrogen receptor in several tissues (Garcia and Rochefort 1979; Thieulant et al. 1981), and exert estrogenic effects there (Handa et al. 2009; Lund et al. 2006).

In collaboration with Jaffe and Gay, the effects of DHT were compared with those of T in regulating LH and FSH secretion after ovariectomy in rats, by measuring the serum levels of these hormones (Beyer et al. 1972, 1974), as well as by assessing the weight and histological characteristics of ovaries and uteri in intact rats (Beyer et al. 1974). Androgens were either injected (100, 500, or 1,000 μg, s.c.) for 15 days or bilaterally implanted into the hypothalamic premammillary arcuate area and mammillary nuclei, immediately after ovariectomy. Blood was collected 24 h after the last injection or 16 days after implantation, and serum levels of LH and FSH were determined by radioimmunoassay. The lowest dose of DHT (100 μg/day) was more effective than that of T in suppressing the postovariectomy rise in serum LH and FSH. At higher doses, both androgens were equally effective. Bilateral implantation of either T or DHT into the premammillary or mammillary areas failed to prevent the postovariectomy increase in serum LH. This result suggested that these androgens may exert their effect on other brain areas or directly on the pituitary in regulating LH secretion (Beyer et al. 1972). When administered to intact female rats, both androgens at the highest doses (500 or 1,000 μg per day) produced ovarian atrophy. DHT was more effective than T in inducing anestrus, as well as uterine and vaginal atrophy, and the highest doses of DHT reduced the number of large ovarian follicles (Beyer et al. 1974). All these results indicated an inhibition of gonadotropin secretion by DHT, which correlated with the inhibition exerted by these doses of DHT on FSH and LH secretion in the ovariectomized rat. These doses of T did not suppress follicular growth. Actually, at the lower dose (100 μg), T significantly stimulated it, induced vaginal cornification, and other signs of estrogenic activity, such as stimulation of the columnar endometrial epithelium. These changes were interpreted as having been elicited by conversion of T to estrogen. The findings that DHT induced a marked, more effective inhibition of LH and FSH secretion than T confirmed previous suggestions that androgen aromatization is not essential for inhibition of gonadotropins (Beyer et al. 1971c, 1972; Swerdloff et al. 1972).

In collaboration with Pérez-Palacios and Lemus, Beyer studied the uptake of androgens by the brain and the pituitary in castrated male rats. By administering H_3-labelled steroids, they obtained evidence of a high uptake of DHT, T, and, to a lesser extent, androstenedione, by the pituitary (Pérez-Palacios et al. 1973). These results were interpreted as due to the presence of androphilic molecules in the pituitary with a higher affinity for DHT than for the other two androgens. Based on those findings, the authors speculated that the high uptake of androgens by the pituitary may influence some of the processes involved in gonadotropin secretion. Indeed, there was a correlation between the levels of radioactivity accumulated by the pituitary after DHT, T, and androstenedione administration, and their antigonadotropic potency (Beyer et al. 1972, 1974) and this, in turn, coincided with reports that androstenedione had a weaker antigonadotropic action than T (Shipley 1962).

1.3.2.2 Effect of 5-Alpha-Dihydrotestosterone and Some of Its Metabolites on Sexual Differentiation of the Brain in the Female Rat

A series of studies were performed (Cruz 1994) to assess the effects of administering T, 5α-DHT, 3α-diol, or 3β-diol during critical periods of the perinatal development of female rats, on functional parameters of the hypothalamo–pituitary–ovarian axis. Their effect on peripheral target tissues, the morphological characteristics of the olfactory bulb (a sexually dimorphic structure) (Roos et al. 1988), and the expression of lordosis and pseudomale behaviors was explored. Steroids were administered either during the prenatal period (to pregnant rats during days 16 to 20 of gestation) or postnatally (to female pups on days 2 and 3 after birth). The effects were analyzed on the age of vaginal opening, and the presence and regularity of vaginal cycles at 15 days after vaginal opening. At 90 days of age, female sexual behavior (LQ) for 15 consecutive days and pseudomale behavior (number of tests with mounting behavioral patterns) across 5 tests performed on alternate days were evaluated. At the end of the test period, rats were ovariectomized, and their feminine and pseudomasculine behavior was again assessed in response to treatment with EB + P. Their behavior was compared to that of control females that received neither perinatal treatment nor behavioral testing before or after ovariectomy and hormonal treatment. In all cases, rats were euthanized after the behavioral testing period and the histological characteristics of the accessory olfactory bulbs were analyzed.

Prenatal treatment with 3α-diol or 3β-diol interfered with vaginal opening, absent in 70% of females and only partially evident in the remaining 30% at 60 days of age. These effects of both forms of androstanediol on vaginal opening, which are coincident with those demonstrated for DHT by Dean et al. (2012), are suggestive of a virilizing effect of these androgens, leading to an inhibition of hypothalamo–pituitary–ovarian activity. This effect contrasted with the lack of a defeminizing effect of 3α- and 3β-diol on the expression of lordosis behavior. LQs were similar but expressed in more of the daily tests compared to those of intact nonandrogenized females. This finding suggested a condition of persistent estrus in the groups prenatally treated with 3α- and 3β-diol, possibly also due to the virilizing effect on the hypothalamic GnRH releasing neurons, shifting the cyclic female gonadotropin secretion pattern to the male tonic secretion pattern (Goy and McEwen 1980) compared with control rats that showed the behavioral fluctuations characteristic of ovarian cycles. Persistent estrus may have accounted for the inability of 3α- and 3β-diol-treated females to become pregnant despite the presence of spermatozoa in the vagina after mating. Pseudomale behavior was not altered by prenatal treatment with the androstanediol compounds, being displayed by only 10% of the females, as in the control females. Histological examination of the accessory olfactory bulbs showed a change to the male phenotype (i.e., a higher neuronal density in the mitral layer) in the T-, DHT-, and 3α-diol-treated females, indicating a virilizing effect of these androgens, but not in the 3β-diol-treated females.

These findings were interpreted as demonstrating a partial virilizing effect of 5α-reduced T metabolites, as demonstrated by interference with the neuroendocrine mechanisms responsible for timely establishment of puberty and ovarian cyclicity, and by altering, to a male phenotype, the morphological characteristics of the accessory olfactory bulbs. Prenatal treatment with DHT or with the androstanediol compounds, however, seemed to exert neither a masculinizing effect on brain mechanisms involved in the expression of pseudomale behavior nor a defeminizing effect on female brain regions involved in the display of female sexual behavior.

Administration of T, DHT, 3α-diol, or 3β-diol to female pups in the postnatal period (days 2 and 3) significantly shortened the latency to vaginal opening (T: 26 days; DHT: 24 days; 3α-diol: 16 days; 3β-diol: 19 days) compared to the control group (38 days). However, only treatment with 3β-diol resulted in shortening the latency to the first vaginal estrus (33 days vs. 40 days in controls). All treatments resulted in irregular vaginal cycles, and 3β-diol treatment, in particular, tended to induce persistent vaginal estrus. Consistent with this result, 3β-diol, and to a lesser degree T, increased the number of tests in which lordosis behavior was expressed (3β-diol: 60%, T: 38%, control rats: 20%). T, 3α-diol, or 3β-diol increased the incidence of pseudomale behavior compared to the control group (tests with mounts, 48%, 36%, and 70%, respectively, vs. 4%).

On the other hand, once ovariectomized, rats of all groups, with the exception of those treated with 3α-diol (LQ = 0.22), showed a response to EB + P administration (LQ: T, 0.90; DHT, 0.67; 3β-diol, 0.80) similar to that of the control rats (LQ = 0.78).

These findings demonstrated that postnatal treatment with 5α-reduced metabolites of T, while exerting effects different from prenatal treatments, also induced partial virilizing effects in hypothalamo–pituitary–ovarian axis function. This was due presumably to increasing the response of GnRH neurons to circulating estrogen, thus advancing puberty, but interfering with the cyclic pattern of gonadotropin secretion and ovarian cyclicity. T and 3β-diol may have exerted a virilizing effect on the female brain, increasing the incidence of pseudomale behavior, but these treatments did not defeminize the neural substrate, that is, they did not block the display of female sexual behavior.

1.4 HORMONAL REGULATION OF MALE SEXUAL BEHAVIOR OF RABBITS AND RATS

1.4.1 Effect of Androgen Replacement on Male Sexual Behavior of Rabbits and Rats; Role of 5-Alpha Reduction

By 1970, available information suggested that even though T was the principal steroid product of the testes in most mammalian species, one of its metabolites, 5α-dihydrotestosterone, was, in fact, the active form of T at several peripheral target organs such as the prostate and the seminal vesicles (Baulieu et al. 1968; Bruchovsky and Wilson 1968). The enzymes needed to convert T to DHT are also present in the hypothalamus of the rat and the dog (Jaffe 1969; Pérez-Palacios et al. 1970), thus suggesting that DHT may also be the active form of T for inducing male sexual behavior. Beyer and McDonald compared the effects of T and DHT for eliciting male sexual behavior in castrated rats. T stimulated the expression of full male sexual behavior including ejaculation patterns, while DHT stimulated only mounting activity (McDonald et al. 1970a). As T can be aromatized in several tissues to estrogenic metabolites (Dorfman and Ungar 1965; Naftolin et al. 1975), and this bioconversion had been suggested to be a mechanism for T to induce sexual behavior (Young 1961), these results suggested that the failure of DHT to induce sexual behavior in castrated male rats could have been due to inability of this androgen to be aromatized. Thus, Beyer and McDonald proposed that aromatization of T was a critical step in the activation of sexual behavior in males and females (the so-called aromatization theory).

Similar results were obtained in a study in male rabbits, in which the capacity of T and DHT for inducing male sexual behavior was compared (Beyer and Rivaud 1973). T propionate and DHT propionate were administered in daily doses of 1, 3, or 9 mg for 30 days to castrated bucks. TP was more effective than DHT propionate at each dose level tested, on several parameters of male copulatory behavior, including the percent of males mounting, latency to initiation of mounts, percentage of tests showing mounts, and number of mounts per test. Even the highest dose of DHT propionate (9 mg/day) failed to stimulate a consistent response, only eliciting a few mounts per test, after a longer treatment period in 45% of rabbits tested and in only 20% of the tests.

1.4.2 Role of Aromatization in the Effect of Testosterone on Male Sexual Behavior of Rabbits and Rats

The effect of 11 natural androgens (precursors or metabolites of T) administered daily (1 mg/day) for 30 days on the initiation of male sexual behavior was studied in inexperienced castrated rats. They included T, androstenedione, androstenediol, dehydroepiandrosterone, 5α-dihydrotestosterone, 5α-androstanedione, 3α,5α-androstanediol (3α-diol), 3α,5β-androstanediol, and 11β-hydroxy-androstenedione. The effect on accessory glands was also determined. As shown in Table 1.2, only those androgens susceptible of being aromatized, that is, T, androstenedione, and androstenediol, were able to stimulate male sexual behavior, including intromission and ejaculation behavioral patterns, while nonaromatizable androgens, though potent in stimulating prostate and seminal vesicle growth, only stimulated mounting activity (Beyer et al. 1973).

Additional support for a role of estrogenic metabolites of T on the initiation of male sexual behavior in castrated male rats was investigated by administering some antiestrogens such as MER-25 (ethamoxytriphetol), cis-clomiphene, or ICI-46,474 (tamoxifen) to inexperienced castrated rats receiving T, 1 mg/day for 21 days. Antiestrogens were administered 24 h before the first T injection and, from then onwards, 1 h before each daily dose of T. Figure 1.5 shows that ICI-46,474, but not the other antiestrogens, exerted a significant inhibitory effect on the sexual behavior induced by T (Beyer et al. 1976).

Failure of the other antiestrogens, in particular MER-25, to inhibit T-induced male sexual behavior was interpreted as due to a possible synergistic effect of the antiestrogen with the androgen to facilitate male sexual behavior, as suggested by its ability to induce mounting in ovariectomized rats (Södersten 1974). Komisaruk and Beyer (1972) had reported that MER-25 alone stimulated significant, but mild, uterine growth

TABLE 1.2
Percent of Castrated Male Rats Showing Mount (M), Intromission (I), or Ejaculation (E) Behavioral Patterns after 21 Days of Treatment with One of the Various Androgens (1 mg/day) or Oil as Control Treatment

Treatment	% Ss with M	% Ss with I	% Ss with E
Testosterone	63	63	55
Androstenedione	75	33	25
Androstenediol	44	33	25
Dehydroepiandrosterone	18	9	0
5α-Dihydrotestosterone	20	0	0
5α-Androstanedione	10	0	0
3α,5α-Androstanediol	22	0	0
3α,5β-Androstanediol	10	0	0
11β-OH-Androstenedione	25	12	0
Oil	22	0	0

Source: Beyer, C. et al., *Horm Behav*, 4, 99–103, 1973.

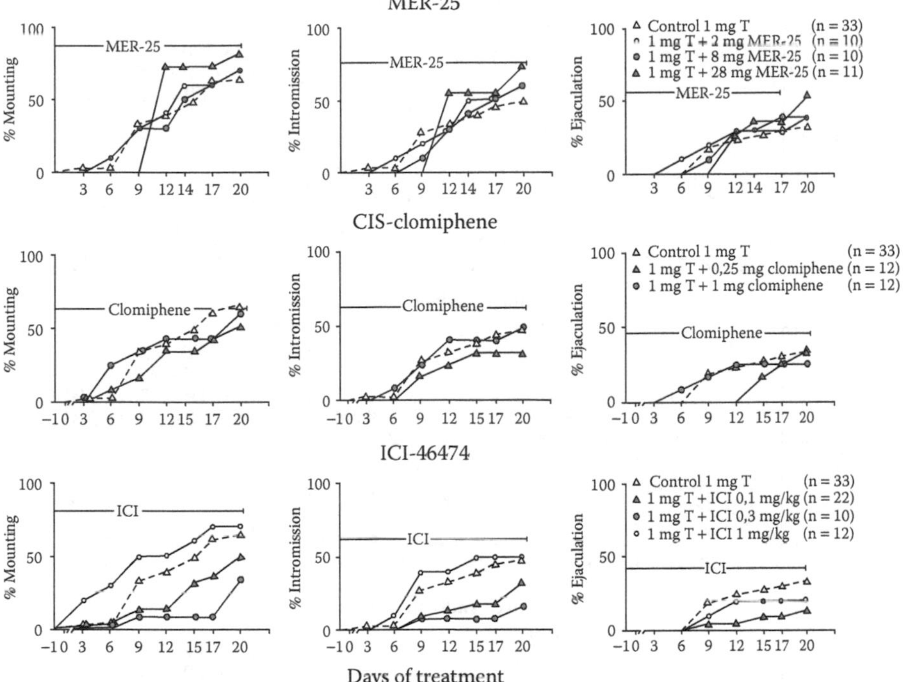

FIGURE 1.5 Effect of antiestrogens (MER-25, cis-clomiphene, and ICI-46474) on the response to testosterone administration. Proportion of rats treated with androgen and antiestrogens displaying at least one mount, intromission or ejaculation. Androgen and antiestrogens were administered for the period indicated by the horizontal line. (Modified from Beyer, C. et al., *Horm Behav*, 7, 353–63, 1976.)

in ovariectomized rats, which led them to conclude that MER-25 is a weak estrogenic agonist, which could account for its anti-estrogenic action. Some aromatase blockers interfering with the metabolic conversion of T to estradiol (e.g., metopirone, aminoglutethimide, or 5α-androstanedione) were administered every 12 h for 96 h to castrated rats receiving a single injection of 6 mg TP and male sexual behavior was daily assessed during 5 days. Aminoglutethimide abolished the sexual behavior induced by TP (Beyer et al. 1976). In a subsequent study, the effects of aromatase blockers including aminoglutethimide, 1,4,6-androstatriene-3,17-dione, and 4-hydroxy-androstenedione were assessed on TP-induced male sexual behavior of sexually inexperienced castrated rats. The males received a single injection of 6 mg TP and twice daily injections of one of the aromatase blockers for 108 h. As shown in Figure 1.6, treatment with the aromatase blockers suppressed ejaculatory behavior in all but one rat and reduced the number of rats displaying intromission and mounting patterns.

Concurrent administration of EB every 12 h counteracted the behavioral inhibitory effect of aromatase blockers, ruling out the possibility of an unspecific effect of these compounds, and supporting the interpretation of estrogen playing an important role on TP induction of male sexual behavior (Moralí et al. 1977). Based on the fact that

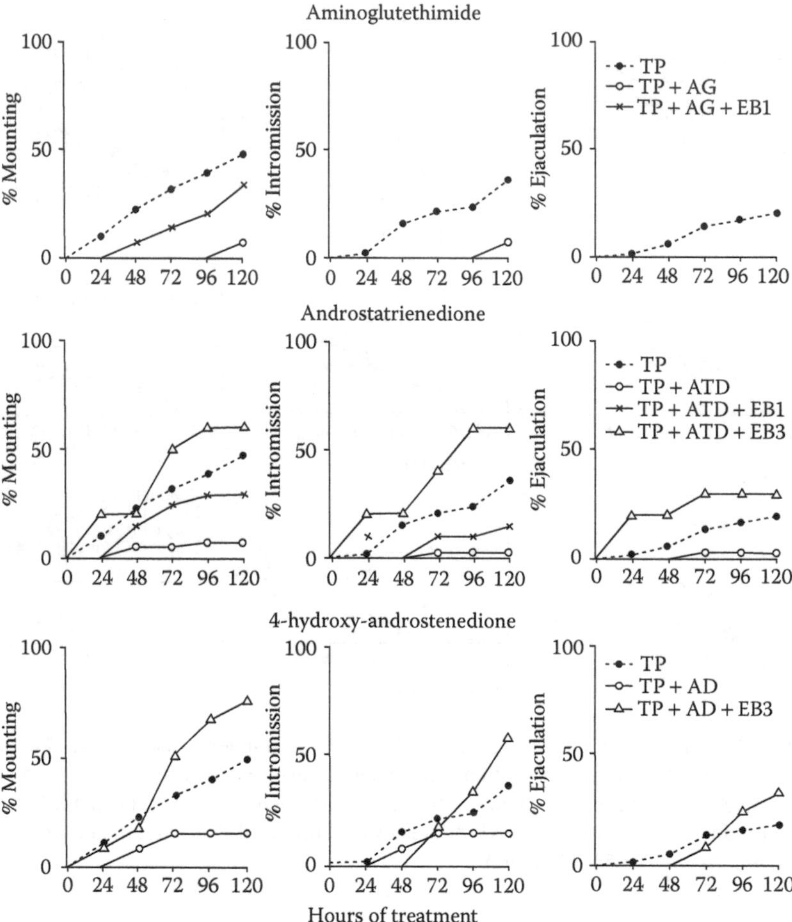

FIGURE 1.6 Effect of aromatase inhibitors (aminoglutethimide, AG; androstatrienedione, ATD; and 4-hydroxy-androstenedione, AD) on the response to a single injection of 6 mg of testosterone propionate (TP). Note prevention of the inhibitory response to aromatase inhibitors by exogenous administration of estradiol benzoate (EB), 1 or 3 μg every 12 h. Data are expressed as cumulative percentages of subjects displaying at least one mount, intromission, or ejaculation. (Modified from Moralí, G. et al., *Horm Behav*, 9, 203–13, 1977.)

simultaneous administration of DHT and EB to castrated rats optimally stimulated their copulatory behavior (Larsson et al. 1973), evidence was obtained that masculine sexual behavior resulting from the concurrent daily administration of DHT + EB for 20 days to castrated rats was not interfered with by aminoglutethimide (Moralí et al. 1977).

The crucial role of T aromatization for the induction of male sexual behavior in rats has been experimentally confirmed in a number of other species, including birds (Södersten 1979; Balthazart and Foidart 1993; Baum 2003).

Controversial results obtained by other researchers demonstrating the ability of DHT to stimulate sexual behavior in castrated rats initiated a discussion of the aromatization hypothesis. Discrepancies may be due to several factors, for example, the effects of sexual experience and the capacity of some DHT metabolites, for example, 3β,5α-androstanediol (3β-diol), to exert estrogenic effects on the brain (Lund et al. 2006; Handa et al. 2009) and to synergize with DHT or other androgens, thus stimulating male sexual behavior (Luttge 1979). Later studies supported this possibility by assessing the capacity of 3α-diol and 3β-diol to restore male copulatory behavior in castrated rats, when administered either alone or concurrently with E_2 or with DHT (Moralí et al. 1994). Full expression of male sexual behavior was indeed restored by the administration of 3β-diol and DHT, comparable in some parameters to that induced by treatment with E_2 and DHT, indicating an estrogen-like behavioral effect of 3β-diol.

Dr. Beyer had an extraordinary talent and great capacity for identifying and focusing on, among the vast bodies of the existing scientific literature, that which was particularly relevant to his concepts and research. In addition, he had a special ability for analyzing and integrating the information and, based on it, precisely interpreting the results, generating new hypotheses, and designing specific projects that invariably provided pioneering information. These attributes were evident in every one of his publications and scientific lectures.

1.5 ELECTROPHYSIOLOGICAL CORRELATES OF NEURAL PROCESSES INVOLVED IN REPRODUCTION-RELATED BEHAVIORAL PHENOMENA: ACTIONS OF SOME NEUROACTIVE STEROIDS ON THE BRAIN

Development of research lines by Dr. Beyer and his group by the late 1960s included the study of the effects of gonadal steroid hormones on the electrical activity of brain areas related to sexual behavior. Although experimental evidence already existed as to sexual behavior being facilitated by a direct effect of sex steroids on the brain (Lisk 1962; Harris and Michael 1964; Palka and Sawyer 1966; Komisaruk 1967), their mechanisms of action and, in particular, the possibility that estrogen effects on sexual behavior involved changes in neuronal excitability in the hypothalamus and other brain structures had not been explored.

1.5.1 Effect of Estrogen on Responsiveness to Sensory or to Sexually Relevant Stimuli

Studying the multiunit (neuronal) activity through the anterior and medial hypothalamus, and the mesencephalic reticular formation of female cats, a differential responsiveness to somatic tactile (lightly touching the back of the subject), visual, auditory, or genital stimuli was shown, as these brain structures were or not under estrogenic influence (Alcaraz et al. 1969). As shown in Figure 1.7, these stimuli may elicit both excitatory and inhibitory multineuronal responses in the hypothalamus and the mesencephalic reticular formation of both anestrous (ovariectomized) and estrous (estrogen treated) cats.

The hypothalamus showed, in ovariectomized cats, clear predominantly inhibitory responses to vaginal stimulation and, under the effect of estrogen, shifted to excitatory responses to sexually relevant stimuli (vaginal and somatic tactile stimulation). On the other hand, an opposite responsiveness to vaginal stimulation was observed

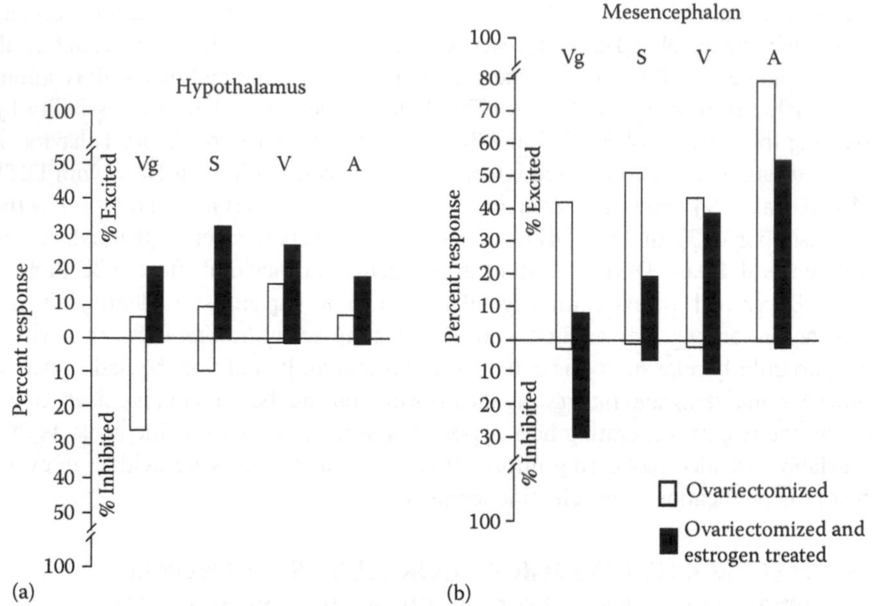

FIGURE 1.7 Histograms showing the percentages of hypothalamic (a) or mesencephalic (b) neuronal pools in ovariectomized (anestrous) and ovariectomized estrogen-treated (estrous) cats (white and black bars respectively), responding with increase or decrease of their firing rate in response to the following stimuli: vaginal probing (Vg), tactile somatic (S), visual (V), and acoustic (A). Note the differential, opposite response of the hypothalamus and the mesencephalic reticular formation to vaginal stimulation, related to the presence or absence of estrogen. (Modified from Alcaraz, M. et al., *Brain Res*, 15, 439–46, 1969.)

in the mesencephalic neurons, as compared to the hypothalamus, and related to the cat's estrogenic conditions: predominant activation of neuronal firing in anestrous, but predominant inhibition in estrous, cats. The particular response of these brain structures was interpreted as related to their differential involvement in processing relevant stimuli for the sexual behavior pattern of the female cat.

Incorporation of Laura De la Torre, Miguel Cervantes, Javier Almanza, and Carlos Kubli to the group directed by Dr. Beyer at the IMSS diversified these studies. The effects of genital stimulation (perineal tapping, vaginal probing, or cervical stimulation) were studied in freely moving, anestrous (ovariectomized) or estrous (ovariectomized, estrogen treated) cats (Beyer et al. 1971b). These stimuli resulted in differential patterns of multiunit neuronal firing in the mesencephalic reticular formation and the medial and lateral hypothalamus, as well as in theta-like EEG synchronization in the parietal cortex. Perineal tapping elicited inhibition of multiunit activity in the mesencephalic reticular formation in both anestrous and estrous cats; moreover, in many cases, multiunit neuronal firing decreased even more during vaginal distention in the estrous cats. On the other hand, as shown in Table 1.3, while cervical stimulation had no effect in anestrous cats, it resulted in a significant increase of multiunit neuronal firing in more than 80% of trials in the mesencephalic reticular formation, and in the medial and lateral hypothalamus in estrous cats.

TABLE 1.3
Effect of Cervical Stimulation on Brain Stem Multiunit Activity in Cats during Anestrus and Estrus

Structure	Anestrus[a] (%)			Estrus[b] (%)		
	+	–	0	+	–	0
Mesencephalic reticular formation	19	19	62	87	0	13
Medial hypothalamus	18	21	61	80	6	14
Lateral hypothalamus	26	14	60	90	0	10

Source: Beyer, C. et al., *Brain Res*, 32, 143–50, 1971b.
Note: Criterion of responsivity was a change of ±30% from the average discharge frequency recorded during a 10-sec control period.
[a] Based on 62 cervical stimulations.
[b] Based on 89 cervical stimulations.

1.5.2 Electrophysiological Expressions of "Relaxation Behavior"

The above data oriented further research toward the relevance of estrogenic actions on the excitability of neuronal circuits involved in neural processes involved in mating behavior, and to the differential electrophysiological expressions of these estrogenic effects on the functioning of specific brain structures.

1.5.2.1 Characterization of "Relaxation Behavior" in Response to Sexually Relevant and Other Pleasant Stimuli

Research focused on characterizing the parietal EEG synchronization elicited by perineal tapping, as this electrographic phenomenon was correlated with "relaxation behavior" (Roth et al. 1967), an emotional condition that could be induced by presumably rewarding stimuli (milk drinking, grooming). Neuronal multiunit activity was significantly decreased in the mesencephalic reticular formation, the ventromedial hypothalamus, and the lateral hypothalamus, while 6–10 Hz EEG synchronization appeared in the parietal cortex, during feeding (*ad libitum* milk drinking or ingestion of tuna fish paste), grooming, petting, and perineal tapping, as exemplified in Figure 1.8 (Beyer et al. 1971a).

Hence, "relaxation behavior" was proposed to be better defined, besides parietal EEG synchronization, on the basis of some other objective data: induction or facilitation by a variety of beneficial or pleasurable stimuli, reduced muscular tone (though not necessarily complete quiescence), appearance of 6–10 Hz EEG synchronization in relatively circumscribed cortical (parietal) and subcortical (hypothalamic) regions, and inhibition of brainstem multiunit firing (Beyer et al. 1971a). A possible correlation of "relaxation behavior" with affective components linked to several modalities of rewarding stimuli, which seem to be involved in the consummatory phenomena of several behavioral patterns, was also suggested.

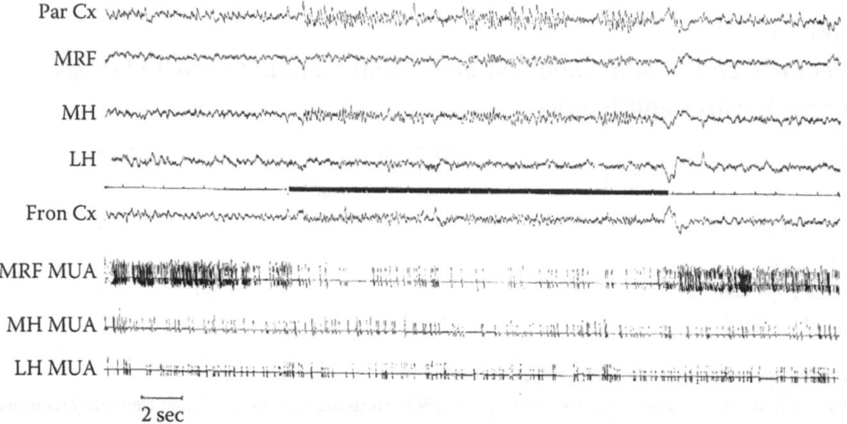

FIGURE 1.8 Effect of perineal tapping on EEG and brain stem multiunit activity. Note EEG spindling in all recorded channels associated with neuronal discharge inhibition. Par Cx, parietal cortex; Fron Cx, frontal cortex; MRF, mesencephalic reticular formation; MH, medial hypothalamus; LH, lateral hypothalamus; MUA, multiunit activity. (Modified from Beyer, C. et al., *Brain Res*, 29, 213–22, 1971a.)

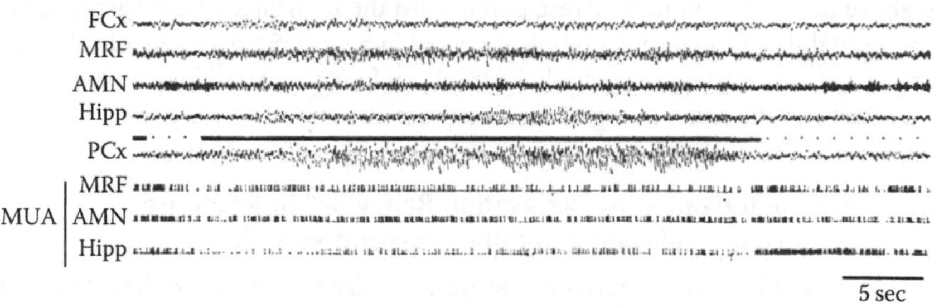

FIGURE 1.9 Changes in EEG and MUA during milk drinking. Note EEG synchronization in the parietal cortex (PCx) and hippocampus (Hipp), and inhibition of MUA in the mesencephalic reticular formation (MRF) and hippocampus. MUA in the amygdala (AMN) was not altered. The mark between channels 4 and 5 indicates the milk drinking period. FCx, frontal cortex. (Modified from Cervantes, M. et al., *Brain Res*, 91, 89–98, 1975.)

As shown in Figure 1.9, when milk drinking elicited a high proportion of parietal EEG synchronization, this electrographic activity also appeared in the dorsal hippocampus, coincident with inhibition of the hippocampal multiunit activity to levels similar to those recorded in the same brain region during slow-wave sleep (Cervantes et al. 1975).

1.5.2.2 Facilitatory Effect of Progesterone and Some Neuroactive Drugs on EEG Synchronization during Milk Drinking

The proportion of parietal 6–10 Hz EEG synchronization, considered to be a reliable electrographic correlate of "relaxation behavior," seemed to be mainly dependent on the level of emotional arousal of the cat, in addition to the rewarding nature of

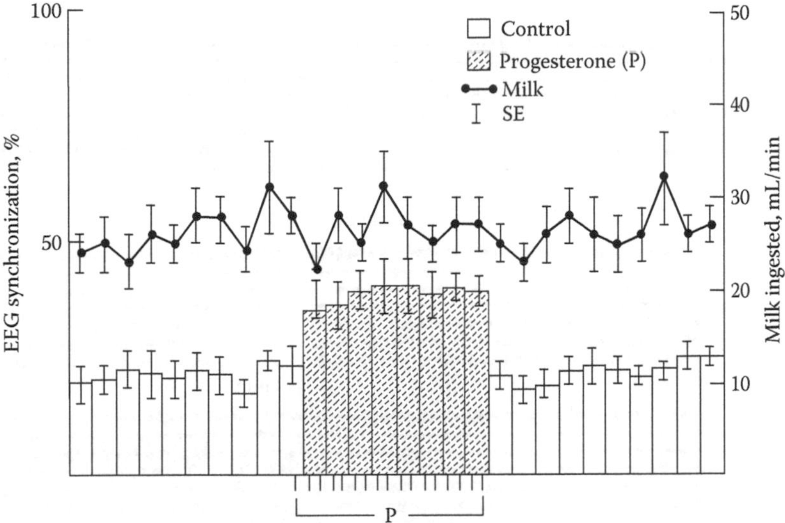

FIGURE 1.10 Average daily values of parieto-occipital synchronization obtained from five cats on successive days before, during, and after progesterone (P) treatment (10 mg/2/day). Values of EEG synchronization were expressed as percentage of the total time of milk drinking during which this electrical phenomenon was recorded. Note the increase of values of EEG synchronization during P treatment and its rapid return to control values at the end of P treatment. Volume of milk (ml/min) ingested daily by the cats was also plotted (Milk). (Modified from Cervantes, M. et al., *Psychoneuroendocrinology*, 4, 245–51, 1979.)

the stimulus. An inverse relationship was usually observed between the amount of time spent by each cat for its habituation to the experimental environment and the duration of parietal EEG synchronization during milk drinking. Thus, those cats showing persistent aversive, hyperactive, or evasive behavior (Brown 1968) inside the sound-proof recording chamber had the lowest values of milk drinking-elicited parietal EEG synchronization. However, in those cats, administration of a single effective dose of chlorpromazine, diazepam, or methocarbamol, as well as long-term progesterone administration, which induced quiescence and behavioral signs of adaptation to the environment, significantly increased the proportion of parietal EEG synchronization during milk drinking (Cervantes et al. 1975, 1979; Cervantes and Ruelas 1985b). Figure 1.10 illustrates the increase of average values of EEG synchronization from about 20% of the time of milk drinking in ovariectomized cats to almost 40% during 8 days of progesterone (10 mg twice daily) treatment (Cervantes et al. 1979).

1.5.2.3 Electrographic Signs of Relaxation Behavior in Lactating Mothers

As shown in Figure 1.11, the electrographic signs of "relaxation behavior" could also be induced in lactating cats by suckling of the nipples by the litter (Cervantes and Ruelas 1985a). Study of this behavior was extended to the human species, by recording the electroencephalographic characteristics of "relaxation behavior" during breast-feeding, which also elicited theta-like EEG synchronization over the parietal

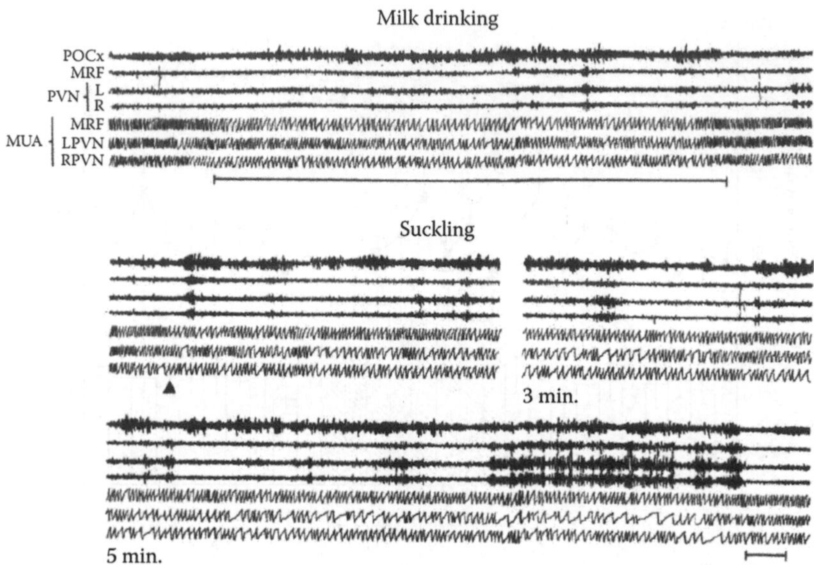

FIGURE 1.11 Low speed EEG recordings during milk drinking (mark below the upper tracings) and suckling (middle and lower tracings). Note the parieto-occipital EEG synchronization shortly after the beginning of suckling (arrowhead) and 3 and 5 min later, and its similarity with EEG synchronization induced by milk drinking. Inhibition of multiunit activity (MUA) from subcortical structures can also be seen. EEG changes induced by milk drinking or suckling can be distinguished readily from the EEG pattern of drowsiness (lower tracing, right). POCx, parieto-occipital cortex; MRF, mesencephalic reticular formation; PVN, paraventricular left and right hypothalamic nuclei. (Modified from Cervantes, M. and R. Ruelas, *Arch Invest Med (Méx)*, 16, 337–48, 1985a.)

cortex in lactating women (Cervantes et al. 1992). In fact, power spectrum analysis of 30 sec samples of EEG recordings, continuously taken from central–parietal and parietal–temporal derivations during breast feeding, showed a shift of the highest EEG power values recorded immediately before suckling at 10–15 Hz toward the highest EEG power at the 6–10 Hz frequency band, shortly after (within 30–60 sec) suckling started and lasting throughout the whole suckling period.

1.5.3 Effect of Progesterone on Brain Electrical Activity; Functional Role of Some Metabolic Pathways

Evaluation of the effect of progesterone and several of its metabolites on the cortical EEG and multiunit activity of some brain structures including the mesencephalic reticular formation, the dorsal hippocampus, and the ventromedial hypothalamus (Kubli-Garfias et al. 1976) was also a subject of interest within this research line. As illustrated in Figure 1.12, of the various progestins whose effects were evaluated, 5β,3α-pregnanolone (epipregnanolone) was the most potent in inhibiting neuronal

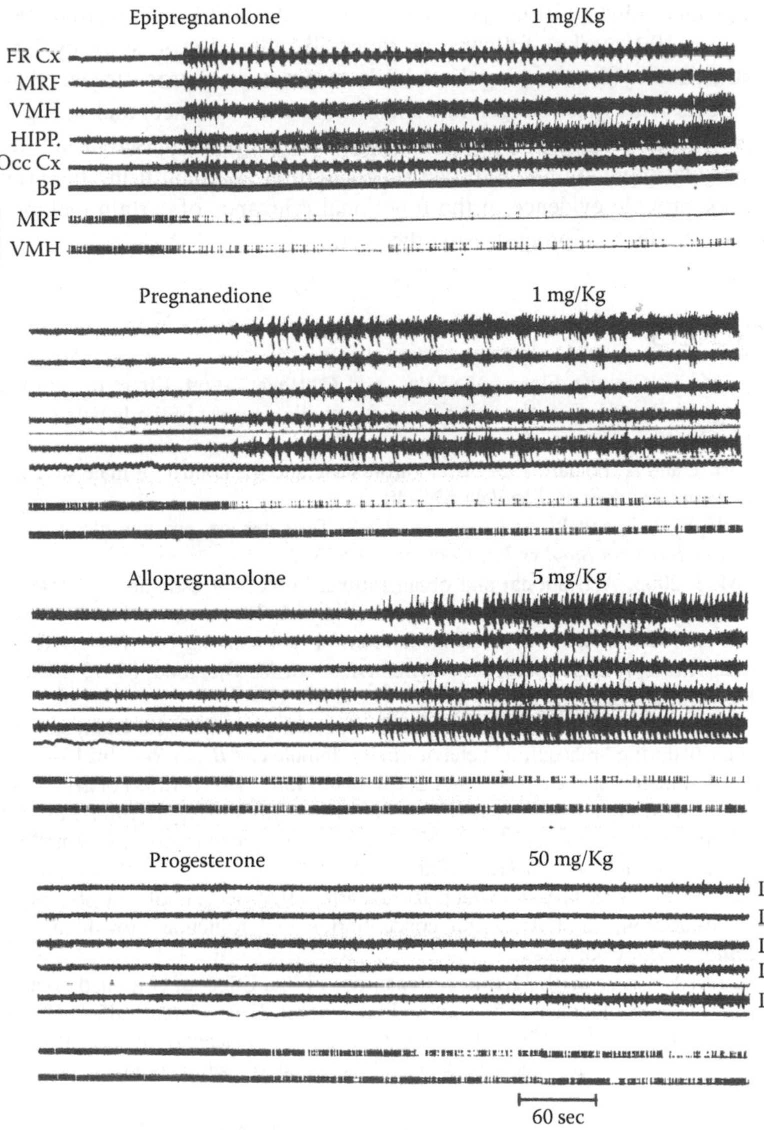

FIGURE 1.12 Typical changes in EEG and MUA induced by several progestins. Bursts of EEG synchronic activity appeared over a wide brain area, including frontal cortex (Fr Cx) and occipital cortex (OccCx), mesencephalic reticular formation (MRF), ventromedial nucleus of the hypothalamus (VMH) and hippocampus (H1PP). Mark in the time record indicates period of injection. Note inhibition of MUA in the MRF, which was more intense than that in the VMH, and differences in latencies and magnitude of the electrophysiological responses (EEG and MUA) among the different progestins. (Modified from Kubli-Garfias, C. et al., *Brain Res*, 114, 71–81, 1976.)

discharge and inducing EEG synchronization, with short latencies (16 to 48 sec) and at low doses (0.1 mg/kg), followed by other 5β-reduced progestins, by 5α-reduced progestins, and by those having the Δ4,3-keto structure as progesterone, that elicited these effects with much longer latencies (more than 300 sec) and at higher doses (50 mg/kg). These results, showing that ring A reduction increased the ability of progesterone and other Δ4,3-keto progestins to inhibit neuronal firing in certain brain structures, provide evidence of the functional relevance of certain metabolic pathways for the effects of progestins on the brain.

REFERENCES

Alcaraz, M., C. Guzmán-Flores, M. Salas, and C. Beyer. 1969. Effect of estrogen on the responsivity of hypothalamic and mesencephalic neurons in the female cat. *Brain Res* 15:439–46.

Balthazart, J. and A. Foidart. 1993. Brain aromatase and the control of male sexual behavior. *J Steroid Biochem Mol Biol* 44:521–40.

Baulieu, E. E., I. Lasnitzki, and P. Robel. 1968. Testosterone, prostate gland and hormone action. *Biochem Biophys Res Commun* 32:575–7.

Baum, M. J. 2003. Activational and organizational effects of estradiol on male behavioral neuroendocrine function. *Scand J Psychol* 44:213–20.

Beach, F. A. 1968. Factors involved in the control of mounting behavior by female mammals. In *Reproduction and Sexual Behavior*, ed. M. Diamond, 83–131. Bloomington, IL: Indiana University Press.

Beyer, C., J. Almanza, L. De la Torre, and C. Guzmán-Flores. 1971a. Brain stem multi-unit activity during "relaxation" behavior in the female cat. *Brain Res* 29:213–22.

Beyer, C., J. Almanza, L. De la Torre, and C. Guzmán-Flores. 1971b. Effect of genital stimulation on the brain stem multi-unit activity of anestrous and estrous cats. *Brain Res* 32:143–50.

Beyer, C., R. Anguiano, and F. Mena. 1961. Oxytocin release in response to stimulation of the cingulate gyrus. *Am J Physiol* 200:625–7.

Beyer, C., M. L. Cruz, V. L. Gay, and R. B. Jaffe. 1974. Effects of testosterone and dihydrotestosterone on FSH serum concentration and follicular growth in female rats. *Endocrinology* 95:722–7.

Beyer, C., M. L. Cruz, and J. Martinez-Manautou. 1970a. Effect of chlormadinone acetate on mammary development and lactation in the rabbit. *Endocrinology* 86:1172–4.

Beyer, C., M. L. Cruz, and N. Rivaud. 1969a. Persistence of sexual behavior in ovariectomized-adrenalectomized rabbits treated with cortisol. *Endocrinology* 85:790–3.

Beyer, C., R. B. Jaffe, and V. L. Gay. 1972. Testosterone metabolism in target tissues: effects of testosterone and dihydrotestosterone injection and hypothalamic implantation on serum LH in ovariectomized rats. *Endocrinology* 91:1372–5.

Beyer, C. and B. Komisaruk. 1971. Effects of diverse androgens on estrous behavior, lordosis reflex, and genital tract morphology in the rat. *Horm Behav* 2:217–25.

Beyer, C., K. Larsson, G. Pérez-Palacios, and G. Moralí. 1973. Androgen structure and male sexual behavior in the castrated rat. *Horm Behav* 4:99–103.

Beyer, C., P. McDonald, and N. Vidal. 1970b. Failure of 5-alpha-dihydrotestosterone to elicit estrous behavior in the ovariectomized rabbit. *Endocrinology* 86:939–41.

Beyer, C. and F. Mena. 1965a. Blockage of milk removal in the cat by periventricular diencephalic lesions. *Am J Physiol* 208:585–8.

Beyer, C. and F. Mena. 1965b. Effect of ovariectomy and barbiturate administration on lactation in the cat and the rabbit. *Bol Inst Estud Med Biol Univ Nac Auton Mex* 23:89–99.

Beyer, C. and F. Mena. 1965c. Induction of milk secretion in the rabbit by removal of the telencephalon. *Am J Physiol* 208:289–92.

Beyer, C. and F. Mena. 1970. Parturition and lactogenesis in rabbits with high spinal cord transection. *Endocrinology* 87:195–7.

Beyer, C., G. Morali, and M. L. Cruz. 1971c. Effect of 5-alpha-dihydrotestosterone on gonadotropin secretion and estrous behavior in the female Wistar rat. *Endocrinology* 89:1158–61.

Beyer, C., G. Moralí, F. Naftolin, K. Larsson, and G. Pérez-Palacios. 1976. Effect of some antiestrogens and aromatase inhibitors on androgen induced sexual behavior in castrated male rats. *Horm Behav* 7:353–63.

Beyer, C. and N. Rivaud. 1969. Sexual behavior in pregnant and lactating domestic rabbits. *Physiol Behav* 4:753–7.

Beyer, C. and N. Rivaud. 1973. Differential effect of testosterone and dihydrotestosterone on the sexual behavior of prepuberally castrated male rabbits. *Horm Behav* 4:175–80.

Beyer, C., N. Rivaud, and M. L. Cruz. 1970c. Initiation of sexual behavior in prepuberally ovariectomized rabbits. *Endocrinology* 86:171–4.

Beyer, C. and N. Vidal. 1971. Inhibitory action of MER-25 on androgen-induced oestrous behaviour in the ovariectomized rabbit. *J Endocrinol* 51:401–2.

Beyer, C., N. Vidal, and P. McDonald. 1969b. Interaction of gonadal steroids and their effect on sexual behaviour in the rabbit. *J Endocrinol* 45:407–13.

Beyer, C., N. Vidal, and A. Mijares. 1970d. Probable role of aromatization in the induction of estrus behavior by androgens in the ovariectomized rabbit. *Endocrinology* 87:1386–9.

Beyer, C., T. Yaschine, and F. Mena. 1964. Alterations in sexual behaviour induced by temporal lobe lesions in female rabbits. *Bol Inst Estud Med Biol (Méx)* 22:379–86.

Brooks, C. Mc. 1937. The role of the cerebral cortex and of various sense organs in the excitation and execution of mating activity in the rabbit. *Am J Physiol* 120:544–53.

Brown, B. B. 1968. Frequency and phase of hippocampal theta activity in the spontaneously behaving cat. *Electroenceph Clin Neurophysiol* 24:53–62.

Bruce, H. M. 1958. Suckling stimulus and lactation. *Proc R Soc Lond B Biol Sci* 149:421–3.

Bruchovsky, N. and J. D. Wilson. 1968. The conversion of testosterone to 5-alpha-androstan-17-beta-ol-3-one by rat prostate in vivo and in vitro. *J Biol Chem* 243:2012–21.

Cervantes, M., L. De La Torre, and C. Beyer. 1975. Analysis of various factors involved in EEG synchronization during milk drinking in the cat. *Brain Res* 91:89–98.

Cervantes, M. and R. Ruelas. 1985a. Effects of diazepam and methocarbamol on EEG signs of "relaxation behavior" induced by milk slurping in cats. *Arch Invest Med (Méx)* 16:337–48.

Cervantes, M. and R. Ruelas. 1985b. Electrophysiological evidence of "relaxation behavior" during suckling in lactating cats. *Arch Invest Med (Méx)* 16:323–36.

Cervantes, M., R. Ruelas, and V. Alcalá. 1992. EEG signs of "relaxation behavior" during breast-feeding in a nursing woman. *Arch Med Res* 23:123–8.

Cervantes, M., R. Ruelas, and C. Beyer. 1979. Progesterone facilitation of EEG synchronization in response to milk drinking in the cat. *Psychoneuroendocrinology* 4:245–51.

Cruz, M. L. 1994. Participación del metabolismo de esteroides en la diferenciación sexual cerebral. (Role of steroid hormone metabolism on brain sexual differentiation). PhD Dissertation. Universidad Autónoma de Tlaxcala, México.

Cruz, M. L. and C. Beyer. 1972. Effects of septal lesions on maternal behavior and lactation in the rabbit. *Physiol Behav* 9:361–5.

Dean, A., L. B. Smith, S. MacPherson, and R. M. Sharpe. 2012. The effect of dihydrotestosterone exposure during or prior to the masculinization programming window on reproductive development in male and female rats. *Int. J. Androl.* 35:330–9.

Dorfman, R. I. and F. Ungar. 1965. *Metabolism of Steroid Hormones.* New York: Academic Press.

Garcia, M. and H. Rochefort. 1979. Evidence and characterization of the binding of two 3H-labeled androgens to the estrogen receptor. *Endocrinology* 104:1797–804.

García-Dalmán, C. and G. González-Mariscal. 2012. Major role of suckling stimulation for inhibition of estrous behaviors in lactating rabbits: acute and chronic effects. *Horm Behav* 61:108–13.

Goy, R. W. and B. S. McEwen. 1980. *Sexual Differentiation of the Brain*. Cambridge, MA: MIT Press.

Greep, R. O. 1961. Physiology of the anterior hypophysis in relation to reproduction. In *Sex and Internal Secretions*, Vol. I, ed. W.C. Young, p. 269. Baltimore, MD: Williams and Wilkins.

Handa, R. J., M. J. Weiser, and D. G. Zuloaga. 2009. A role for the androgen metabolite, 5α-androstane-3β,17β-diol, in modulating oestrogen receptor β-mediated regulation of hormonal stress reactivity. *J Neuroendocrinol* 21:351–8.

Harris, G. W. and R. P. Michael. 1964. The activation of sexual behaviour by hypothalamic implants of oestrogen. *J Physiol* 171:275–301.

Hilliard, J., D. Archibald, and C. H. Sawyer. 1963. Gonadotropic activation and preovulatory synthesis and release of progestin in the rabbit. *Endocrinology* 72:59–66.

Jaffe, R. B. 1969. Testosterone metabolism in target tissues. Hypothalamic and pituitary tissues of the adult rat and human fetus, and the immature rat epiphysis. *Steroids* 14:483–98.

Klüver, H. and P. C. Bucy. 1939. Preliminary analysis of functions of the temporal lobes in monkeys. *Arch Neurol Psychiat (Chic)* 42:979–1000.

Komisaruk, B. R. 1967. Effects of local brain implants of progesterone on reproductive behavior in ring doves. *J Comp Physiol Psychol* 64:219–224.

Komisaruk, B. R. and C. Beyer. 1972. Differential antagonism, by MER-25, of behavioral and morphological effects of estradiol benzoate in rats. *Horm Behav* 3:63–70.

Kubli-Garfias, C., M. Cervantes, and C. Beyer. 1976. Changes in multiunit activity and EEG induced by the administration of natural progestins to flaxedilimmovilized cats. *Brain Res* 114:71–81.

Larsson, K., P. Södersten, and C. Beyer. 1973. Sexual behavior in male rats treated with estrogen in combination with dihydrotestosterone. *Horm Behav* 4:289–99.

Lisk, R. D. 1962. Diencephalic placement of estradiol and sexual receptivity in the female rat. *Am J Physiol* 203:493–6.

Lund, T. D., L. R. Hinds, and R. J. Handa. 2006. The androgen 5alpha-dihydrotestosterone and its metabolite 5alpha-androstan-3beta, 17beta-diol inhibit the hypothalamo-pituitary-adrenal response to stress by acting through estrogen receptor beta-expressing neurons in the hypothalamus. *J Neurosci* 26:1448–56.

Luttge, W. G. 1979. Endocrine control of mammalian male sexual behavior: An analysis of the potential role of testosterone metabolites. In *Endocrine Control of Sexual Behavior*, ed. C. Beyer, 341–63. New York: Raven Press.

McDonald, P. G., C. Beyer, F. Newton et al. 1970a. Failure of 5α-dihydrotestosterone to initiate sexual behavior in the castrated male rat. *Nature (Lond)* 227:964–5.

McDonald, P. G., N. Vidal, and C. Beyer. 1970b. Sexual behavior in the ovariectomized rabbit after treatment with different amounts of gonadal hormones. *Horm Behav* 1:161–72.

Mena, F., R. Anguiano, and C. Beyer. 1961. Relase of oxytocin by stimulation of the caudal part of the hypothalamus. *Bol Inst Estud Med Biol (Mex)* 19:119–24.

Mena, F. and C. Beyer. 1963. Effect of high spinal section on established lactation in the rabbit. *Am J Physiol* 205:313–16.

Moralí, G., K. Larsson, and C. Beyer. 1977. Inhibition of testosterone-induced sexual behavior in the castrated male rat by aromatase blockers. *Horm Behav* 9:203–13.

Moralí, G., M. V. Oropeza, A. E. Lemus, and G. Pérez-Palacios. 1994. Mechanisms regulating male sexual behavior in the rat: Role of 3α- and 3β-androstanediols. *Biol Reprod* 51:562–71.

Naftolin, F., K. J. Ryan, I. J. Davies et al. 1975. The formation of estrogens by central neuro-endocrine tissues. *Rec Prog Horm Res* 31:295–319.

Palka, Y. S. and C. H. Sawyer. 1966. Induction of estrous behavior in rabbits by hypothalamic implants of testosterone. *Am J Physiol* 211:225–8.

Pérez-Palacios, G., E. Castañeda, F. Gómez-Pérez et al. 1970. In vitro metabolism of androgens in dog hypothalamus, pituitary, and limbic system. *Biol Reprod* 3:205–13.

Pérez-Palacios, G., A. E. Pérez, M. L. Cruz, and C. Beyer. 1973. Comparative uptake of (3 H) androgens by the brain and the pituitary of castrated male rats. *Biol Reprod* 8:395–9.

Roos, J., M. Roos, C. Schaeffer, and C. Aron. 1988. Sexual differences in the development of accessory olfactory bulbs in the rat. *J Comp Neurol* 270:121–31.

Roth, S. R., M. B. Sterman, and C. D. Clemente. 1967. Comparison of EEG correlates of reinforcement, internal inhibition and sleep. *Electroenceph Clin Neurophysiol* 23:509–20.

Shealey, C. N. and T. N. Peele. 1957. Studies on amygdaloid nucleus of cat. *J Neurophysiol* 20:125–39.

Shipley, R. A. 1962. The gonads and aging. *J Am Geriatr Soc* 10:26–34.

Södersten, P. 1974. Effects of an estrogen antagonist, MER-25, on mounting behavior and lordosis behavior of the female rat. *Horm Behav* 5:111–21.

Södersten, P. 1979. Role of estrogen in the display and development of sexual behavior in male rats. In *Endocrine Control of Sexual Behavior*, ed. C. Beyer, 305–15. New York: Raven Press.

Swerdloff, R. S., P. C. Walsh, and W. D. Odell. 1972. Control of LH and FSH secretion in the male: Evidence that aromatization of androgens to estradiol is not required for inhibition of gonadotropin secretion. *Steroids* 20:13–22.

Thieulant, M. L., S. Samperez, and P. Jouan. 1981. Evidence for 5 alpha-androstane-3 beta, 17 beta-diol binding to the estrogen receptor in the cytosol from male rat pituitary. *Endocrinology* 108:1552–60.

Tindall, J. S., C. Beyer, and C. Sawyer. 1963. Milk ejection reflex and maintenance of lactation in the rabbit. *Endocrinology* 72:720–4.

Uyldert, I. E. 1946. Mammary development and lactation in rats. *Acta Brev Neerl Physiol* 14:86–94.

Yaschine, T., F. Mena, and C. Beyer. 1967. Gonadal hormones and mounting behavior in the female rabbit. *Am J Physiol* 213:867–72.

Young, W. C. 1961. The hormones and mating behavior. In *Sex and Internal Secretions*, Vol. II, ed. W.C. Young, p. 1196. Baltimore, MD: Williams and Wilkins.

2 Neuroendocrine and Behavioral Role of the Medial Preoptic Area in Rabbits

Recollections of Collaboration with Carlos Beyer

Anders Ågmo and Knut Larsson

CONTENTS

2.1 The Rabbit as Subject in Studies of Sexual Behavior....33
2.2 Neural Control of Male Sexual Behavior: The Preoptic Area....35
2.3 A Note on Neural Control of Lordosis in the Female Rabbit....38
2.4 Endocrine Role of the Female Rabbit Preoptic Area....39
2.5 Other Functions of the Preoptic Area....41
2.6 Conclusion....41
References....41

2.1 THE RABBIT AS SUBJECT IN STUDIES OF SEXUAL BEHAVIOR

The overwhelming majority of experimental studies of sexual behavior and its neuroendocrine control have been performed in rodents, the most popular species being the rat, hamster, and guinea pig. Many years ago Frank Beach expressed concern about the concentration of scientific effort in these few species (Beach 1950), which he poetically called the ramstergig (*Iurasequens sexualis*), a species of the order *Myostrichomorpha* (Beach 1971). Even though the guinea pig has been little used during the past few decades, rats and hamsters remain in demand. The most evident change in the choice of research animal is probably the increasing popularity of the mouse. Nevertheless, rodents in general are still most widely used, and it would appear that Beach's much cited 1950 paper did not provoke any dramatic shift in the choice of experimental subjects.

Among nonrodent mammals, only primates have had a long, continuous, but modest presence in laboratory studies of sexual behavior. There are, of course, a couple of studies of more exotic mammalian species, but most of them are descriptive rather than experimental. The heavy concentration on rodents certainly has many explanations, but it might be worthwhile to mention that the earliest experiments on the mechanisms controlling sexual behavior were made on frogs (Spallanzani 1784) and the discovery of the role of gonadal secretions were made in the fowl (Berthold 1849). Perhaps the first to employ rats in studies of the endocrine control of sexual behavior was Eugen Steinach (1894, 1910). The earliest studies of the central nervous control of that behavior were made in dogs (von Bechterew 1911) and roosters (Ceni 1917). Spallanzani's (1784) much earlier studies of the effects of decapitation of frogs were not specifically concerned with the role of central nervous processes and can be ignored in this context. Even though the studies in dogs and roosters employed such rudimentary procedures that any meaningful conclusion was impossible, the precarious data were interpreted as showing the existence of a cortical sexual center. The first useful experimental data concerning the role of the central nervous system were obtained in rabbits (Stone 1925a, 1925b, 1926). A section of the olfactory bulbs, as well as extensive cortical lesions, failed to modify male sexual behavior, thereby disproving the proposal of a cortical sexual center.

Despite Stone's elegant experiments, there was little interest in pursuing studies of rabbit sexual behavior. In Figure 2.1, the number of publications on rat and rabbit sexual behavior from 1941 to 2010 is shown. As evident from the figure, the rabbit was almost completely ignored until the 1960s. In fact, it was not until Carlos Beyer and his associates started to publish a long series of papers on several aspects of the neuroendocrine control of male and female sexual behavior that the rabbit again was given the opportunity to contribute to this field. It is not likely that this lagomorph was used in order to expand the comparative aspects in studies of sexual behavior in the spirit of Frank Beach. Rather, it was because Carlos Beyer had used rabbits in his many studies of lactation, in collaboration with Flavio Mena, at the Instituto de Investigaciones Biomedicas in Mexico City and in the laboratory of Charles H.

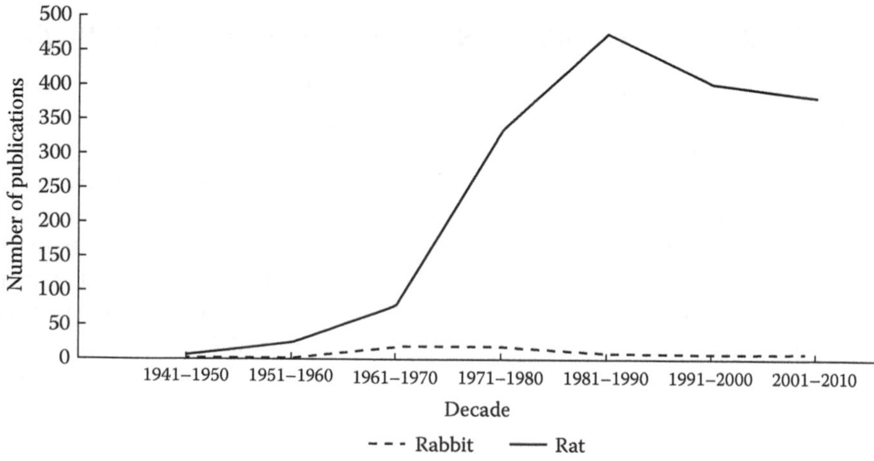

FIGURE 2.1 The total number of publications per decade obtained from the Web of Science by crossing the search terms rabbit and sexual behavior or rat and sexual behavior.

Sawyer at the University of California, Los Angeles. It must also be observed that Carlos Beyer's contribution to the neuroendocrinology of sexual and maternal behaviors was not limited to studies in rabbits. A substantial proportion of the publications from the Beyer laboratories were based on data from other species, mainly the rat.

2.2 NEURAL CONTROL OF MALE SEXUAL BEHAVIOR: THE PREOPTIC AREA

After the pioneering studies by Stone, scattered efforts were made to elucidate the importance of the cerebral cortex for male sexual behavior (e.g., Beach 1940). An interest in the role of subcortical structures also emerged (e.g., Soulairac and Soulairac 1956a; Soulairac 1963). An often-ignored but extremely important paper appeared in 1954 (Hillarp et al. 1954). In the course of a study of the hypothalamic—pituitary system, small electrolytic lesions were placed in an area extending from the mammillary bodies to the preoptic area in male and female rats. It was found that animals lesioned in the preoptic area started to display intense male sexual behavior within 20 to 30 min of awakening from ether anesthesia. Lesions in other areas did not have this effect. Moreover, the effect was similar in males and females, and in intact and gonadectomized animals. Although the mechanism behind the stimulation of copulatory behavior was unknown, the authors suggested that nervous structures in the vicinity of the lesion might have been stimulated. It was concluded that "The fact that such a specific behavior can be produced from a very limited brain region gives strong support to the assumption that either a centre is situated here for the integration of the nervous mechanisms for the male mating behavior, or that there are in this region direct connections with such a centre" (Hillarp et al. 1954, pp. 224–225). A few years later, it was shown that the electrolytic lesion led to the deposit of ferric and/or ferrous ions, exciting adjacent neurons (Everett and Radford 1961), exactly as suggested by Hillarp and colleagues. Their observation concerning the stimulatory action of small electrolytic lesions in the medial preoptic area was replicated in a more detailed study some time ago (Paredes and Ågmo 1992).

The importance of the preoptic area was largely overlooked during the decade following the Hillarp et al. (1954) study. The reports from André Soulairac's laboratory in Paris (Soulairac and Soulairac 1956a, 1956b) were essentially ignored, probably because they were published in French and in journals little read outside of France. It was not until the publication of a series of papers by Knut Larsson and Lennart Heimer (Heimer and Larsson 1966; Larsson and Heimer 1964) that it started to become generally accepted that the medial preoptic area is crucial for male rat sexual behavior. In the following years, all kinds of evidence accumulated in support of the notion that the medial preoptic area is an essential structure for male rat sexual behavior (see Paredes 2003; Paredes and Baum 1997 for reviews). However, in the early 1970s, very little was known about its role in other species. It was not surprising, then, that Carlos Beyer, at the instigation of one of the present authors (KL), planned a study of the effects of preoptic lesions on male rabbit sexual behavior. By some strange coincidence, the other of the present authors (AÅ) happened to be charged with that study in the summer of 1976. Shortly before, Carlos Beyer had moved the main part of his laboratory from the Department of Scientific Research at the

Mexican Institute of Social Security to the newly founded Autonomous Metropolitan University, Iztapalapa campus. The lab was located in a brand-new building and full of new equipment, like a radiofrequency lesion generator.

After determining the appropriate parameters to achieve complete destruction of the medial part of the preoptic area, a group of sexually experienced, intact male rabbits was lesioned and another group subjected to the same procedure except that the radiofrequency generator was not turned on. All subjects were tested for sexual behavior with receptive females every 10 days for at least 2 months. As can be seen in Figures 2.2 and 2.3, the lesion caused a drastic reduction of sexual behavior that seemed to be irreversible. At least there was no sign of recovery during the observation period. The results obtained in these rabbits were very similar to what had been previously reported in rats.

Several unexpected incidents occurred between the end of behavioral observations and the histological analysis of exact location and size of the lesions. In fact, complete histology was obtained from only a few of the rabbits in the lesioned group.

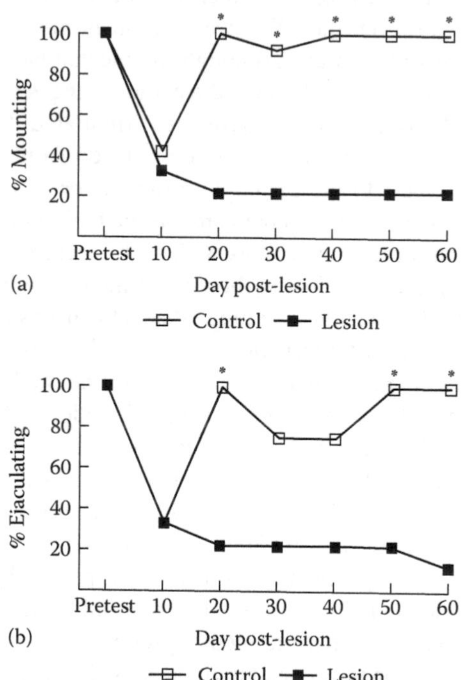

FIGURE 2.2 The proportion of rabbits performing at least one mount (a) or one ejaculation (b) in 10-min tests with a sexually receptive female performed shortly before radiofrequency lesion of the medial preoptic area or a sham lesion and every 10 days after lesion for 60 days. There were 9 animals in the lesion group and 12 animals in the sham group. The reduced sexual behavior in both groups at Day 10 post-lesion can be attributed to incomplete recovery from the surgical procedure. At later tests, the sham group returned to prelesion levels of behavior, whereas the lesioned group remained at a very low level. *, significantly different from sham as determined by the Fisher exact probability test, $p < 0.05$.

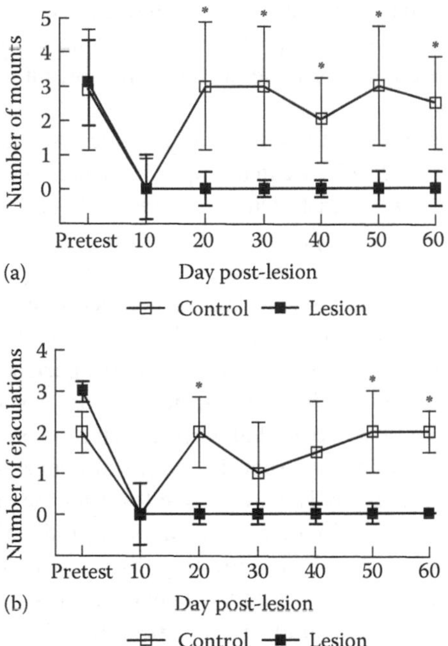

FIGURE 2.3 Median ± semi-interquartile range of the number of mounts (a) or ejaculations (b) performed by male rabbits with or without lesion in the medial preoptic area. *, different from sham lesion according to the Mann–Whitney *U* test, $p < 0.05$. For further experimental details, see Figure 2.2.

They all had massive destruction of the medial preoptic area, as intended, with damage extending to parts of the lateral preoptic area. Nevertheless, because of the incomplete histology, it was decided not to publish these data, but to wait for a future opportunity to repeat the experiment. However, studies from other species showing that the preoptic area was important for male copulatory behavior rapidly accumulated, for example, frogs (Schmidt 1968), chicks (Gardner and Fisher 1968), cats (Hart et al. 1973), dogs (Hart 1974), and rhesus monkeys (Slimp et al. 1978). Because of that, it was not considered worthwhile to invest time and effort in the repetition of a study adding very little new information, and the replication of the rabbit experiment never materialized. Nevertheless, even the somewhat preliminary data obtained makes it possible to add the rabbit to the long list of species in which the preoptic area is crucial for the display of male copulatory behavior.

Although lesion studies can provide evidence for the need of an intact preoptic area for the display of masculine copulatory behavior, they do not offer any clear-cut evidence as to the site of action of the gonadal hormone(s) essential to that behavior. In male rats, several studies have reported complete restoration of copulatory behavior after implants of testosterone in the medial preoptic area (e.g., Smith et al. 1977). This kind of observation has made it possible to suggest that hormone action within the preoptic area is necessary and sufficient for male sexual behavior. However, the behavior in animals carrying preoptic testosterone implants is often of an intensity inferior to that found in intact animals or in animals given testosterone systemically. This fact is often interpreted as suggesting that brain areas outside of the preoptic area participate in the control of male sexual behavior, although they are neither necessary nor sufficient.

Strangely enough, a study of the effects of preoptic testosterone implants in rabbits was not reported until 2008. As can be expected, it stemmed from Carlos Beyer's laboratory (Melo et al. 2008). Castrated males were bilaterally implanted with cannulae filled with testosterone propionate. Three weeks after implantation, the males were subjected to a 5-min test for sexual behavior in the presence of a receptive female. None of the males copulated. The authors suggested that the unexpected absence of sexual activity could be attributed to the lack of sexual experience. In fact, the 5-min test may have been too short for the inexperienced subjects to initiate mounting. Another possibility is that the testosterone dose was insufficient. For some reason, the 22-gauge cannulae were filled with testosterone propionate diluted 1:5 in cholesterol. This may have led to a subeffective local concentration or to a limited diffusion around the cannula tip. However, such testosterone propionate implants in the medial preoptic area did stimulate scent marking, indicating a clear biological activity of the implanted steroid, and suggesting that scent marking has a lower hormonal response threshold than copulatory behavior. It must be considered unlikely that the male rabbit is an exception to what has been found in all vertebrates studied thus far. The negative results in the Melo et al. (2008) study are, therefore, difficult to explain.

2.3 A NOTE ON NEURAL CONTROL OF LORDOSIS IN THE FEMALE RABBIT

In the early part of his career, Carlos Beyer, together with his long-time student, colleague, and friend, Flavio Mena, performed a most interesting study of the effects of temporal lobe lesions in female rabbits (Beyer et al. 1964). Ovariectomized, estradiol-treated rabbits displayed intense mounting after ablation of the basal part of the temporal lobes, including most of the entorhinal cortex and parts of the amygdala. The lesioned females even mounted males of other species, that is, cats. After withdrawal of estrogen, mounting disappeared. Control subjects did not mount at all. The effects of the ablation of the temporal lobes in rabbits were, then, partly similar to what has been observed in the famous Klüver–Bucy syndrome (e.g., Green et al. 1957). For some reason, the Beyer et al. (1964) paper did not attract much attention, not even from Carlos Beyer himself, because he never pursued this line of inquiry. Instead he looked into the role of the temporal lobes in milk ejection (Mena and Beyer 1968).

In female rats, an intact ventromedial nucleus of the hypothalamus is necessary for the expression of sexual behavior in response to estrogen treatment (reviewed in Blaustein 2009). In the 1960s, it was shown that estradiol implants into the ventromedial hypothalamus–premamillary region also activated lordosis in ovariectomized rabbits (Palka and Sawyer 1966). Many years later, Carlos Beyer's group replicated and extended this observation (Melo et al. 2008). In addition to showing that ventromedial hypothalamic estradiol implants activate lordosis, they also reported that these implants stimulate chinning, a behavior consisting of rubbing the chin on solid objects, which deposits secretions from the submandibular glands. This behavior is estrogen dependent (Hudson et al. 1990) and supposedly is performed to attract males (González-Mariscal et al. 1990). Chinning might, then, be an expression of sexual motivation as is female rat proceptive behavior. It is noteworthy that in female rats and mice, activation of ventromedial estrogen receptors is essential to both lordosis and expression of the motivational components of sexual behavior (Musatov et al. 2006; Spiteri et al. 2010). Medial preoptic implants of estradiol facilitated chinning, and to a minor degree lordosis, in the Melo et al. (2008) study. In female rats, the preoptic area is thought to be involved in some motivational aspects of female sexual behavior (see Spiteri et al. 2012 for a discussion). Taken together, these data suggest that the neural control of female sexual behavior in lagomorphs is similar to what has been reported in rodents. However, there might be some subtle differences, probably related to the fact that rabbits and other lagomorphs are induced ("reflex") ovulators, whereas rodents ovulate spontaneously. This issue has been brilliantly discussed elsewhere (Beyer et al. 2007).

2.4 ENDOCRINE ROLE OF THE FEMALE RABBIT PREOPTIC AREA

One of the characteristics of lagomorphs is reflex ovulation, normally initiated by copulation. Within minutes of the male's ejaculation, the female releases gonadotropin-releasing hormone (GnRH), and shortly thereafter, serum LH concentration increases and remains elevated for several hours (Yang et al. 1996). Ovulation usually occurs 10–12 h after copulation (Ramirez and Beyer 1988). GnRH is mainly released in the arcuate nucleus-median eminence area, but mating-induced release has also been reported in the anterior hypothalamus (Kaynard et al. 1990) and activation of GnRH neurons extends rostrally into the preoptic area (Caba et al. 2000b). Recent data show that mating causes enhanced c-fos expression in the preoptic area in estrous but not in lactating females (González-Mariscal et al. 2015). Since the latter do not ovulate in response to mating, this observation indirectly supports the importance of the preoptic area in mating-induced ovulation.

The mechanisms controlling the enhanced GnRH release are only partly known, but a tentative picture has started to emerge. Findings from a microdialysis study with the probe located in the area of the arcuate nucleus-median eminence showed that noradrenaline release increased shortly before the surge of GnRH following copulation (Yang et al. 1996). An adrenergic α_1 antagonist administered into the 3rd ventricle blocked mating-induced release of both GnRH and LH, whereas the drug blocked only GnRH release when infused into the arcuate nucleus-median eminence. Serum LH concentrations were not affected in the latter case (Yang et al. 1998), suggesting that noradrenergic actions outside the arcuate nucleus-median eminence somehow contribute to enhanced LH release. Moreover, another α_1 antagonist, dibenamine, blocked ovulation in rabbits (Sawyer et al. 1947). Thus, noradrenaline release is evidently necessary for mating-induced gonadotropin release and the ensuing ovulation.

An important question concerns the origin of the noradrenergic neurons promoting GnRH release after copulation. It has been reported that the immediate early gene c-fos is activated by mating in female (and male) rabbits (Reyna-Neyra et al. 2000), and an elegant study showed that c-fos mRNA is enhanced in cells immunolabeled for dopamine-β-hydroxylase in the noradrenergic cell groups A1 (lateral tegmentum of the medulla oblongata) and A2 (the nucleus of the solitary tract) but not in A6 (locus ceruleus) (Caba et al. 2000a). This observation is extremely interesting since the solitary nucleus responds to vaginocervical stimulation, relayed through the vagus, hypogastric, and pelvic nerves (Guevara-Guzmán et al. 2001), and the vagus nerves in women, based on functional MRI (Komisaruk et al. 2004). Somatosensory information from the genitals may also reach the medullary lateral tegmentum and activate the noradrenergic A1 cells. In rabbits, it is not known whether the noradrenergic neurons originating in A1 or A2 project to GnRH neurons in the preoptic area/hypothalamus. In rats, there are data showing that estradiol activates noradrenergic A2 neurons, leading to increased noradrenaline release in the preoptic area, contributing to the LH surge (Szawka et al. 2013). A similar arrangement in rabbits with regard to the mating-induced GnRH release is not unlikely.

In addition to noradrenaline as mediator of the mating-induced GnRH release, there is evidence suggesting that neuropeptide Y (NPY) may participate in this process. In rabbits, infusion of NPY into the mediobasal hypothalamus increases release of GnRH, and this effect is blocked by adrenergic α_1 antagonists, whereas α_2 antagonists are ineffective (Berria et al. 1991). This observation coincides nicely with the observations mentioned above that α_1 antagonists block coitus-induced release of GnRH and LH as well as ovulation in rabbits. Moreover, NPY and dopamine-β-hydroxylase are co-localized in brain areas activated by copulation in female rabbits, notably the lateral medullary tegmentum (A1) and the nucleus of the solitary tract (A2) (Pau et al. 1997). This offers an anatomical substrate for the suggested involvement of both NPY and noradrenaline in mating-induced ovulation.

Whether the female rabbit's preoptic area is necessary for mating-induced ovulation is not known, for neither lesion studies nor the effects of local drug administration have been reported. There is an abstract from the annual meeting of the American Physiological Society held in Memphis, TN, in 1937, in which it is mentioned that electrical stimulation of an area above and anterior to the optic chiasm induced ovulation (Haterius and Derbyshire 1937). That area might correspond to the ventral part of the preoptic area. The abstract was not followed by a full-length paper, making it difficult to judge the importance of these data. Sawyer (1959) reported that lateral preoptic lesions failed to block mating-induced ovulation, but he could not determine the effects of medial lesions because of high mortality among the lesioned subjects. Nevertheless, and as summarized in the preceding paragraphs, there is substantial indirect evidence for the participation of preoptic GnRH neurons in mating-induced ovulation. It is not clear, however, whether these neurons are indispensable. However, González-Mariscal et al. (2015) recently reported an approximately eightfold increase in the number of c-fos immunoreactive cells following mating in the preoptic area of estrous rabbits. Further studies are clearly needed to reveal the complex neuroendocrine cascade involved in mating-induced ovulation in rabbits.

2.5 OTHER FUNCTIONS OF THE PREOPTIC AREA

The Beyer laboratories have been engaged in studies of the role of the rabbit preoptic area in maternal behavior (see, e.g., González-Mariscal et al. 2005; Olazábal et al. 2013). These elegant studies will be reviewed in other chapters and are not, therefore, mentioned here.

2.6 CONCLUSION

The rabbit has been marginal as an experimental subject in neuroendocrine research and in research on sexual behavior. It has attracted some attention, however, as the female is a reflex ovulator. Most of the studies of rabbit sexual behavior come from Carlos Beyer and his colleagues, and so do many of those on the induction of ovulation. Both areas of inquiry have provided valuable knowledge, not only concerning the unique characteristics of rabbits, but also concerning basic neuroendocrine mechanisms.

We will not make a general statement concerning the vast contributions of Carlos Beyer to the understanding of the neuroendocrinology of several basic behavior patterns; this has been done by others. Instead we emphasize how Carlos used his extensive knowledge of this field and his sensitive intuition to generate ideas for himself and for those surrounding him, and how elegantly he exploited the possibilities in his local environment and how he avoided most of the related inconveniences. For example, a generous supply of rabbits, combined with available space and excellent caretakers, made it possible to perform studies requiring large number of animals at reasonable cost. We believe that many young scientists can develop successful scientific careers by cultivating Carlos Beyer's capacity of turning obstacles into opportunities.

REFERENCES

Beach, F.A. 1940. Effects of cortical lesion upon the copulatory behavior of male rats. *Journal of Comparative Psychology* 29:193–244.

Beach, F.A. 1950. The snark was a boojum. *American Psychologist* 5:115–24.

Beach, F.A. 1971. Hormonal factors controlling the differentiation, development and display of copulatory behavior in the ramstergig and related species. In *The Biopsychology of Development,* eds. E. Tobach, L.R. Aronson, and E. Shaw, 249–96. New York: Academic Press.

Berria, M., Pau, K.Y.F., and Spies, H.G. 1991. Evidence for a_1-adrenergic involvement in neuropeptide Y-stimulated GnRH release in female rabbits. *Neuroendocrinology* 53:480–6.

Berthold, A.A. 1849. Transplantation der Hoden. *Archiv für Anatomie, Physiologie und Wissenschaftliche Medizin* 16:42–6.

Beyer, C., Hoffman, K.L., and González-Flores, O. 2007. Neuroendocrine regulation of estrous behavior in the rabbit: Similarities and differences with the rat. *Hormones and Behavior* 52:2–11.

Beyer, C., Yaschine, T., and Mena, F. 1964. Alterations in sexual behaviour induced by temporal lobe lesions in female rabbits. *Boletín del Instituto de Estudios Médicos y Biológicos, Universidad Nacional Autonoma de Mexico* 22:379–86.

Blaustein, J.D. 2009. Feminine reproductive behavior and physiology in rodents: Integration of hormonal, behavioral, and environmental influences. In *Hormones, Brain and Behavior (2nd Ed.), Part 1*. eds. D.W. Pfaff, A.P. Arnold, S.E. Fahrbach, A.M. Etgen, and R.T. Rubin, 67–108. San Diego, CA: Academic Press.

Caba, M., Bao, J.Z., Pau, K.Y.F., and Spies, H.G. 2000a. Molecular activation of noradrenergic neurons in the rabbit brainstem after coitus. *Molecular Brain Research* 77:222–31.

Caba, M., Pau, K.Y.F., Beyer, C., Gonzalez, A., Silver, R., and Spies, H.G. 2000b. Coitus-induced activation of c-fos and gonadotropin-releasing hormone in hypothalamic neurons of female rabbits. *Molecular Brain Research* 78:69–79.

Ceni, C. 1917. The influence of cerebral excitation on the male sexual glands. Experimental and anatomico-pathologic researches on the cortical genetic centres. *Alienist and Neurologist* 38:359–92.

Everett, J.W. and Radford, H.M. 1961. Irritative deposits from stainless steel electodes in preoptic rat brain causing release of pituitary gonadotropin. *Proceedings of the Society for Experimental Biology and Medicine* 108:604–9.

Gardner, J.E. and Fisher, A.E. 1968. Induction of mating in male chicks following preoptic implantation of androgen. *Physiology and Behavior* 3:709–12.

González-Mariscal, G., Chirino, R., Rosenblatt, J.S., and Beyer, C. 2005. Forebrain implants of estradiol stimulate maternal nest-building in ovariectomized rabbits. *Hormones and Behavior* 47:272–79.

González-Mariscal, G., García Dalmán, C., and Jiménez, A. 2015. Biostimulation and nursing modify mating-induced c-FOS immunoreactivity in the female rabbit forebrain. *Brain Research* 1608:66–74.

González-Mariscal, G., Melo, A.I., Zavala, A., and Beyer, C. 1990. Variations in chin-marking behavior of New Zealand female rabbits throughout the whole reproductive cycle. *Physiology and Behavior* 48:361–65.

Green, J.D., Clemente, C.D., and De Groot, J. 1957. Rhinencephalic lesions and behavior in cats. An analysis of the Klüver–Bucy syndrome with particular reference to normal and abnormal sexual behavior. *Journal of Comparative Neurology* 108:505–45.

Guevara-Guzmán, R., Buzo, E., Larrazolo, A., de la Riva, C., Da Costa, A.P., and Kendrick, K.M. 2001. Vaginocervical stimulation-induced release of classical neurotransmitters and nitric oxide in the nucleus of the solitary tract varies as a function of the oestrus cycle. *Brain Research* 898:303–13.

Hart, B.L. 1974. Medial preoptic-anterior hypothalamic area and sociosexual behavior of male dogs: A comparative neuropsychological analysis. *Journal of Comparative and Physiological Psychology* 86:328–49.

Hart, B.L., Haugen, C.M., and Peterson, D.M. 1973. Effects of medial preoptic-anterior hypothalamic lesions on mating behavior of male cats. *Brain Research* 54:177–91.

Haterius, H.O. and Derbyshire, Jr., A.J. 1937. Ovulation in the rabbit following upon stimulation of the hypothalamus. *American Journal of Physiology* 119:329–30.

Heimer, L. and Larsson, K. 1966. Impairment of mating behavior in male rats following lesions in the preoptic-anterior hypothalamic continuum. *Brain Research* 3:248–63.

Hillarp, N.Å., Olivecrona, H., and Silfverskiöld, W. 1954. Evidence for the participation of the preoptic area in male mating behavior. *Experientia* 10:224–5.

Hudson, R., González-Mariscal, G., and Beyer, C. 1990. Chin marking behavior, sexual receptivity, and pheromone emission in steroid-treated, ovariectomized rabbits. *Hormones and Behavior* 24:1–13.

Kaynard, A.H., Pau, K.Y.F., Hess, D.L., and Spies, H.G. 1990. Gonadotropin-releasing hormone and norepinephrine release from the rabbit bediobasal and anterior hypothalamus during the mating-induced luteinizing hormone surge. *Endocrinology* 127:1176–85.

Komisaruk, B.R., Whipple, B., Crawford, A., Grimes, S., Liu, W.-C., Kalnin, A., and Mosier, K. 2004. Brain activation during vaginocervical self-stimulation and orgasm in women with complete spinal cord injury: fMRI evidence of mediation by the Vagus nerves. *Brain Research* 1024:77–88.

Larsson, K. and Heimer, L. 1964. Mating behaviour of male rats after lesions in the preoptic area. *Nature* 202:413–14.

Melo, A.I., Chirino, R., Jiménez, A., Cuamatzi, E., Beyer, C., and González-Mariscal, G. 2008. Effect of forebrain implants of testosterone or estradiol on scent-marking and sexual behavior in male and female rabbits. *Hormones and Behavior* 54:676–83.

Mena, F. and Beyer, C. 1968. Induction of milk secretion in the rabbit by lesions in the temporal lobe. *Endocrinology* 83:618–20.

Musatov, S., Chen, W., Pfaff, D.W., Kaplitt, M.G., and Ogawa, S. 2006. RNAi-mediated silencing of estrogen receptor a in the ventromedial nucleus of the hypothalamus abolishes female sexual behaviors. *Proceedings of the National Academy of Sciences of the United States of America* 103:10456–60.

Olazábal, D.E., Pereira, M., Agrati, D. et al. 2013. Flexibility and adaptation of the neural substrate that supports maternal behavior in mammals. *Neuroscience and Biobehavioral Reviews* 37:1875–92.

Palka, Y.S. and Sawyer, C.H. 1966. Effects of hypothalamic implants of ovarian steroids on oestrous behaviour in rabbits. *Journal of Physiology (London)* 185:251–69.

Paredes, R.G. 2003. Medial preoptic area/anterior hypothalamus and sexual motivation. *Scandinavian Journal of Psychology* 44:203–12.

Paredes, R.G. and Ågmo, A. 1992. Facilitation of sexual behavior shortly after electrolytic lesion of the medial preoptic area. What does it mean? *Brain Research Bulletin* 29:125–8.

Paredes, R.G. and Baum, M.J. 1997. Role of the medial preoptic area/anterior hypothalamus in the control of masculine sexual behavior. In *Annual Review of Sex Research*, eds. R.C. Rosen, C.R. Davis, and H.J. Ruppel Jr., 68–101. Allentown, PA: Society for the Scientific Study of Sexuality.

Pau, K.Y.F., Ma, Y.J., Yu, J.H., Yang, S.P., Airhart, N., and Spies, H.G. 1997. Topographic comparison of the expression of norepinephrine transporter, tyrosine hydroxylase and neuropeptide Y mRNA in association with dopamine β-hydroxylase neurons in the rabbit brainstem. *Molecular Brain Research* 48:367–81.

Ramirez, V.D. and Beyer, C. 1988. The ovarian cycle of the rabbit: Its neuroendocrine control. In *The Physiology of Reproduction,* vol. 2, eds. E. Knobil and J.D. Neill, 1873–92. New York: Raven Press.

Reyna-Neyra, A., Camacho-Arroyo, I., Cerbón, M.A., and González-Mariscal, G. 2000. Mating modifies c-fos expression in the brain of male and female rabbits. *Neuroscience Letters* 284:1–4.

Sawyer, C.H. 1959. Effects of brain lesions on estrous behavior and reflexogenous ovulation in the rabbit. *Journal of Experimental Zoology* 142:227–46.

Sawyer, C.H., Markee, J.E., and Hollinshead, W.H. 1947. Inhibition of ovulation in the rabbit by the adrenergic-blocking agent dibenamine. *Endocrinology* 41:395–402.

Schmidt, R.S. 1968. Preoptic activation of frog mating behavior. *Behaviour* 30:239–57.

Slimp, J.C., Hart, B.L., and Goy, R.W. 1978. Heterosexual, autosexual and social behavior of adult male rhesus monkeys with medial preoptic-anterior hypothalamic lesions. *Brain Research* 142:105–22.

Smith, E.R., Damassa, D.A., and Davidson, J.M. 1977. Plasma testosterone and sexual behavior following intracerebral implantation of testosterone propionate in the castrated male rat. *Hormones and Behavior* 8:77–87.

Soulairac, A. and Soulairac, M.L. 1956a. Effets des lésions hypothalamiques sur le comportement sexuel et le tractus génital du rat mâle. *Annales d'Endocrinologie (Paris)* 17:731–45.

Soulairac, A. and Soulairac, M.L. 1956b. Modifications du comportement sexuel et du tractus génital du rat mâle après lésions hypothalamiques. *Comptes Rendus des Séances de la Société de Biologie et de ses Filiales* 150:1097–100.

Soulairac, M.L. 1963. *Etude expérimentale des régulations hormono-nerveuses du comportement sexuel du rat mâle.* Paris: Masson.

Spallanzani, L. 1784. *Expériences pour servir à l'histoire de la génération.* Genève, Switzerland: Barthelemi Chirol.

Spiteri, T., Musatov, S., Ogawa, S., Ribeiro, A., Pfaff, D.W., and Ågmo, A. 2010. Estrogen-induced sexual incentive motivation, proceptivity and receptivity depend on a functional estrogen receptor α in the ventromedial nucleus of the hypothalamus but not in the amygdala. *Neuroendocrinology* 91:142–54.

Spiteri, T., Ogawa, S., Musatov, S., Pfaff, D.W., and Ågmo, A. 2012. The role of the estrogen receptor α in the medial preoptic area in sexual incentive motivation, proceptivity and receptivity, anxiety, and wheel running in female rats. *Behavioural Brain Research* 230:11–20.

Steinach, E. 1894. Untersuchungen zur vergleichenden Physiologie der männlichen Geschlechtsorgane insbesondere der accessorischen Geschlechtsdrüsen. *Pflügers Archiv für die gesamte Physiologie des Menschen und der Tiere* 56:304–38.

Steinach, E. 1910. Geschlechtstrieb und echt sekundäre Geschlechtsmerkmale als Folge der innersekretorischen Funktion der Keimdrüsen. *Zentralblatt für Physiologie* 24:551–66.

Stone, C.P. 1925a. The effects of cerebral destruction on the sexual behavior of rabbits: I. The olfactory bulbs. *American Journal of Physiology* 71:430–5.

Stone, C.P. 1925b. The effects of cerebral destruction on the sexual behavior of rabbits: II. The frontal and parietal regions. *American Journal of Physiology* 72:372–85.

Stone, C.P. 1926. The effects of cerebral destruction on the sexual behavior of male rabbits. III. The frontal, parietal, and occipital regions. *Journal of Comparative Psychology* 6:435–48.

Szawka, R.E., Poletini, M.O., Leite, C.M. et al. 2013. Release of norepinephrine in the preoptic area activates anteroventral periventricular nucleus neurons and stimulates the surge of luteinizing hormone. *Endocrinology* 154:363–74.

von Bechterew, W. 1911. *Die Funktionen der Nervencentra. Heft 3. Hemisphären des Grosshirns.* Jena, Germany: Gustav Fischer.

Yang, S.P., Pau, K.Y.F., Airhart, N., and Spies, H.G. 1998. Attenuation of gonadotropin-releasing hormone reflex to coitus by α_1-adrenergic receptor blockade in the rabbit. *Proceedings of the Society for Experimental Biology and Medicine* 218:204–9.

Yang, S.P., Pau, K.Y.F., Hess, D.L., and Spies, H.G. 1996. Sexual dimorphism in secretion of hypothalamic gonadotropin-releasing hormone and norepinephrine after coitus in rabbits. *Endocrinology* 137:2683–93.

3 Hormonal Regulation of the Copulatory Motor Pattern in Mammals

Gabriela Moralí, Knut Larsson, José Luis Contreras, and Miguel Cervantes

CONTENTS

3.1 Introduction 46
3.2 Approaches for Describing the Motor and Genital Components of Male Copulation 46
3.2.1 Analysis of Some Parameters of Motor and Genital Components of Copulation 46
3.2.2 The Polygraphic and Accelerometric Technique 47
3.3 Characteristics of the Masculine Copulatory Motor Pattern of Some Laboratory Species 47
3.3.1 Rabbit 47
3.3.1.1 Description of Homotypical (Male) Sexual Behavior 47
3.3.1.2 Motor Characteristics of Heterotypical (Pseudomale) Sexual Behavior 49
3.3.2 Rat 50
3.3.2.1 Description of Homotypical (Male) Sexual Behavior 50
3.3.2.2 Motor Characteristics of Heterotypical (Pseudomale) Sexual Behavior 52
3.3.3 Golden Hamster 52
3.3.4 Guinea Pig 54
3.3.5 Mouse 55
3.4 Hormonal Regulation of the Masculine Copulatory Motor Pattern: Effects of Castration and Hormone Replacement 56
3.4.1 Rabbit 56
3.4.1.1 Males 56
3.4.1.2 Females 57
3.4.2 Rats 58
3.4.3 Golden Hamsters 60
3.4.4 Guinea Pigs 61
References 62

3.1 INTRODUCTION

Male sexual behavior includes a series of precopulatory, copulatory, and postejaculatory behavioral patterns. Compared to classical studies, more attention has been given recently to expressions of precopulatory behavior as indicative of the level of sexual motivation, as well as to postejaculatory behavior as indicative of sexual satiety in mammals. Copulatory behavior has been the most extensively studied, regarding its behavioral patterns, their temporal sequence, and the hormonal, neural, social, and environmental factors involved in its expression (Dewsbury 1979; Larsson 1979; Meisel and Sachs 1994; Hull and Rodríguez-Manzo 2009). Nevertheless, there is a dearth of studies providing a detailed description of the morphology of the various copulatory behavioral patterns, including the characteristics of their motor and genital components.

Male copulation in mammals involves the activation of three interacting components: (1) a motor component involving the contraction and relaxation of the various muscles of the lower trunk participating in the performance of pelvic thrusting; (2) an external genital component allowing the penile responses leading to erection and intravaginal insertion; and (3) an internal genital component including the pattern of contraction of the various sex accessory organs participating in seminal emission and ejaculation (Moralí and Beyer 1992). Precise information of the interactions among these three components is required for a full understanding of copulatory behavior.

The accelerometric and polygraphic technique developed in Dr. Carlos Beyer's laboratory and applied by us in several subsequent studies enabled the recording of signals generated by the motor and genital responses displayed at copulation. This has led to precise descriptions of the male copulatory pattern of several species, intact or subjected to castration, hormonal replacement, penile desensitization, or pharmacological treatments, and their electrographic correlates. In this review, we present our data on the characteristics of motor and genital responses of intact male rabbits, rats, hamsters, guinea pigs, and mice. We also present our data on the characteristics of pseudomale behavior spontaneously displayed by female rabbits and rats, and on the hormonal regulation of the copulatory motor pattern of males and females.

3.2 APPROACHES FOR DESCRIBING THE MOTOR AND GENITAL COMPONENTS OF MALE COPULATION

3.2.1 Analysis of Some Parameters of Motor and Genital Components of Copulation

Some studies were performed before the 1980s to describe the morphology of copulation, including high-speed cinematography or videotape recording of motor copulatory responses with subsequent analysis in slow motion (Stone and Ferguson 1940; Bermant 1965; Sachs and Barfield 1976). An electric circuit was also used, closing when moist contact between male and female partners occurred at copulation, thus allowing investigators to determine the precise duration of penile insertions during intromission and ejaculation, in rabbits and rats (Peirce and Nuttal 1961; Carlsson and Larsson 1962; Rubin and Azrin 1967). The rate of pelvic thrusting of rabbits was also estimated (Rubin and Azrin 1967). However, no information was available, for any species, on the intensity, vigor, or rhythmicity of pelvic movements of copulation. More recent approaches have included electromyographic recordings of the bulbospongiosus and ischiocavernosus muscles during copulation in the male rat as parameters indicative of their role in penile responses occurring during mount, intromission, and ejaculation (Holmes et al. 1991).

3.2.2 The Polygraphic and Accelerometric Technique

It is clear from a casual observation of male copulation that there are gross differences between rabbits and rats in their stereotypic movements, temporal sequence, duration of mounting trains, and rate of pelvic thrusting (faster in rats than in rabbits). Based on these observations, Dr. Beyer, José Luis Contreras, and Prof. Knut Larsson (in several of his fruitful annual collaborations in Dr. Beyer's laboratory in Mexico City) became interested in describing these patterns quantitatively. Dr. Miguel Cervantes and Engineer Javier Almanza suggested using an accelerometer—a transducer commonly used in neurology for evaluating the intensity of tremor in patients—to be attached to the male's pelvis, and the signal recorded on an oscilloscope or polygraph. This provided precise images of pelvic thrusting as a series of wave-like signals whose amplitude was related to the strength of each movement, and enabled measurement of the duration of individual thrusts, mounting trains, and frequency, rhythmicity, and vigor of pelvic thrusting. A pressure transducer recorded seminal vesicle pressure (SVP) and an electronic circuit recorded moist contact as intromissions. These initial studies were extended by their students and colleagues, Gabriela Moralí, Magdalena Olmedo, Laura Carrillo, Guadalupe Hernández, Martha Reynoso, María Pía Soto, Marcela Artcaga, Dolores González-Vidal, and Marisela Hernández. See review of these studies by Moralí and Beyer (1992), Beyer and González-Mariscal (1994), and Moralí et al. (2003).

3.3 CHARACTERISTICS OF THE MASCULINE COPULATORY MOTOR PATTERN OF SOME LABORATORY SPECIES

3.3.1 Rabbit

3.3.1.1 Description of Homotypical (Male) Sexual Behavior

Copulation in male rabbits involves the display of a mount with a variable period of pelvic thrusting, its duration depending on lordosis by the doe, which may culminate in intromission, at which time ejaculation invariably occurs. Male rabbits may display "effective" mounts (i.e., those culminating in intromission and ejaculation) predominantly and variable numbers of "ineffective" mounts (Rubin and Azrin 1967; Contreras and Beyer 1979; Dewsbury 1979). The relative number of "effective" vs. "ineffective" mounts changes across tests that allow males to copulate *ad libitum* to reach sexual satiety (Jiménez et al. 2012).

Recording of accelerometric signals in copulating New Zealand white rabbits allowed establishing the intra- and interindividual variability of the male motor copulatory pattern. Differences were observed among and within males on different occasions in the number, duration, and dynamic characteristics of the mounts (Contreras and Beyer 1979). Intromittive mounts were slightly shorter in duration than nonintromittive mounts (Mean ± SD: 2.61 ± 1.50 sec vs. 3.08 ± 2.16 sec), and pelvic thrusting frequency was significantly higher in intromittive mounts (13.54 ± 1.11 vs. 12.08 ± 0.98). The latter were also more rhythmic, periodic, and regular in their vigor, as shown by the amplitude of the signals generated by thrusts. These differences were interpreted as having relevance for eliciting the lordosis posture in the doe (Contreras and Beyer 1979). Based on this methodology, as shown in Figure 3.1, sexual experience did not influence the copulatory pattern in the male rabbit; first responses displayed by naïve males were similar to those of experienced males. By contrast, as also shown in Figure 3.1, female rabbit pseudomale mounting patterns were clearly different from the mounting pattern of males (Contreras and Beyer 1979).

As shown in Figure 3.2, no changes in pressure of the seminal vesicles were recorded during pelvic thrusting. However, shortly after intromission, during which pelvic thrusting ceases, a rise in pressure was recorded, evidence of contraction of the

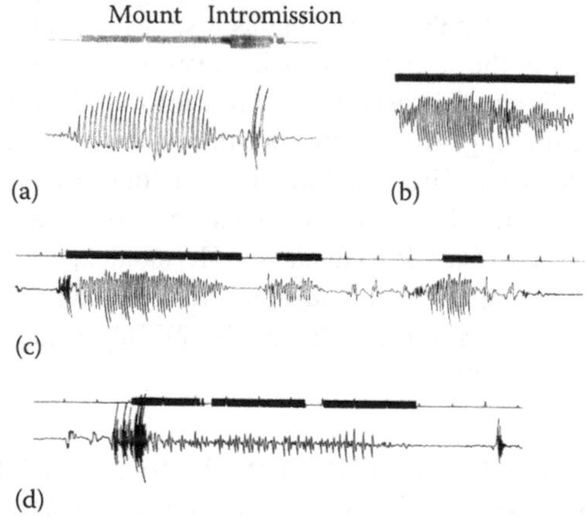

FIGURE 3.1 (a) Typical record of a copulatory response of an experienced male rabbit including ejaculation. Lower trace shows the pelvic thrusting movements recorded with the use of the accelerometer. Upper trace carries a time mark (1 sec) and a signal (horizontal bar) that was operated by an observer. Narrow signal indicates time during which mounting occurred; wide signal indicates the moment in which the observer recorded intromission. Note that during intromission, pelvic thrusting stopped. (b) Mounting train by an experienced rabbit. (c) Mounting train of an inexperienced rabbit. Note that the mounting pattern is similar to that of experienced rabbits. (d) Record of a mounting train performed by a female rabbit, interrupted by pauses, and lacking the rhythmicity and vigor of male rabbit thrusting trains. (Reprinted with permission from Contreras, J. L. and C. Beyer, *Physiol. Behav.*, 23, 939–43, 1979.)

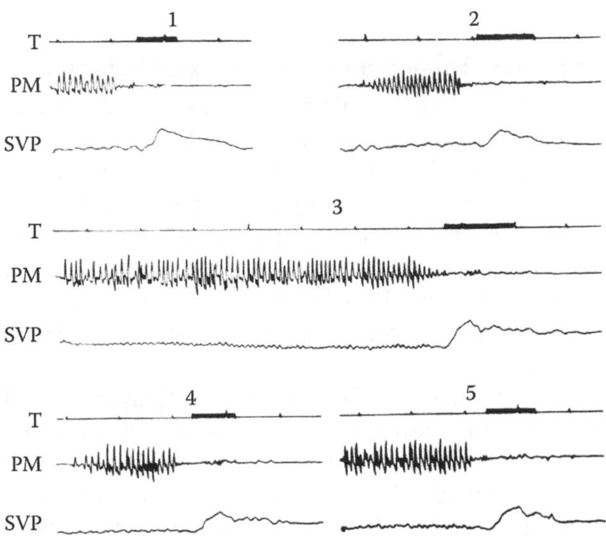

FIGURE 3.2 Recordings of five successive copulations displayed by an adult intact male rabbit. Upper trace shows the 1-sec time (T) signal and a mark, introduced by an observer, to indicate the occurrence of intromission. Middle trace shows the pelvic movements (PM) recorded with the use of the accelerometer. Lower trace shows the changes in seminal vesicle pressure (SVP) occurring during intromission and outlasting the duration of the copulatory response. (Reprinted with permission from Contreras, J. L. and C. Beyer, *Physiol. Behav.*, 23, 939–43, 1979.)

seminal vesicles and the occurrence of seminal emission and ejaculation. Thereafter, pressure of the seminal vesicles gradually declined, outlasting copulation (Contreras and Beyer 1979).

3.3.1.2 Motor Characteristics of Heterotypical (Pseudomale) Sexual Behavior

Female rabbits commonly display pseudomale (mounting) sexual behavior toward other females as an expression of their estrous behavior (Brooks 1937; Yaschine et al. 1967; Albonetti and Dessi-Fulgheri 1990). This pseudomale behavior may appear to be similar to male copulation; however, accelerometric analysis of the temporal and dynamic characteristics of mounting trains performed by female rabbits showed clear differences relative to males (Contreras and Beyer 1979; Soto et al. 1984). As exemplified in Figure 3.1, mounting trains of females are usually shorter, with shallow, weak movements, generating signals of lower amplitude, lacking rhythmicity. Sexual differences in the dynamic organization of pelvic thrusting displayed by male and female rabbits are also evident via power frequency spectrum analysis. While pelvic thrusting by males shows a predominant peak around 14–15 Hz, female mounts usually generate signals of lower amplitude, disorganized in their periodicity, generally lacking a predominant peak frequency (Soto et al. 1984) (See Section 3.4.1.2).

3.3.2 Rat

3.3.2.1 Description of Homotypical (Male) Sexual Behavior

In contrast to rabbits, copulating male rats display brief mounting responses, in which climbing onto the female is accompanied by his fast forelimb palpation of her flanks and the display of rhythmic pelvic thrusting against her rump. Some mounts (intromission responses) are terminated by a deep thrust, usually corresponding to intravaginal penile insertion, followed by an abrupt dismount. A series of intromission responses is normally required by male rats to ejaculate. Ejaculatory responses are characterized by mounts terminated by a deep pelvic thrust, which is prolonged for 2 or 3 sec, while repeated irregular flexion of the hind limbs occurs (Larsson 1956, 1979; Beyer 1979).

Accelerometric recordings of pelvic thrusting displayed by male rats during mounts, intromissions, and ejaculations reveal their temporal characteristics and dynamic organization. As shown in Figure 3.3, mounts consist of a series of 6–12 pelvic thrusts of similar duration but of gradually increasing amplitude up to a maximum and decreasing thereafter, giving the mounting train a fusiform appearance (Beyer et al. 1981).

Duration of mounting trains may vary from 0.3 to 0.6 sec, with an average of 0.38 ± 0.08 sec (Mean ± SD), and frequency of pelvic thrusting is relatively regular (Mean ± SD: 20.95 ± 0.82 thrusts/sec). During mounts, none or only occasional brief contacts occur between the penis of the male and the vaginal orifice of the female (Moralí et al. 1983), and a slight increase in SVP occurs in 52% of mounts as a single wave (Beyer et al. 1982), in association with pelvic thrusting.

Pelvic thrusting trains during intromissions are of shorter duration than during mounts (Mean ± SD: 0.31 ± 0.07 sec). As can also be seen in Figure 3.3, they have the same appearance as during mounts in their initial component, but differ in showing a final period of irregular broad signals associated with penile insertion and abrupt withdrawal. Establishment of a genital contact during penile insertion, as recorded by the intromission detection circuit, results in the interruption of pelvic thrusting and is maintained for 0.41 ± 0.15 sec (Mean ± SD). As in mounts, a rise in SVP occurs during pelvic thrusting at the intromission responses; then, coinciding with penile insertion, a steep further rise in SVP occurs. The duration of the pressure rise is longer for intromissions than for mounts.

Pelvic thrusting trains during ejaculatory behavioral responses are longer than those at mounts or intromissions. As shown in Figure 3.3, the accelerometric technique differentiated two types of ejaculatory responses: long and short (Beyer et al. 1981, 1982; Moralí et al. 1983). These two patterns differ not only in the duration of the thrusting trains (long: 1.04 sec, and short: 0.68 sec as averages) but also in their dynamic organization. Pelvic thrusting was not interrupted by penile insertion as was the case for intromissions, but continued as a period of intravaginal thrusting at both ejaculatory responses. At long ejaculations, clearly identifiable pelvic thrusting trains were recorded before and after intromission, which coincides with a reduction in the amplitude of thrusts. By contrast, during short ejaculations, extra- and intravaginal thrusting periods proceeded in immediate succession as a single phase of gradually increasing amplitude, culminating in a series of irregular pelvic

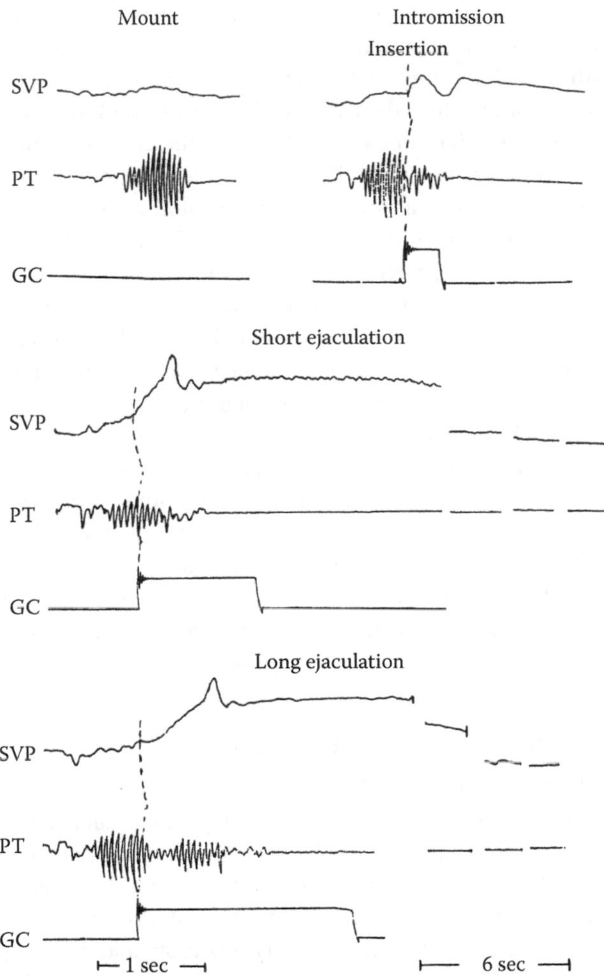

FIGURE 3.3 Polygraphic recordings of changes in seminal vesicle pressure (SVP), of the signals generated by the accelerometer in relation to pelvic thrusting movements (PT), and of genital contacts (GC), occurring during a typical mount, intromission, and a short and a long ejaculation behavioral pattern of intact male rats. Note the fusiform organization of the accelerometric record of the mounting train, and the similar appearance of the intromission and ejaculation thrusting trains until penile insertion is achieved, as indicated by the dotted line and by the plateau signal generated during genital contact. Note the differences between short and long ejaculation patterns in the duration and dynamic organization of the pelvic thrusting train, and in the duration of the period of intravaginal thrusting that precedes ejaculation, as revealed by this methodology. Note the minimal, gradual, rise in SVP during the mount and the sharp increase coinciding with penile insertion. Note that the SVP rises during penile insertion and intravaginal thrusting at ejaculation responses, culminating in a further steep rise associated with seminal emission and remaining above the baseline for several seconds. (Reprinted with permission from Moralí, G. et al., *Scand. J. Psychol.*, 44, 279–88, 2003.)

movements associated with seminal emission and ejaculation. It should be noted that the intravaginal thrusting period of short ejaculation responses is brief (0.31 ± 0.08 sec), usually involving only a few pelvic thrusts, while in long ejaculations, a longer period of intravaginal thrusting (0.54 ± 0.08 sec) seems to be required for ejaculation to occur (Moralí et al. 2003). As in the case of mounting and intromission, a smooth gradual rise in SVP occurs at the beginning of mounting in the ejaculatory trains, followed by a steeper rise associated with penile insertion. After a variable period of intravaginal thrusting (around 0.31 sec in short ejaculations and 0.54 sec in long ejaculations), a further rise in SVP occurs, associated with seminal emission. Thereafter, the SVP falls to its previous level and remains above baseline for several seconds (Beyer et al. 1982; Moralí et al. 2003). Factors such as hormonal condition, age, sexual excitation, and phenomena related to the sequential display of successive ejaculatory series seem to contribute to the display of short vs. long ejaculatory patterns (Moralí et al. 2003).

Suppression of thrusting movements by penile insertion at intromissions, and transient reduction of their amplitude in long ejaculations, suggests that activation of penile receptors may stimulate interneurons inhibitory to the spinal motoneurons, forming the neural circuits involved in the generation of the motor copulatory pattern in male rats (Beyer and González-Mariscal 1994). By contrast, the increasing amplitude of thrusts preceding seminal emission and ejaculation seems to provide further penile stimulation that elicits ejaculation.

3.3.2.2 Motor Characteristics of Heterotypical (Pseudomale) Sexual Behavior

Particularly during estrus, female rats may display mount and intromission motor patterns similar to those of males, including the forward movement associated with penile insertion followed by the abrupt dismount shown by males (Beach 1942; Moralí and Beyer 1979). The accelerometric recordings show a great similarity between males and females in the temporal organization, and in the vigor and rhythmicity of the mounting and intromission motor patterns, although with a longer duration of mounts in the females (Moralí et al. 1985). In addition, as shown in Figure 3.4, frequency spectrum analysis revealed similar frequency (20 to 22 Hz), rhythmicity, and vigor of pelvic thrusting during mounts and intromissions in males and females.

3.3.3 Golden Hamster

Four types of copulatory behavioral patterns are displayed by male golden hamsters: mounts, intromissions, ejaculations, and long intromissions (Bunnell et al. 1976). The polygraphic analysis of pelvic thrusting and genital contacts during these copulatory responses showed a series of regular rhythmical extravaginal pelvic thrusts, with frequencies between 14 and 15.5 thrusts/sec. As in rabbits, mounting trains not resulting in penile insertion were usually longer (1.28 ± 0.15 vs. 0.87 ± 0.10 sec) and had slightly lower thrusting frequencies (14.78 ± 0.28 vs. 15.20 ± 0.30 thrusts/sec) than those at intromission and ejaculation (Arteaga and Moralí 1997; Moralí et al. 2003).

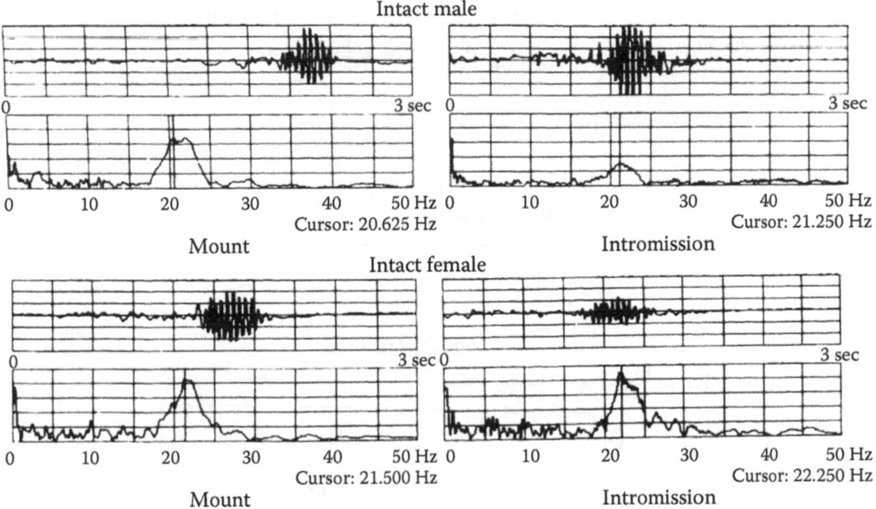

FIGURE 3.4 Frequency spectrum analysis of pelvic thrusting performed by intact male and female rats during typical mount and intromission patterns. Upper trace shows the signal generated by the accelerometer during the behavioral responses. Note the similar fusiform organization of male and female mounting trains. Lower trace corresponds to the power frequency spectrum analysis (range: 0 to 50 Hz) of the signals generated during an 8-sec period within which individual responses occurred. Note that the peak values (indicated by cursor) corresponding to the predominant frequency of pelvic thrusting, and the narrow dispersion of the spectrum as an indicator of rhythmicity, are similar between male and female responses. (Modified from Moralí, G. et al., *Physiol. Behav.*, 34, 267–75, 1985, reprinted with permission.)

As shown in Figure 3.5, the pelvic thrusting train at intromission ceased when penile insertion occurred, and this period (lasting 2.21 ± 0.16 sec) was followed by a slow dismount.

During ejaculation, the extravaginal pelvic thrusting train also ceased at penile insertion; however, after a brief pause (50–100 msec), a short train of intravaginal pelvic thrusting, generating polygraphic signals of lower amplitude and higher frequency (16 thrusts/sec) than those of the extravaginal train, occurred (see Figure 3.5). This intravaginal thrusting train, like that of rats, seems to be associated with ejaculation. The duration of this period (0.45 sec) is highly consistent among and within individuals, and occurs with a precise latency (1.25 ± 0.08 sec) after the onset of penile insertion. As described by Bunnell et al. (1976), after several ejaculatory series, when a male is approaching sexual satiety, it presents "long" intromissions, in which penile insertion is held for 30 sec or more, while displaying intravaginal pelvic thrusting of slow frequency. Accelerometric recordings showed that extravaginal fast pelvic thrusting trains at long intromissions are of similar frequency, rhythmicity, and vigor as those of intromissions and ejaculations, and, in response to penile insertion, they shift to slow intravaginal thrusting (around 2 thrusts/sec), generating low-amplitude signals (see Figure 3.5) (Arteaga and Moralí 1997).

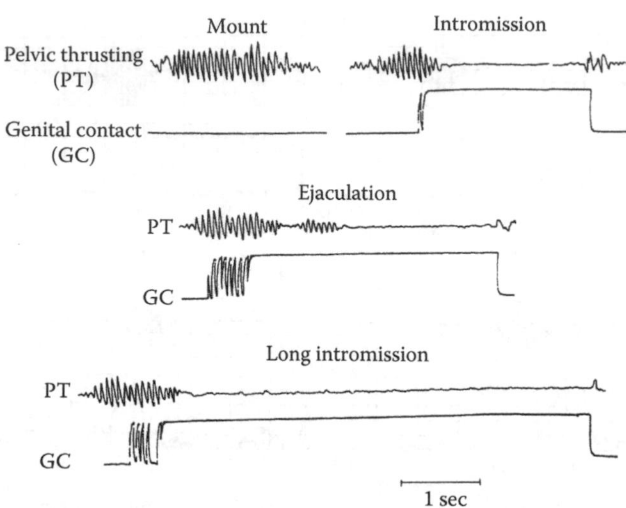

FIGURE 3.5 Polygraphic records of signals generated by the accelerometer in relation to pelvic thrusting (PT), and of the intromission detection circuit indicating the occurrence and duration of genital contact (GC) associated with penile insertion, during mount, intromission, ejaculation, and long intromission patterns displayed by male hamsters. Note that when penile insertion occurs at intromission and ejaculation responses, the vigorous extravaginal pelvic thrusting ceases, and it is followed, at ejaculation responses, by a well-defined period of intravaginal thrusting associated with ejaculation. At long intromission patterns, slow intravaginal thrusting (about 2/sec) is displayed during the prolonged period of penile insertion. (Modified from Arteaga, M. and G. Moralí, *J. Physiol (Paris)*, 91, 311–16, 1997, reprinted with permission.)

3.3.4 Guinea Pig

The copulatory behavior of the male guinea pig may involve the performance of a variable number of mounts and intromissions preceding ejaculation. However, as in rabbits, guinea pigs are capable of ejaculating during a single penile insertion, at least on some occasions (Dewsbury 1979). As shown in Figure 3.6, the polygraphic analysis of copulatory responses of adult Hartley albino male guinea pigs has revealed the display of rhythmic pelvic thrusting trains with frequencies of 11 to 12 thrusts/sec at mounting, before penile insertion (González-Vidal 1995; Moralí et al. 2003).

When penile insertion occurred, fast pelvic thrusting was interrupted and it shifted to a slow pattern of about 1.5 thrusts/sec, with a duration of several seconds. At the ejaculatory behavioral responses, after a brief period of slow thrusting performed during penile insertion, a train of fast intravaginal pelvic thrusting was displayed, generating signals of lower amplitude and slightly higher frequency than those of preinsertion pelvic thrusting. It was associated with ejaculation, had a characteristic duration (1.14 ± 0.05 sec), and usually occurred with a precise latency (2.5 sec) after the onset of penile insertion. Thus, as in hamsters and rats, triggering of ejaculation requires a period of intravaginal fast pelvic thrusting with highly predictable characteristics.

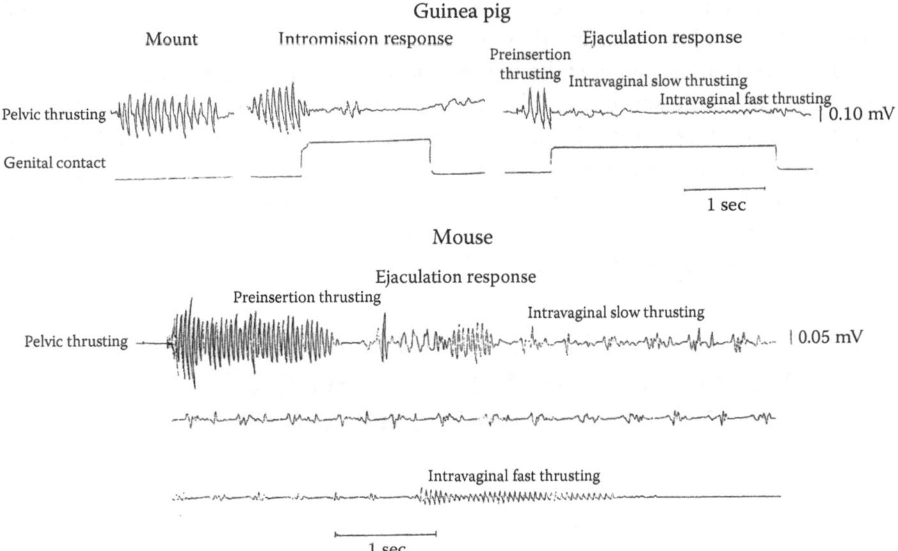

FIGURE 3.6 Representative polygraphic records of signals generated in relation to pelvic thrusting (PT), and to genital contact (GC) established at penile insertion, during mount, intromission, and ejaculation displayed by a male guinea pig; and to pelvic thrusting during an ejaculation response displayed by a mouse. In both species, fast pelvic thrusting occurring before penile insertion shifts during insertion to a slow pattern (that in the mouse may last for 20 sec or more) that is followed by a characteristic period of intravaginal fast thrusting associated with ejaculation. (Reprinted from Moralí, G. et al., *Scand. J. Psychol.*, 44, 279–88, 2003, with permission.)

3.3.5 Mouse

Similar to rats and hamsters, mice display several mounts and intromissions before ejaculation. However, by contrast, mice display prolonged periods (several seconds) of slow intravaginal pelvic thrusting during penile insertion at intromission and ejaculation (Dewsbury 1979). Accelerometric recordings of the motor copulatory pattern of a small number of B6D2F1/J male mice (Wang et al. 1989) provided clear images of the signals generated by extravaginal and intravaginal pelvic thrusting (see Figure 3.6). Mounts, characterized by fast (22–25 thrusts/sec) rhythmic thrusting trains, may have a variable duration. When penile insertion was achieved, fast pelvic thrusting ceased and shifted to a slow thrusting pattern (2 thrusts/sec), similar to guinea pigs and the long intromissions of hamsters, but of longer duration. When the mouse lost penile insertion, fast pelvic thrusting was resumed until insertion was regained. After variable periods (several seconds) of slow intravaginal pelvic thrusting, males either dismounted or maintained penile insertion and ejaculated. At ejaculation, as in guinea pigs, the intravaginal slow thrusting period, which in mice may last for 20 sec or more, is followed by a characteristic new period of fast pelvic thrusting of similar frequency to that of extravaginal thrusting (22 thrusts/sec) and low amplitude, presumably associated with ejaculation (see Figure 3.6). After this period, the male may maintain genital contact for several seconds and then withdraw (Moralí et al. 2003).

The above findings support the interpretation that differences exist among these species in the penile sensory requirements for ejaculation: A brief penile insertion with no pelvic thrusting is adequate to elicit brief latency ejaculation in the rabbit. In the other species, whether they display a prolonged (several seconds) period of intravaginal slow pelvic thrusting (guinea pigs and mice) or not (rats and hamsters), a period of intravaginal fast pelvic thrusting, of similar frequency to the extravaginal train, leads to ejaculation (Moralí et al. 2003).

3.4 HORMONAL REGULATION OF THE MASCULINE COPULATORY MOTOR PATTERN: EFFECTS OF CASTRATION AND HORMONE REPLACEMENT

3.4.1 Rabbit

Based on the temporal and spatial analysis of the male copulatory patterns of rabbits and rats, and an analysis of the literature on rhythmic motor patterns both in vertebrates and invertebrates, proposals about several characteristics of the neural organization of this behavior were postulated by Dr. Beyer (Beyer and González-Mariscal 1994). The alternate, forward–backward rhythmic nature of pelvic thrusting when mounting the female reveals that different, antagonistic sets of motoneurons discharge phasically and then become inhibited, in a reciprocal manner. The generation of periodic behavior involves command neurons that trigger the motor pattern, a set of interneurons responsible for the generation of rhythmic activity (the so-called Central Pattern Generator), and the motoneurons innervating the muscles of the trunk and pelvis involved in the motor pattern (Delcomyn 1980; Harris-Warrick and Johnson 1989). As noted by Beyer and González-Mariscal (1994), evidence supports the concept that some of the command neurons related to copulation are located in the medial preoptic area (MPOA), while central pattern generator neurons for pelvic thrusting are located within the spinal cord.

Based on this information, two models were proposed by Beyer and González-Mariscal (1994) for the neuroendocrine regulation of male copulation in mammals: (1) a single-site model in which testosterone or its metabolites would only need to act, directly or indirectly, on the brain command neurons, which would activate the central pattern generator to elicit pelvic thrusting; and (2) a multiple-site model in which steroid hormones, besides acting on the brain, also modulate the excitability of the spinal cord neural circuits associated with the generation of the motor copulatory pattern. Rat copulation appears to belong to the first model and rabbit copulation to the second. Evidence supporting this proposal is described below.

3.4.1.1 Males

Castrated male rabbits may continue displaying mounts for several months after surgery. However, only a low proportion of mounts stimulate lordosis in the female and culminate in ejaculation (Beyer et al. 1980). As shown in Figure 3.7, the accelerometric recordings of mounts performed after castration show periods of weak,

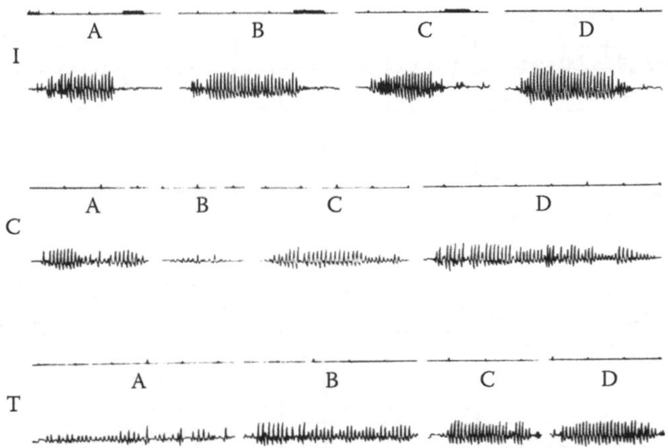

FIGURE 3.7 Accelerometric records of thrusting movements at the first four mounts performed by a New Zealand white rabbit when intact (I), 30 days after castration (C), and 15 days after the initiation of testosterone propionate administration (T). Upper tracings in each record carry a time mark (1 sec) and a signal operated by the observer when detecting intromission. Note that castration results in a diminution of the amplitude of the thrusting movements and in the appearance of trains having weak thrusts interspersed with periods of more vigorous activity. Note that T administration tended to restore the normal mounting pattern. (Reprinted from Beyer, C. et al., *Horm. Behav.*, 14, 179–90, 1980, with permission.)

low-amplitude movements, with interthrust intervals that disrupt the rhythm of thrusting and cause a reduction of its frequency (Beyer et al. 1980).

Thus, reduction of thrusting frequency is not due to a slowing of pelvic movements, but to intercalation of pauses between individual thrusts. These alterations occurred earlier than the decrease in the number of mounts displayed per test. As mentioned above, these data suggest that in the rabbit, gonadal hormones act both at the brain level to stimulate the command neurons triggering copulation and at the level of the spinal cord, influencing the excitability and/or the organization of the spinal cord circuitry involved in vigor and rhythm of pelvic thrusting. Their findings suggested that these characteristics are more sensitive to androgen deprivation than is sexual motivation (Beyer and González-Mariscal 1994). Administration of testosterone propionate (TP, 10 mg/day) for several days to castrated male rabbits gradually restored their mounting activity, but more prolonged treatments were required for their copulatory motor pattern to regain its normal characteristics (Beyer et al. 1980) (see Figure 3.7).

3.4.1.2 Females

Additional support for rabbit copulation as belonging to the multiple-site model for the effect of steroid hormones comes from data on female rabbits. Administration of TP to either intact or ovariectomized rabbits results in a marked increase in the vigor of pelvic thrusting (generating signals of higher amplitude), and in its rhythmicity, with thrusting frequencies similar to those of males (approximately 13 thrusts/sec). Treatment with estradiol benzoate (EB) for several weeks resulted in mounting trains

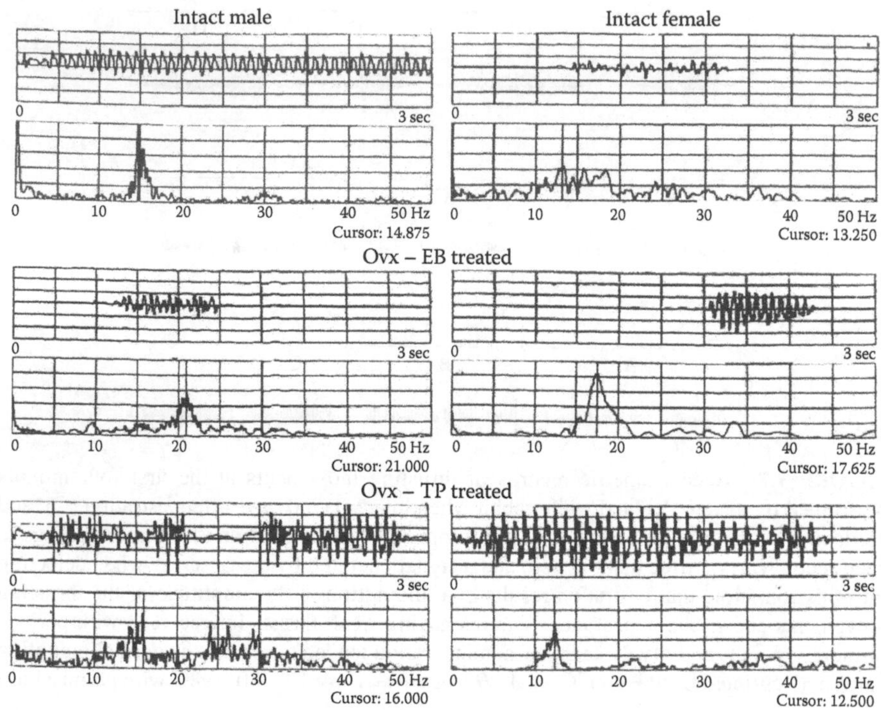

FIGURE 3.8 Accelerometric records (upper tracings) and frequency spectrum analysis graphs (Range: 1–50 Hz) (lower tracings) of pelvic movements in typical mounts displayed by intact male and female rabbits, and by ovariectomized (ovx) rabbits receiving estradiol benzoate (EB) or testosterone propionate (TP) treatments. The frequency analysis of the male mount shows a single component with a peak frequency value (F) of 14.875 Hz, as indicated by the cursor. In contrast, that of the female mount does not show a dominant frequency but rather a series of ill-defined components. Note sharp, single components in the spectra of the EB-treated, ovariectomized rabbits, also showing high thrusting frequencies (F = 21.00 and 17.625 Hz). Frequency analysis of TP-stimulated mounts reveals two patterns: one with a well-defined component (F = 12.500 Hz), and the other showing both a dominant frequency (F = 16.000 Hz) and an ill-defined band of higher frequencies. (Modified from Soto, M. A. et al., *Horm. Behav.*, 18, 225–34, 1984; Reprinted from Moralí, G. et al., *Scand. J. Psychol.*, 44, 279–88, 2003, with permission.)

with highly rhythmic pelvic thrusting characterized by significantly higher frequencies than those of males (Soto et al. 1984). Representative accelerometric recordings and frequency spectrum analysis of mounts displayed by ovariectomized rabbits under treatment with EB or with TP are shown in Figure 3.8.

3.4.2 Rats

In contrast to the rabbit, the motor copulatory pattern of the male rat is not altered by castration. As shown in Figure 3.9, the amplitude, frequency, or rhythmicity of

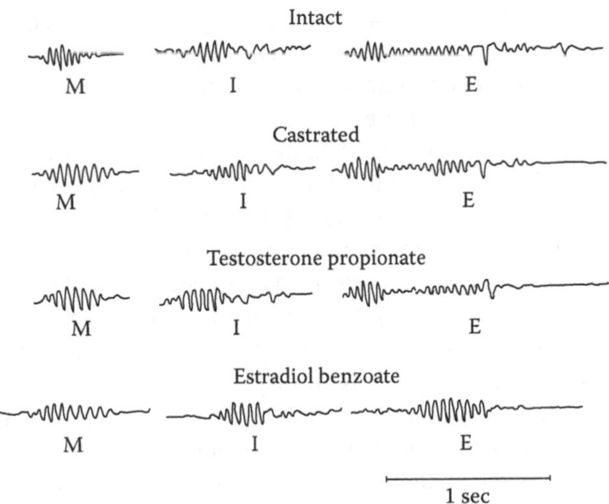

FIGURE 3.9 Accelerometric records of pelvic thrusting at mounts (M), intromission responses (I), and ejaculation responses (E) of intact, castrated, testosterone-propionate-treated, and estradiol-benzoate-treated male rats. Note that similar patterns of pelvic thrusting in the respective behavioral responses are displayed under the various hormonal conditions. (Modified from Beyer, C. et al., *Physiol. Behav.*, 27, 727–30, 1981, reprinted with permission.)

copulatory pelvic thrusting displayed by rats several weeks after castration did not differ from those of intact rats (Beyer et al. 1981).

Only the duration of the mounting trains was slightly longer in castrated rats than in intact rats (0.54 ± 0.02 vs. 0.38 ± 0.08 sec). This was interpreted as due to a failure either of penile erection or orientation of the penis to the vaginal region (Moralí and Beyer 1992) as had also been described in castrated cats, in addition to an increase in mount duration (Rosenblatt and Aronson 1958). Administration of TP to castrated rats restored their full copulatory behavior, while EB stimulated the display of mounts and intromissions, but the ejaculatory pattern on only two occasions. As shown in Figure 3.9, under either TP or EB treatment, copulatory pelvic thrusting had motor characteristics that were identical to those shown before castration (Beyer et al. 1981). These findings are consistent with the proposal by Beyer and González-Mariscal (1994) that copulation of the male rat corresponds to a single-site model for the effects of androgen on the neural tissue: these would be exerted on the command neurons in the brain, which, in turn, would activate the central pattern generator at the spinal cord level to elicit pelvic thrusting. As further support for this proposal, implantation of TP in the mPOA restored copulation in castrated rats with characteristics of vigor, frequency, and temporal organization similar to those of intact rats, thus providing evidence that activation of a single brain site (the mPOA) by androgen enables the display of a normal copulatory motor pattern in the rat (Moralí et al. 1986). These findings can be interpreted as indicative that spinal cord circuits related to pelvic thrusting in the rat do not require sexual steroids for their proper functioning. However, as an alternative possibility (Moralí and Beyer 1992;

Beyer and González-Mariscal 1994), rats may stop copulating before alterations in the motor copulatory pattern are detected. This would be the case if motivation to copulate requires a higher concentration of sex steroids than that needed for displaying a normal copulatory motor pattern.

As mentioned above (Section 3.3.2.2), the dynamic characteristics and temporal organization of pelvic thrusting at mounting and intromission behavioral patterns of female rats are similar to those of males (Moralí et al. 1985). Furthermore, neonatal castration of male rats or androgen administration to female rats, though affecting the incidence of copulatory responses performed by subjects at adulthood after receiving sex steroid treatment, do not modify the characteristics of copulatory pelvic thrusting. Thus, neonatally castrated male rats receiving TP as adults showed full copulatory behavior, and those receiving EB only displayed mounts and intromission patterns. On the other hand, neonatally androgenized female rats displayed a higher number of mounts and intromission patterns than control females and some showed the ejaculation motor pattern. As shown in Figure 3.10, pelvic thrusting in all cases had similar characteristics to those of control male rats (Moralí et al. 1985).

3.4.3 Golden Hamsters

Castration of hamsters, as previously described by other authors (Lisk and Heimann 1980), resulted in a progressive reduction of their copulatory activity, evidenced by a progression of slowing behavioral responses with a consequent reduction in

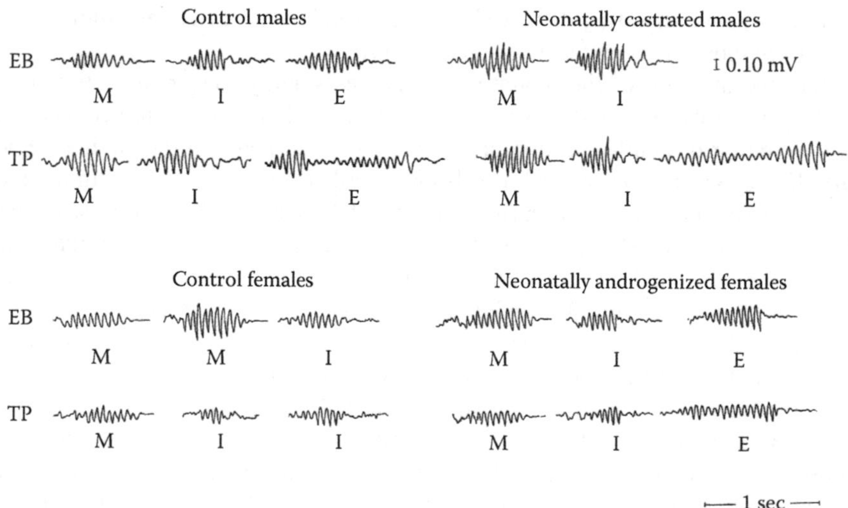

FIGURE 3.10 Accelerometric records of pelvic thrusting at mount (M), intromission (I), and ejaculation (E) behavioral responses of control (post-puberally castrated) and neonatally castrated male rats, and by control and neonatally androgenized female rats, after ovariectomy and treatment with estradiol benzoate (EB) or testosterone propionate (TP). Dynamic organization and rhythmicity of pelvic thrusting are similar in the respective behavioral responses under the various experimental conditions. (Reprinted with permission from Moralí, G. et al., *Scand. J. Psychol.*, 44, 279–88, 2003.)

the number of copulatory series per test (Arteaga 1995; Moralí et al. 2003). Three weeks after castration, no ejaculation patterns were shown. However, coinciding with castrated rats, copulatory pelvic thrusting displayed by hamsters after castration had similar vigor, frequency, and rhythmicity as those of intact hamsters. An effect of castration on the copulatory pattern of the hamster was a lengthening of extravaginal pelvic thrusting trains. Thus, duration of thrusting trains at mounts (Mean ± SEM: 1.75 ± 0.16 sec), intromissions (1.03 ± 0.14 sec), and ejaculation (1.10 ± 0.16 sec) was longer than that of behavioral responses displayed before castration (1.28 ± 0.15; 0.83 ± 0.10; and 0.87 ± 0.05 sec, respectively). As in the case of castrated rats, this effect may be due to a failure either of penile erection or of the orientation of the penis to the vagina. This interpretation is supported by the finding that genital contacts at intromission and ejaculation of castrated hamsters (1.45 ± 0.22 and 2.75 ± 0.22 sec, respectively) were significantly shorter than those of intact subjects (2.21 ± 0.16 and 3.15 ± 0.10 sec, respectively) (Arteaga and Moralí 1997; Moralí et al. 2003). On the other hand, when the ejaculatory pattern was displayed after castration, duration (0.42 ± 0.02 sec), vigor, frequency, and rhythmicity of the intravaginal thrusting period in these responses were similar to those displayed before castration.

These findings can be interpreted, as in rats, to indicate that spinal cord circuits related to pelvic thrusting in the hamster do not require sexual steroids for their characteristic functioning, provided androgen acts on a single site, that is, on brain command neurons that regulate both sexual motivation and the copulatory motor pattern. The possibility also exists, as proposed for rats (Beyer and González-Mariscal 1994), that motivation to copulate requires higher levels of sex steroids than those needed for displaying a normal copulatory motor pattern. If so, hamsters would stop copulating before alterations in the copulatory motor pattern are detected.

Sexual behavior of castrated hamsters was optimally restored by the administration of TP (0.5 mg/day) and 5α-dihydrotestosterone propionate (0.3 mg/day) for eight weeks. Hormonal treatment restored the duration of genital contacts and of extravaginal pelvic thrusting trains to similar values as those of intact hamsters (Arteaga 1995; Moralí et al. 2003).

3.4.4 Guinea Pigs

Similar to descriptions by other researchers (Grunt and Young 1953), castration results in an abrupt reduction in the ability of male guinea pigs to achieve penile insertion and ejaculation; mounts are displayed for variable periods differing among subjects, gradually reducing their incidence until they cease 10 weeks after castration (González-Vidal 1995; Moralí et al. 2003). During that period, pelvic thrusting trains were shorter in duration than those of intact animals (Mean ± SEM: 0.77 ± 0.05 sec vs. 1.18 ± 0.07 sec) and showed lower vigor (0.08 ± 0.02 mV vs. 0.12 ± 0.03 mV), but had similar frequency and rhythmicity as those of intact subjects. The fact that frequency and rhythmicity of pelvic thrusting were not altered by castration may indicate that guinea pig copulation corresponds, as in rats and hamsters, to the single-site model of action of androgen, on brain command neurons regulating both sexual motivation and the copulatory motor pattern (Beyer and González-Mariscal 1994).

However, in relation to vigor of pelvic thrusting, guinea pigs seem to correspond to the multiple-site model, in which androgen, besides stimulating sexual motivation at the brain level, also modulates the excitability of the spinal cord neural circuits and the synchronous activity of motoneurons that innervate the muscles involved in the copulatory motor pattern. Synchronization of firing in these groups of motoneurons has been proposed (Beyer and González-Mariscal 1994) to rely on rearrangements of their afferents or of their interconnections, possibly through the formation of gap junctions between these motoneurons in response to androgen, as has been described for other nuclei of the lumbosacral spinal cord of the rat (Matsumoto et al. 1988). Depletion of androgen after castration would interfere with these structural rearrangements and with the synchronous firing of motoneurons, essential for the vigorous contraction of muscles involved in pelvic thrusting.

Under TP treatment, castrated guinea pigs gradually recover mounting and intromission, and then, ejaculation. These responses have similar motor characteristics to those of intact subjects (González-Vidal 1995; Moralí et al. 2003).

As reviewed here, precise descriptions of the male copulatory pattern of several species were obtained by using the accelerometric and polygraphic technique developed in Dr. Carlos Beyer's laboratory. These descriptions enabled recognizing variations among species, differences between males and females, and analyzing the effects of castration and hormonal replacement on motor and genital responses of copulation. The findings enabled our group to derive conclusions on the neuroendocrine regulation of these events and have constituted a starting point for determining the influence of several other factors on their occurrence and electrographic correlates.

REFERENCES

Albonetti, M. E. and F. Dessi-Fulgheri. 1990. Female-female mounting in the European rabbit. *Z. Säugetierk.* 55:128–38.

Arteaga, M. 1995. Description of the motor and genital characteristics of copulation of the male hamster (*Mesocricetus auratus*): Effects of castration and treatment with androgens. MSc. diss., National Autonomous University of Mexico (UNAM), Mexico.

Arteaga, M. and G. Moralí. 1997. Characteristics of the motor and genital copulatory responses of the male hamster. *J. Physiol. (Paris)* 91:311–16.

Beach, F. A. 1942. Execution of the complete masculine copulatory pattern by sexually receptive female rats. *J. Genet. Psychol.* 60:137–42.

Bermant, G. 1965. Rat sexual behavior: Photographic analysis of the intromission response. *Psychon. Sci.* 2:65–6.

Beyer, C. (Ed.) 1979. *Endocrine Control of Sexual Behavior*. New York: Raven Press.

Beyer, C., J. L. Contreras, K. Larsson, M. Olmedo, and G. Moralí. 1982. Patterns of motor and seminal vesicle activities during copulation in the male rat. *Physiol. Behav*. 29:495–500.

Beyer, C., J. L. Contreras, G. Moralí, and K. Larsson. 1981. Effects of castration and sex steroid treatment on the motor copulatory pattern of the rat. *Physiol. Behav.* 27:727–30.

Beyer, C. and G. González-Mariscal. 1994. Effects of sex steroids on sensory and motor spinal mechanisms. *Psychoneuroendocrinology* 19:517–27.

Beyer, C., J. Velázquez, K. Larsson, and J. L. Contreras. 1980. Androgen regulation of the motor copulatory pattern in the male New Zealand white rabbit. *Horm. Behav.* 14:179–90.

Brooks, C. 1937. The role of the cerebral cortex and of various sense organs in the excitation and execution of mating activity in the rabbit. *Am. J. Physiol.* 120:544–53.

Bunnell, B. N., B. D. Boland, and D. A. Dewsbury. 1977. Copulatory behaviour of golden hamsters (*Mesocricetus auratus*). *Behaviour* 61:244–50.

Carlsson, S. and K. Larsson. 1962. Intromission frequency and intromission duration in the male rat mating behavior. *Scand. J. Physiol.* 3:189–91.

Contreras, J. L. and C. Beyer. 1979. A polygraphic analysis of mounting and ejaculation in the New Zealand white rabbit. *Physiol. Behav.* 23:939–43.

Delcomyn, F. 1980. Neural bases of rhythmic behavior in animals. *Science* 210:492–8.

Dewsbury, D. A. 1979. Description of sexual behavior in research on hormone behavior interactions. In *Endocrine Control of Sexual Behavior*, ed. C. Beyer, 1–32, New York: Raven Press.

González-Vidal, M. D. 1995. Characteristics of motor and genital responses of the copulatory activity of the male guinea pig (*Cavia porcellus*): Effects of castration and testosterone treatment. 1995. MSc. diss., National Autonomous University of Mexico (UNAM), Mexico.

Grunt, J. A. and W. C. Young. 1953. Differential reactivity of individuals and the response of the male guinea pig to testosterone propionate. *Endocrinology* 51:237–48.

Harris-Warrick, R. M. and B. R. Johnson. 1989. Motor pattern networks: Flexible foundations for rhythmic pattern production. In *Perspectives in Neural Systems and Behavior,* eds. T. J. Carew and D. B. Kelley, 51–71, New York: Alan R. Liss Inc.

Holmes, G. M., W. D. Chapple, R. E. Leipheimer, and B. D. Sachs. 1991. Electromyographic analysis of male rat perineal muscles during copulation and reflexive erections. *Physiol. Behav.* 49:1235–46.

Hull, E. M. and G. Rodríguez-Manzo. 2009. Male sexual behavior. In *Hormones, Brain, and Behavior, 2nd Ed*, ed. D. W. Pfaff, 5–66, Oxford: Elsevier.

Jiménez, P., Serrano-Meneses, M. A., Cuamatzi, E., and González-Mariscal, G. 2012. Analysis of copulatory parameters in male rabbits across successive tests leading to sexual exhaustion. *World Rabbit Sci.* 20:13–23.

Larsson, K. 1956. *Conditioning and Sexual Behavior in the Male Albino Rat.* Stockholm, Sweden: Almqvist & Wiksell.

Larsson, K. 1979. Features of the neuroendocrine regulation of masculine sexual behavior. In *Endocrine Control of Sexual Behavior*, ed. C. Beyer, 77–163, New York: Raven Press.

Lisk, R. D. and J. Heimann. 1980. The effects of sexual experience and frequency of testing on retention of copulatory behavior following castration in the male hamster. *Behav. Neural Biol.* 28:156–71.

Matsumoto, A., A. P. Arnold, G. S. Zampighi, and P. E. Micevych. 1988. Androgenic regulation of gap junctions between motoneurons in rat spinal cord. *J. Neurosci.* 8:4177–83.

Meisel, R. I. and B. D. Sachs. 1994. The physiology of male sexual behavior. In *The Physiology of Reproduction,* Vol 2, eds. E. Knobil and J. D. Neill, 5–105, New York: Raven Press.

Moralí, G. and C. Beyer. 1979. Neuroendocrine control of mammalian estrous behavior. In *Endocrine Control of Sexual Behavior,* ed. C. Beyer, 33–75, New York: Raven Press.

Moralí, G. and C. Beyer. 1992. Motor aspects of masculine sexual behavior in rats and rabbits. In *Advances in the Study of Behavior*, Vol. 21, eds. P. J. R. Slater, J. S. Rosenblatt, C. Beer and M. Milinski, 201–38, New York: Academic Press.

Moralí, G., L. Carrillo, and C. Beyer. 1983. A method for assessing intravaginal thrusting during copulation in rats. *Conf. Reprod. Behav.* 15th, p. 54. Medford, Massachusetts.

Moralí, G., L. Carrillo, and C. Beyer. 1985. Neonatal androgen influences sexual motivation but not the masculine copulatory motor pattern in the rat. *Physiol. Behav.* 34:267–75.

Moralí, G., G. Hernández, and C. Beyer. 1986. Restoration of the copulatory pelvic thrusting pattern in castrated male rats by the intracerebral implantation of androgen. *Physiol. Behav*. 36:495–9.

Moralí, G., M. A. Soto, J. L. Contreras, M. Arteaga, M. D. González-Vidal, and C. Beyer. 2003. Detailed analysis of the male copulatory motor pattern in mammals: Hormonal bases. *Scand. J. Psychol.* 44:279–88.

Peirce, J. T. and R. L. Nuttal. 1961. Duration of sexual contacts in the rat. *J. Comp. Physiol. Psychol.* 5:585–7.

Rosenblatt, J. S. and L. R. Aronson. 1958. The decline of sexual behavior in male cats after castration with special reference to the role of prior sexual experience. *Behaviour* 12:285–338.

Rubin, H. B. and N. H. Azrin. 1967. Temporal patterns of sexual behavior in rabbits as determined by an automatic recording technique. *J. Exp. Anal. Behav.* 10:219–31.

Sachs, B. D. and R. J. Barfield. 1976. Functional analysis of masculine behavior in the rat. *Adv. Study Behav.* 7:91–154.

Soto, M. A., M. Reynoso, and C. Beyer. 1984. Sexual dimorphism in the motor mounting pattern of the New Zealand white rabbit: steroid regulation of vigor and rhythmicity of pelvic thrusting. *Horm. Behav*. 18:225–34.

Stone, C. P. and L. W. Ferguson. 1940. Temporal relationships in the copulatory acts of adult male rats. *J. Comp. Psychol.* 30:419–33.

Yaschine, T., F. Mena, and C. Beyer. 1967. Gonadal hormones and mounting behavior in the female rabbit. *Am. J. Physiol.* 213:867–72.

4 Neuronal and Neurochemical Correlates of Copulatory Motor Patterns in Male Rats

Marisela Hernández González and Miguel Ángel Guevara

CONTENTS

4.1 Introduction 65
4.2 Computerized Capture and Analysis of Pelvic Thrusting 67
4.3 Spinal Neurochemical Modulation of Pelvic Copulatory Movements 70
4.4 Ventral Tegmental Area and Pedunculopontine Nucleus in the Control of Pelvic Thrusting 72
4.5 Conclusion 76
References 76

4.1 INTRODUCTION

Sexual interaction is a motivated behavior that, like other goal-directed behaviors, is manifested through a complex sequence of fine and coarse motor patterns. In male sexual behavior in particular, the fine motor components include penile responses and the contraction of various organs associated with erection, penile insertion, seminal emission, and ejaculation. The coarse motor components involve the contraction and relaxation of the muscles responsible for the stereotypic copulatory movements of mount (M), intromission (I), and ejaculation (E).

These copulatory responses constitute the most general manner in which males interact with females and, hence, play a basic role in the expression of sexual behavior. In the case of the male rat, all copulatory responses are characterized by the performance of repetitive and alternating pelvic thrusting (17–22 cycles/sec) against the female's rump (Sachs & Barfield, 1976; Beyer, Moralí & Larsson, 1981) that likely results from alternating periods of excitation and inhibition of spinal motoneurons that innervate specific groups of muscles (Moralí, Komisaruk & Beyer, 1989). These repetitive motor patterns have a voluntary component controlled by specific cerebral structures, as well as an involuntary component that is generated in spinal neuronal networks. In general terms, we know that the neural substrate responsible for rhythmic motor patterns consists of (1) brain structures that contain command neurons (CNs) that trigger the motor pattern itself (Armstrong, 1988), (2) excitatory and inhibitory interneurons located in the thoracolumbar level of the spinal cord involved in generating rhythmic activity (Central Pattern Generator: CPG) (Pratt & Jordan, 1987), and (3) motoneurons that innervate the skeletal muscles involved in the motor pattern (Grillner & Wallen, 1985; Grillner & Dubuc, 1988; Harris-Warrick & Johnson, 1989). This neural organization allows a set of motoneurons to discharge phasically during the forward movement of the male rat's pelvic thrusting (the basic unit of copulatory movements) toward the female's perineum, while in the backward movement away from this region, another set of antagonistic motoneurons is activated, while the initial set is inhibited.

In relation to the supraspinal modulation of pelvic thrusting, it has been suggested that structures such as the medial preoptic area (MPOA) (Oomura, Yoshimatsu & Aou, 1983; Horio, Shimura, Hanada & Shimokochi, 1986; Paredes, Highland & Karam, 1993; Shimura, Yamamoto & Shimokochi, 1994; Paredes, Tzschentke & Nakach, 1998), the ventral tegmental area (VTA), and the pedunculopontine nucleus (PPN) (Hernández-González Guevara, Moralí & Cervantes, 1997a) could all play essential roles in both the motivational and motor components of male sexual behavior. Other structures of the limbic system, such as the prefrontal cortex (PFC) (Fernández-Guasti, Omana-Zapata, Luján & Condés, 1994; Agmo, Villalpando, Picker & Fernández, 1995; Hernández-González, Guevara, Cervantes, Moralí & Corsi-Cabrera, 1998; Hernández-González, Prieto-Beracoechea, Arteaga-Silva & Guevara, 2007), the accumbens nucleus (Liu, Sachs & Salamone, 1998; Guevara, Martínez-Pelayo & Hernández-González, 2008), and the amygdala (Wood, 1998; Maras & Petrulis, 2006; Hernández-González, Robles-Aguirre, Guevara, Quirarte & Haro-Magallanes, 2014), could be responsible for processing the sexually relevant incentive stimuli that induce the sexual motivation involved in searching for a nearby receptive female.

Neurochemical and hormonal factors modulate the cerebral and spinal circuits involved in the performance of the rhythmic motor components of sexual behavior. The cerebral areas control the neurons located at the spinal level, thereby triggering the trains of pelvic thrusting (Oomura et al., 1983; Hernández-González et al., 1997a), while the connectivity among the several interneuron types that constitute the CPG and motoneurons is responsible for the thrusting frequency and rhythmicity (Pratt & Jordan, 1987; Grillner & Dubuc, 1988).

In 1979, Contreras and Beyer developed an innovative accelerometric and polygraphic technique that made it possible to obtain a precise, detailed description of the temporal and dynamic characteristics of pelvic thrusting that male rabbits perform during copulation. Subsequently, this technique was used to analyze pelvic thrusting in male rats, and several studies were carried out to investigate the precise temporal association between the fine motor components and the occurrence of copulatory responses (Beyer, Contreras, Larsson, Olmedo & Moralí, 1982; Moralí & Beyer, 1992), the sensory control (Moralí, Contreras & Beyer, 1982; Contreras & Agmo, 1993) as well as the hormonal (Moralí, Carrillo & Beyer, 1985; Moralí,

Hernández & Beyer, 1986; Moralí & Beyer, 1992; Beyer & González-Mariscal, 1994), and neurochemical regulation (Moralí & Larsson, 1984; Moralí et al., 1989; Ågmo, Contreras & Paredes, 1991; Hernández-González, Oropeza, Guevara, Cervantes & Moralí, 1994) of the rhythmic patterns that characterize M, I, and E responses in both rabbits and rats.

By applying this technique, the parameters of duration, frequency (number of pelvic thrusts per second), and amplitude, or vigor, of copulatory pelvic thrusting in male rats have been described (Beyer et al., 1981, 1982; Moralí & Beyer, 1992; Hernández-González, Guevara, Oropeza & Moralí, 1993). In general terms, the mount response has a duration of 400–600 msec and a frequency of 16–22 pelvic thrusts/sec (Hz), the duration of intromission is approximately 200–450 msec with a frequency of 18–23 Hz, and the duration of ejaculation can be about 500 msec in the case of short ejaculations, or up to 2,000 msec in the case of long ejaculations.

The accelerometric–polygraphic technique allowed researchers to determine certain features of pelvic copulatory movements where previous methodologies had failed (Beyer et al., 1981, 1982). However, parameters such as rhythmicity and regularity, as well as quantitative analyses of vigor or potency (determined by the amplitude or "area under the curve") of pelvic-thrusting trains in each response, represented the greatest challenge for manual quantification. The introduction of computers in the 1980s led to the development of automatic analysis techniques that significantly improved the methods by which bioelectrical signals could be observed and interpreted.

Computerized systems for analyzing bioelectrical signals—in this particular case, the signals generated when transducing the pelvic thrusting trains performed by males during copulatory responses—have proven to be useful, as they help clarify the spinal mechanisms involved in the integration and expression of those responses. Also, the simultaneous recording and analysis of accelerometric and electroencephalographic (EEG) signals has made it possible to probe more precisely the participation of specific brain structures in both the motivational and performance components of the male rat's sexual behavior.

This chapter describes the computerized programs that were developed to assess more easily and accurately the characteristic parameters of pelvic thrusting performed by the male rat during copulatory patterns. It also reviews how the application of the accelerometric–polygraphic technique allows researchers to study the influence of the neurochemical modulation of copulatory movements at the spinal level. Finally, it discusses the functionality of the cortical and subcortical areas that have been implicated in the modulation of sexual motivation and copulatory responses in male rats.

4.2 COMPUTERIZED CAPTURE AND ANALYSIS OF PELVIC THRUSTING

Before explaining the computer programs that were developed in the Pascal programming language to capture and analyze the copulatory pelvic thrusting of male rats, we first describe the traditional system used by Beyer (Contreras & Beyer, 1979; Beyer et al., 1981). This procedure utilized a cloth harness, adjusted to the pelvic region of the rat, which was equipped with a strain gauge transducer to measure maximum acceleration in one plane (Grass SPA-1 accelerometer). The accelerometer output was connected to a DC Grass preamplifier coupled to a Grass 7B polygraph, enabling the traces of pelvic thrusting to be recorded directly on paper, while the experimenter indicated each occurrence of M, I, or E with a pencil.

Capture and storing accelerometric signals in specific computer files required connecting the polygraph outlet to an analog-digital converter (PCL 812) that operated as an interface to the microcomputer, and to a unit board equipped with buttons corresponding to each response, so that the amplified and filtered accelerometric signals (0–30 Hz) could be saved for subsequent analysis. For this task, a computerized program called CAPTUMAR was used. Using a 50 mV calibration pulse generated by the polygraph as a reference, all values were directly converted to millivolts (mV), and the moment at which the female rat was introduced into the cage and the test began was set as zero time on the timer integrated into the program, such that it was possible to record the latency of each response and, simultaneously, the sequence in which those responses occurred. Thus, as copulatory interaction proceeded, the tester pushed the appropriate buttons to capture the three seconds prior to the moment that the button was pushed, which included the response performed by the male. CAPTUMAR also made it possible to distribute the responses to create four files containing the data of M, I, E, and the number and uninterrupted sequence of all the responses and times at which the rat performed each pattern.

To analyze accelerometric records, each 3-sec segment captured was examined to select and delimit the signal interval that corresponded to a specific response. This was achieved by means of the CHECAMAR program, which displayed the recorded signal on the screen and determined the exact duration of each response with the help of a pair of cursors operated by the user (see Figure 4.1). By then applying

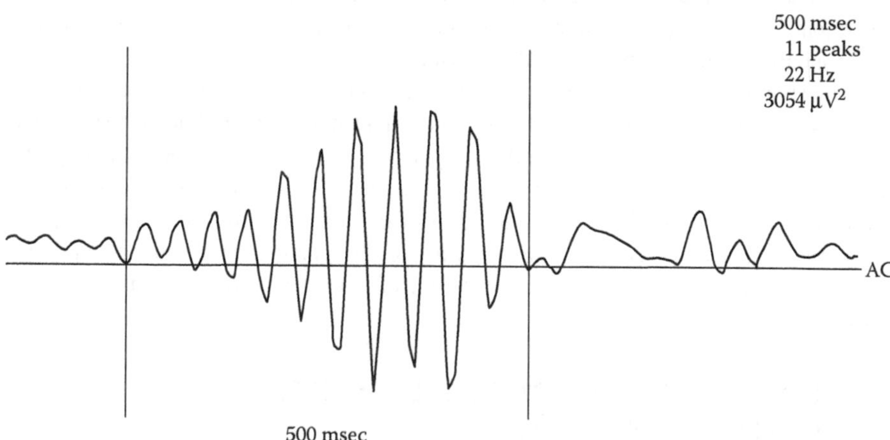

FIGURE 4.1 Accelerometric recording (AC) from a short ejaculation on the computer monitor after delimitation by means of cursors that made it possible to select the signal interval that corresponds to this copulatory response. Upper right margin: the duration value (msec), peak number, frequency (Hz), and potency–or amplitude (mV²)–of the pelvic-thrusting train are indicated.

Fast Fourier Analysis, CHECAMAR calculated the frequency and amplitude—or vigor—parameters of the pelvic-thrusting train for each response.

A third program, SPECTRO, made it possible to obtain the frequency spectra in relation to the amplitude for each separate response, or for the sum of all responses in one file, whether for M, I, or E. This program also allowed the determination of the rhythmicity or regularity of the responses, as well as the amplitude of the frequency spectra that represented the vigor of each response. These programs made it possible to define the large amplitude difference that exists between the extravaginal and the intravaginal pelvic movements of long ejaculations. Based on this methodology, the first phase, which corresponds to extravaginal movements, was found to have a greater amplitude (about 1184 mV^2) than the second (about 810 mV^2) (Hernández-González et al., 1993) (Figure 4.2).

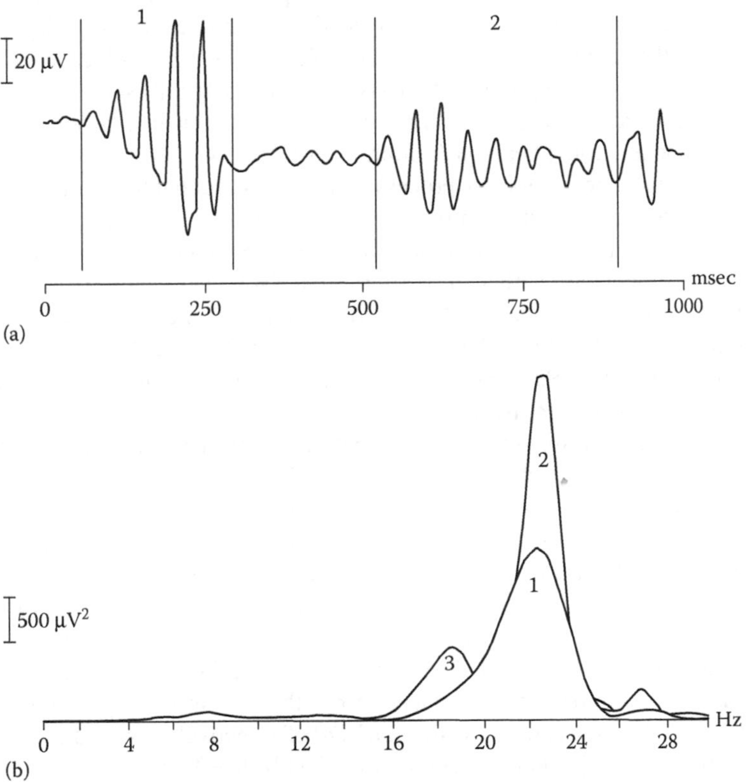

FIGURE 4.2 Monitor images representing: (a) accelerometric recording from a long ejaculation after delimitation by means of cursors that made it possible to select the signal intervals that correspond to (1) extra-vaginal and (2) intra-vaginal, thrusting phases; (b) spectrum of frequencies in relation to amplitude of the same long ejaculation, indicating the amplitude of (1) the complete response; (2) the extra-vaginal thrusting phase; and (3) the intra-vaginal thrusting phase.

4.3 SPINAL NEUROCHEMICAL MODULATION OF PELVIC COPULATORY MOVEMENTS

The alternating, rhythmic pelvic movements that male rats perform during the copulatory responses of M, I, and E result from the activation–inhibition of interneurons and motoneurons located at the thoracolumbar level. These spinal networks of interconnected neurons are termed "CPG" and are capable of generating a rhythmic pattern of motor activity independent of phasic sensory input from peripheral receptors. CPGs have been identified and analyzed in more than 50 rhythmic motor systems, including those that control such diverse behaviors as walking, swimming, feeding, respiration, and flying (Edgerton, Grillner, Sjostrom & Langger, 1976; Grillner & Kashin, 1976).

Most CPGs produce a complex temporal pattern of activation of different groups of motoneurons. The sequencing of activity in those motoneurons is regulated by a number of mechanisms, perhaps the simplest one of which is mutual inhibition. Interneurons that fire in consistent, but different, phases with each other are usually reciprocally coupled by inhibitory connections. Another rhythmicity-inducing mechanism is the latency of recovery from inhibition, which can influence the relative time of the onset of activity in two neurons simultaneously released from inhibition. Finally, mutual excitation can generate synchronous firing in a group of neurons (Marder & Bucher, 2001; Pearson & Gordon, 2013).

The excitation and inhibition processes that occur sequentially in these CPGs are mediated by different neurotransmitters and hence, by the potential interaction of a variety of receptors present in the spinal cord. Spinal cord neurons that control rhythmic movements are activated by supraspinal input. Descending glutamatergic, dopaminergic, serotonergic, and GABAergic pathways exert complex effects on the spinal network (Grillner, Deliagina, Ekebuerg, el Manira, Hill, Lansner et al., 1995; Katz & Frost, 1995; Harris-Warrik, Johnson, Peck, Kloppenburg, Ayali & Skarbinski, 1998; Lillvis & Katz, 2013).

One of the neurotransmitters that affects male sexual behavior is serotonin. Several studies have suggested that a decrease in 5-HT neurotransmission facilitates, whereas an increase in 5-HT neurotransmission inhibits, male sexual behavior (Ahlenius, Larsson, Svensso, Hjorth, Carlsson, Lindberg et al., 1981; Olivier, Chan, Snoeren, Olivier, Veening, Vinkers et al., 2011). The systemic administration of the postsynaptic 5-HT receptor agonist 8-hydroxy-2-(di-*n*-propylamino) tetralin (8-OH-DPAT) in male rats elicited a dose-dependent decrease in the number of mounts and intromissions, and a decrease in the ejaculation latency (Ahlenius et al., 1981), but did not modify the thrusting pattern of the copulatory responses (Moralí & Larsson, 1984). Thus, treatment with 8-OH-DPAT selectively lowered the threshold for elicitation of the ejaculation response, but had little effect on the motor copulatory pattern.

Gamma-aminobutyric acid (GABA) inhibits male copulatory behavior and genital reflexes. The concentration of GABA in the cerebrospinal fluid samples from the cisterna magna of male rats increased dramatically during the postejaculatory interval (Qureshi & Södersten, 1986). There is evidence that different inhibitory interneurons of the CPGs use GABA as a neurotransmitter (Jovanović, Petrov & Stein, 1999). When administered intraperitoneally, two GABA transaminase inhibitors, γ-acetylene GABA (GAG) and sodium valproate, which increase GABAergic function, reduced intromission behavior and increased the postejaculatory interval in male rats, but did not affect mounting behavior (Agmo, Paredes & Fernández, 1987). However, while neither of these GABA transaminase inhibitors affected the duration of pelvic thrusting, GAG did induce a significant, although minor, reduction of thrust frequency during mount and intromission responses (Agmo & Contreras, 1990). A similar lack of effect of these GABA transaminase inhibitors on copulatory thrusting was also demonstrated in rabbits (Agmo et al., 1991).

In order to exert a more direct effect of the drugs at the spinal level, other studies have used intrathecal administration of agonists and antagonists at the lumbosacral level. There, a combination of the accelerometric technique and spinal administration of different drugs was used to study the possible role of glycine and GABA in modulating pelvic thrusting in male rats. Whether the animals were treated with the glycine antagonist, strychnine, or the GABA antagonist bicuculline, they showed sensory and motor disturbances, although the frequency, duration, and rhythmicity of the copulatory thrusting movements performed during mounts, intromissions, or ejaculations were not affected (Moralí et al., 1989). This suggests that glycine and GABA are not critically involved in regulating these rhythmic motor patterns.

These earlier studies were interpreted as indicating that the duration and rhythmicity of copulatory movements in male rats are controlled by synapses that are insensitive, or inaccessible, to the various drugs administered. Another possible reason for the lack of effect of those agents is that the excitatory and inhibitory interneurons localized at the thoracolumbar level of the spinal cord involved in the generation of rhythmic pelvic thrusting are controlled by other than serotonergic, glycinergic, and/or GABAergic neurotransmitter systems.

Following this lead, Hernández-González et al. (1994) evaluated the possible role of the adrenergic pathways in modulating the spinal mechanisms involved in the control of rhythmic pelvic copulatory thrusting. Several different adrenergic receptors participate in the neural mechanisms that control rhythmic movement (Wiesenfeld-Hallin, 1987; Fischer, Merrywest & Sillar, 2001). Administration of adrenergic drugs induces "fictive" locomotion in spinal animals (Forssberg & Grillner, 1973), as well as rhythmic swimming in spinalized frogs (McLean & Sillar, 2003). Thus, the computerized recording and analysis of pelvic thrusting made it possible to determine the effect of different adrenergic agonists administered intrathecally at the lumbosacral level. Noradrenaline (a nonselective α-adrenoceptor agonist) increased the frequency of pelvic thrusting in the M and I responses, whereas clonidine (a selective α2-adrenoceptor agonist) and isoproterenol (a nonselective β-adrenoceptor agonist) decreased the frequency of pelvic thrusting without affecting the duration or amplitude of the thrusting trains. The differential effects of these adrenergic agonists on the frequency of thrusting in M and I responses could result from their interaction with different receptors that exert a direct action on motoneurons or interneurons, or from a blockage of the facilitation provided by excitatory pathways that alter the normal pattern of motoneuron rhythmic activity.

4.4 VENTRAL TEGMENTAL AREA AND PEDUNCULOPONTINE NUCLEUS IN THE CONTROL OF PELVIC THRUSTING

Few studies have been conducted to determine the brain structures that contain the CNs involved in triggering rhythmic copulatory patterns. There is evidence that some of these CNs are located in the brainstem and hypothalamic nuclei (including the MPOA, the VTA, and the PPN). While these brain structures may not project to the spinal neurons directly, it is possible that indirect synaptic connections could trigger and modulate the rhythmic patterns.

The MPOA may contain CNs, for its destruction interferes with copulation, and sex steroid implants in this area restore copulation in castrated males (see Larsson, 1979). CNs trigger motor events but do not necessarily participate in their patterning and duration. Thus, Oomura et al. (1983) demonstrated in male monkeys an increase in the firing rate of single-unit activity in the dorsomedial hypothalamic nucleus related precisely to mounting and intromission thrusting, together with an increased single-unit activity of the MPOA neurons, immediately before initiation of copulation that decreased and remained silent during copulation itself. Similar results were reported by recording multiunit activity (MUA) in the MPOA of male rats (Horio et al., 1986), where the maximum firing rate was obtained during pursuit-mounting, but was immediately inhibited when mounting was followed by intromission and ejaculation.

The possible role of other brain structures located in the brainstem also has been analyzed. The VTA, which is the main source of the mesocorticolimbic dopaminergic pathway (Lindvall & Bjorklund, 1974; Fallon & Moore, 1978; Balfour et al., 2004), plays an important role in the initiation, execution, and sensory-motor integration of driven/species-typical behaviors, as well as in reward-related behavior patterns (Yim & Mogenson, 1980; Swanson & Mogenson, 1981). Hence, it is evidently involved in motivational and/or motor aspects of male copulatory behavior (Eibergen & Caggiula, 1973; Hull, Bazzett, Warner, Eaton & Thompson, 1990; Hull, Weber, Eaton, Dua, Markowski, Lumley et al., 1991). The PPN is another structure of the brainstem that has been shown to be involved in the motor regulation of rewarding behaviors; its primary role is the final pathway of the circuit for transitions from motivation to action, as proposed by Mogenson (Mogenson, Jones & Yim, 1980; Mogenson & Yong, 1991). The PPN, located in the caudal mesencephalic tegmentum, is a major component of the mesencephalic locomotor region (MLR), which includes the cuneiform nucleus (Skinner & García-Rill, 1984; García-Rill, 1986). The PPN sends descending fibers to neuronal components of the spinal cord pattern generators involved in the control of rhythmical limb movements for locomotion (Grillner & Shik, 1973). The accelerometric recordings of copulatory pelvic thrusting performed by male rats has been precisely correlated temporally with MUA recordings (Hernández-González et al., 1997a,b), showing that both VTA and MLR present a characteristic increase of the neuronal firing rate associated with several motor copulatory patterns (see Figure 4.3). The firing rate of the VTA neurons increased 500 ms before, and continued during, the execution of pelvic thrusting in mount, intromission, and ejaculation responses, compared to baseline. In the MLR, the highest firing rates were obtained during execution of pelvic thrusting in each

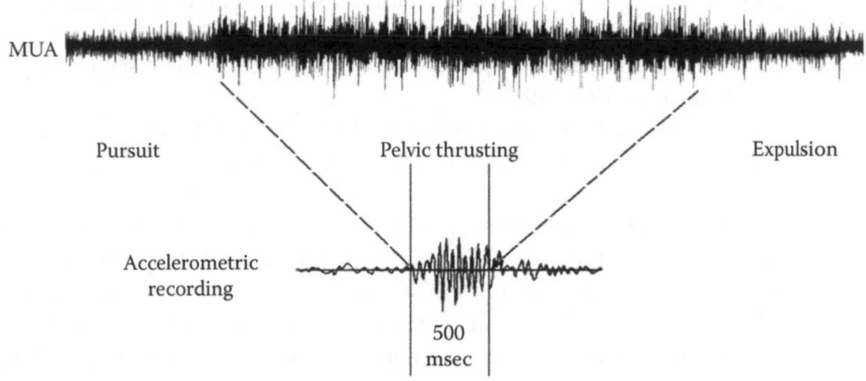

FIGURE 4.3 Example of the simultaneous recording of MUA from the pedunculopontine nucleus and accelerometric signal during a short ejaculation (monitor image). Delimitation of the accelerometric recording with cursors also made it possible to delimit and analyze the firing rate of the pedunculopontine nucleus in the 500 msec before, during, and after the pelvic-thrusting train of the ejaculation response. Note the sequential increase of the MUA in relation to the pelvic thrusting and how the increase is maintained during the expulsion phase of the short ejaculation.

copulatory response, whereas during post-intromission and post-ejaculatory genital grooming, its activity decreased to basal values at the onset of the post-ejaculatory interval. Considering that the largest changes in the MUA of the VTA and MLR were related to the onset and performance of rhythmic pelvic thrusting in the M, I, and E responses, it is possible that both brainstem structures contain CNs involved in executing motor copulatory responses.

In the case of the cerebral cortex, the cortical areas may play only a minor role in modulating sexual behavior, and an even less important role in the regulation of copulatory pelvic thrusting. However, several studies suggest that the PFC, defined as the cortical region where the largest number of projections from the mediodorsal nucleus of the thalamus (MD) are received (Rose & Woolsey, 1948; Uylings & Van Eden, 1990), plays a role in the processing of all sensory modalities. Due to its special position in the circuitry of the forebrain, the PFC has been implicated in the motivation, planning, monitoring, and modifying of sequential behavior (Goldman-Rakic, 1996; Fuster, 1997), as well as in the modulation of motor activity (Avila-Costa, Colin-Barenque, Fortoul, Espinosa-Villanueva, Rugerio-Vargas, Borgonio et al., 2001). Moreover, it has been suggested that due to its projections to the MPOA, this frontal area could modulate the motivation and performance of sexual behavior (Balfour et al., 2006).

The PFC of the rat has been divided into three functionally distinct areas: (1) the medial prefrontal region (mPFC) that includes the following areas of the cingulate cortex: Cg1, Cg2, and Cg3 (i.e., prelimbic and infralimbic areas); (2) the orbital prefrontal region (oPFC) including the ventral, lateral, ventrolateral, medial orbital, and agranular insular areas; and (3) a region that is likely analogous to the frontal eye fields of primates (Fr2) (Kolb, 1990; Uylings & Van Eden, 1990; Uylings,

Groenewegen & Kolb, 2003). This latter area has also been considered part of the mPFC, and it has been reported that dorsal lesions in the mPFC, including the Fr2 area, alter motor tasks that involve spatial memory and the adequate sequencing of goal-directed behaviors (Kolb, 1990).

Sexual activity of the male rat follows a well-organized temporal sequence, in which voluntary goal-directed movements, stereotypic responses, and consummatory acts can be distinguished.

EEG activity, defined as a mixture of rhythmic, sinusoidal-like fluctuations in voltage generated by the brain, represents the global activity of the pyramidal cells of the cortex and the activity of neurons in subcortical structures. Thus, quantitative analysis of EEGs has allowed researchers to investigate the simultaneous functioning of several brain structures in a precise temporal relationship with specific physiological states, behavior, and sensory processing.

A few studies have evaluated the correlation of EEG of different cortical areas with motor execution of copulatory responses in male rats (Kurtz & Adler, 1973; McIntosh, Barfield & Thomas, 1984). One of the most recent experiments reported that the changes in the EEG of the Fr2 prefrontal subregion correlated precisely on time with well-defined elements of male rat copulation (Hernández-González et al., 1998). By applying principal component analysis, it was possible to identify three distinct frequency bands (4–16, 18–24, and 26–32 Hz), and to calculate absolute power (AP), defined as the power density of each frequency band expressed in microvolts squared ($\mu V^2/Hz$). This revealed that approach behavior patterns, such as orientation to, and pursuit of, the female rat correlated with a decreased AP of the fast frequencies but an increase in the slow frequencies. In the copulatory sequence itself, the AP of the 4–16 Hz band increased in the 500-ms periods before, during, and after the execution of pelvic thrusting in M, I, and E responses; the AP of the 18–24 Hz band increased selectively during execution of pelvic thrusting at the three copulatory responses, while the AP of the 26–32 Hz band increased only during the pelvic movements of M and I responses (see Figure 4.4).

The increase in the slow frequencies (4–16 Hz) during pursuit and orientation (index of sexual motivation), and before, during, and after the three copulatory responses, probably represents the manifestation of the appetitive process of male sexual behavior. Other studies suggest that the PFC participates in sexually motivated approach behavior or the appetitive component. For example, Febo's (2011) recordings of single-unit activity in the PFC showed that the firing rate of the PFC neurons increased during the expression of approach responses toward a sexually receptive female. Another possible explanation could be that the increase in AP in the 4–16 Hz band represents the orientation of the male's movements toward the female because the cells that constitute the theta hippocampal generator change their firing rate in relation to the rat's direction of movement, speed, and turning angle (Wiener, Paul, and Eichenbaum 1989). This suggestion is supported by the different studies made by Komisaruk and cols. (1970, 1977), who demonstrated that the theta frequency (7–10 Hz) in limbic structures is characteristic of approach/exploratory/appetitive behavior such as the vibrissa twitching during the exploratory sniffing behavior in rats. These low frequencies are also related to the synchrony of the unitary activity that is recorded in limbic structures during exploratory behavior (Komisaruk & Olds, 1968)

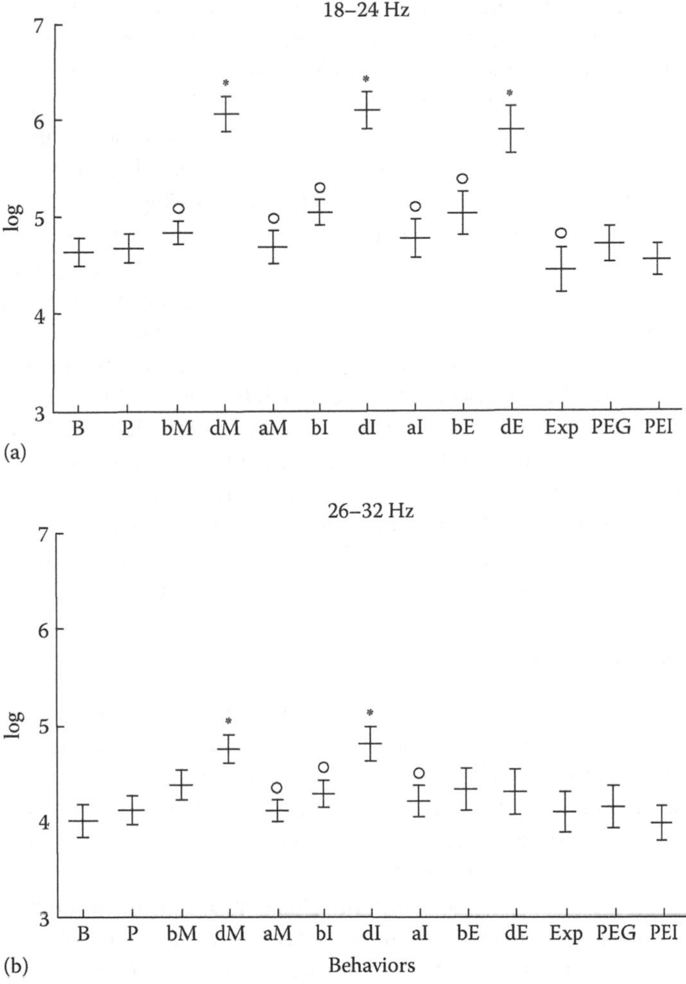

FIGURE 4.4 Mean ± SEM of the log-transformed AP of the (a) 18–24 Hz, and (b) 26–32 Hz bands, as a function of different behavioral situations during sexual interaction of male rats. B, basal condition in awake-quiet state; P, pursuit; bM, before pelvic thrusting at mount; dM, during pelvic thrusting at mount; aM, after pelvic thrusting at mount; bI, before pelvic thrusting at intromission; dI, during pelvic thrusting at intromission; aI, after pelvic thrusting at intromission; bE, before pelvic thrusting at ejaculation; dE, during pelvic thrusting at ejaculation; Exp, seminal expulsion and withdrawal; PEG, post-ejaculatory genital grooming; PEI, post-ejaculatory interval. °P < 0.01 as compared to the basal situation. *P < 0.01 as compared to periods during the execution of pelvic thrusting in M, I or E responses.

and with the rhythmical thalamic MUA bursts that were recorded during a fine tremor of the jaw and/or vibrissae in rats (Semba, Szechtman & Komisaruk, 1980).

The increase limited to the 18–24 Hz band during execution of pelvic-thrusting trains in M, I, and E could indicate that this band is specifically related to the neuromotor aspects that modulate the rhythmic, alternating movements of pelvic thrusting.

Considering that the PFC plays a role in anticipation of reward and goal orientation (Kolb, 1990), it is probable that the increase in the 26–32 Hz band during M and I, but not E, may represent the appetitive component of the copulatory sequence but not performance of the consummatory act of ejaculation (Hernández-González et al., 1998). Although it has been reported that lesions in the PFC only affect sexual motivation and not the motor performance of copulatory responses (Fernández-Guasti et al., 1994; Agmo et al., 1995), the possible role of the PFC in modulating general motor activity has been analyzed. In a model of oxidative stress (induced by acute ozone exposure), Avila-Costa et al. (2001) found alterations in motor behavior and a significant reduction of dendritic spines in the PFC. Thus, it is likely that these EEG changes in the PFC in relation to sexual motor activity could influence the neuronal activity of MPOA and other brain structures of the mesolimbic system involved in modulating the male rat's sexual behavior, as other authors have proposed (Balfour et al., 2004, 2006).

As the above description of electrical activity makes clear, PFC manifests a characteristic functionality in relation to the motivational and motor components of sexual behavior. It is still an open question as to which aspect of PFC functionality is essential to the regulation of the motor performance in sexual behavior. Nevertheless, some studies suggest that the PFC participates in the evaluation of sexually relevant stimuli and, probably, also in modulating motor components.

4.5 CONCLUSION

The importance and usefulness of the accelerometric–polygraphic technique developed by Carlos Beyer and colleagues constituted a significant advance in research into the hormonal, neurophysiological, and sensory regulation of pelvic thrusting performed during copulatory activity of both rabbits and rats. The computerized adaptation of this technique made it possible to readily and accurately analyze the various parameters of frequency, duration, and amplitude of each copulatory pattern. This chapter presents a review of various studies conducted to investigate neurochemical modulation at the spinal level and the role of several brain structures as possible CNs of the copulatory pelvic thrusting. There is a significant influence of noradrenergic innervation at the spinal level on the frequency of pelvic thrusting, in contrast to the apparent nonparticipation of the serotoninergic, GABAergic, and glycinergic pathways. At the supraspinal level, there is evidence that CNs are involved in triggering the rhythmic copulatory patterns in VTA and PPN, as determined by computer analysis of accelerometric and brain electrical signals recorded simultaneously in male rats. There is also evidence of PFC participation in the motivational and motor components of male rat sexual behavior. However, the precise participation of other neurotransmitters and brain structures in modulating copulatory pelvic thrusting remains to be explored.

REFERENCES

Ågmo, A. & Contreras, J.L. (1990). Copulatory thrusting pattern in the male rat after acute treatment with GABA transaminase inhibitors. *Physiology & Behavior, 47* (2), 311–314.

Ågmo, A., Contreras, J.L. & Paredes, R. (1991). Sexual behavior and copulatory thrusting patterns in male rabbits treated with GABA transaminase inhibitors. *Physiology & Behavior, 49* (1), 73–78.

Ågmo, A., Paredes, R. & Fernández, H. (1987). Differential effects of GABA transaminase inhibitors on sexual behavior, locomotor activity, and motor execution in the male rat. *Pharmacology, Biochemistry and Behavior, 28* (1), 47–52.

Ågmo, A., Villalpando, A., Picker, Z. & Fernández, H. (1995). Lesions of the medial prefrontal cortex and sexual behavior in the male rat. *Brain Research, 696*, 177–186.

Ahlenius, S., Larsson, K., Svensson, L., Hjorth, S., Carlsson, A., Lindberg, P., Wikstrom, H., Sanchez, D., Arvidsson, L.E., Hacksell, U. & Nilsson, J.L.G. (1981). Effects of a new type of 5-HT receptor agonist on male rat sexual behavior. *Pharmacology Biochemestry Behavior, 15*, 785–792.

Armstrong, D.M. (1988). The supraspinal control of mammalian locomotion. *The Journal of Physiology, 405*, 1–37.

Avila-Costa, M.R., Colin-Barenque, L., Fortoul, T.I., Machado-Salas, J.P., Espinosa-Villanueva, J., Rugerio-Vargas, C., Borgonio, G., Dorado, C. & Rivas-Arancibia, S. (2001). Motor impairments in an oxidative stress model and its correlation with cytological changes on rat striatum and prefrontal cortex. *International Journal of Neuroscience, 108*, 193–200.

Balfour, M.E., Brown, J.L., Yu, L. & Coolen, L.M. (2006). Potential contributions of efferents from medial prefrontal cortex to neuralactivation following sexual behavior in the male rat. *Neuroscience, 137*, 1259–1276.

Balfour, M.E., Yu, L. & Coolen, L.M. (2004). Sexual behavior and sex-associated environmental cues activate the mesolimbic system in male rats. *Neuropsychopharmacology, 29*, 718–730.

Beyer, C., Contreras, J.L., Larsson, K., Olmedo, M. & Moralí, G. (1982). Patterns of motor and seminal vesicle activities during copulation in the male rat. *Physiology and Behavior, 29* (3), 495–500.

Beyer, C., Contreras, J.L., Moralí, G. & Larsson, K. (1981). Effects of castration and sex steroid treatment on the motor copulatory pattern of the rat. *Physiology & Behavior, 27*, 727.

Beyer, C. & González-Mariscal, G. (1994). Effects of sex steroids on sensory and motor spinal mechanisms. *Psychoneuroendocrinology, 19* (5–7), 517–527.

Contreras, J.L. & Agmo, A. (1993). Sensory control of the male rat's copulatory thrusting patterns. *Behavioral and Neural Biology, 60*, 234–240.

Contreras, J.L. & Beyer, C. (1979). A polygraphic analysis of mounting and ejaculation in the New Zealand white rabbit. *Physiology & Behavior, 23*, 939–943.

Edgerton, V.R., Grillner, S., Sjostrom, A. & Langger, P. (1976). Central generation of locomotion in vertebrates. In R.M. Herman (Ed.), *Neural Control of Locomotion* (p. 439). New York, EE.UU: Plenum Press.

Eibergen, R.D. & Caggiula, A.R. (1973). Ventral midbrain involvement in copulatory behavior of the male rat. *Physiology & Behavior, 10*, 435–441.

Fallon, J. & Moore, R. (1978). Catecholamine innervation of the basal forebrain. IV Topography of the dopamine projection to the basal forebrain and neostriatum. *Journal of Comparative Neurology, 180*, 545–580.

Febo, M. (2011). Prefrontal cell firing in male rats during approach towards sexually receptive female: Interactions with cocaine. *Synapse, 65*, 271–277.

Fernández-Guasti, A., Omana-Zapata, I., Luján, M. & Condés-Lara, M. (1994). Actions of sciatic nerve ligature on sexual behavior of sexually experienced and inexperienced male rats: effects of frontal pole decortication. *Physiology & Behavior, 55*, 577–581.

Fischer, H., Merrywest, S.D. & Sillar, K.T. (2001). Adrenoreceptor-mediated modulation of the spinal locomotor pattern during swimming in Xenopuslaevis tadpoles. *European Journal of Neuroscience, 13* (5), 977–986.

Forssberg, H. & Grillner, S. (1973). The locomotion of the acute spinal cat injected with clonidine i.v. *Brain Research, 50*, 184–186.

Fuster, J.M. (1997). The prefrontal cortex: Anatomy, physiology, and neuropsychology of the frontal lobe. New York, EE.UU: Lipincott-Raven.

García-Rill, E. (1986). The basal ganglia and the locomotor regions. *Brain Research Reviews, 11*, 47–63.

Goldman-Rakic, P.S. (1996). The prefrontal landscape: Implications of functional architecture for understanding human mentation and the central executive. *Philosophical Transactions of the Royal Society London, 351*, 1445–1453.

Guevara, M.A., Martínez-Pelayo, M., Arteaga-Silva, M., Bonilla-Jaime, H. & Hernández-González, M. (2008). Electrophysiological correlates of the mesoaccumbens system during male rat sexual behaviour. *Physiology & Behavior, 95*, 545–552.

Grillner, S., Deliagina, T., Ekebuerg, O., el Manira, A., Hill, R.H., Lansner, A., Orlovsky, G.N. & Wallen, P. (1995). Neural networks that coordinate locomotion and body orientation in lamprey. *Trends in Neurosciences, 18*, 270–280.

Grillner, S. & Dubuc, R. (1988). Control of locomotion in vertebrates: spinal and supraspinal mechanisms. In S.G. Waxman (Ed.), *Advances in Neurology: Functional Recovery in Neurological Disease* (pp. 425–453). New York, EE.UU: Raven Press.

Grillner, S. & Kashin, S. (1976). On the generation and performance of swimming in fish. In R.M. Herman (Ed.), *Neural Control of Locomotion* (p. 181). New York, EE.UU: Plenum Press.

Grillner, S. & Shik, M.I. (1973). On the descending control of the lumbosacral spinal cord from the mesencephalic locomotor region. *Acta Physiologica Scandinavica, 87*, 320–333.

Grillner, S. & Wallen, P. (1985). Central pattern generators for locomotion with special reference to vertebrates. *Annual Review of Neuroscience, 8*, 233–261.

Harris-Warrick, R.M. & Johnson, B.R. (1989). Motor pattern networks: Flexible foundations for rhythmic pattern production. In T.J. Carew and D.B. Kelley (Eds.), *Perspectives in Neural Systems and Behavior* (pp. 51–71). New York, EE.UU: Alan R. Liss, Inc.

Harris-Warrick, R.M., Johnson, B.R., Peck, J.H., Kloppenburg, P., Ayali, A. & Skarbinski, J. (1998). Distributed effects of dopamine modulation in the crustacean pyloric network. *Annals of the New York Academy of Sciences, 860*, 155–67.

Hernández-González, M., Guevara, M.A., Cervantes, M., Moralí, G. & Corsi-Cabrera, M. (1998). Characteristic frequency bands of the cortico-frontal EEG during the sexual interaction of the male rat as a result of factorial analysis. *Journal of Physiology, 92*, 43–50.

Hernández-González, M., Guevara, M.A., Moralí, G. & Cervantes, M. (1997a). Subcortical multiple unit activity changes during male sexual behavior of the rat. *Physiology & Behavior, 61* (2), 285–291.

Hernández-González, M., Guevara, M.A., Moralí, G. & Cervantes, M. (1997b). Computer programs to analyze brain electrical activity during copulatory pelvic thrusting in male rats. *Physiology & Behavior, 62* (4), 701–708.

Hernández-González, M., Guevara, M.A., Oropeza, M.V. & Moralí, G. (1993). Male rat pelvic copulatory movements: Computarized analysis of accelerometric data. *Archives of Medical Research, 24* (2), 155–160.

Hernández-González, M., Oropeza, M.V., Guevara, M.A., Cervantes, M. & Moralí, G. (1994). Effects of intrathecal administration of adrenergic agonists on the frequency of copulatory pelvic thrusting of the male rat. *Archives of Medical Research, 25* (4), 419–425.

Hernández-González, M., Prieto-Beracoechea, C.A., Arteaga-Silva, M. & Guevara, M.A. (2007). Different functionality of the medial and orbital prefrontal cortex during a sexually motivated task in rats. *Physiology & Behavior, 90*, 450–458.

Hernández-González, M., Robles-Aguirre, F.A., Guevara, M.A., Quirarte, G.L. & Haro-Magallanes, P. (2014). Basolateral amygdala inactivation reduces sexual motivation in male rats during performance of a T-maze task with a sexual reward. *Journal of Behavioral and Brain Science, 4* (5), 223–232.

Horio, T., Shimura, T., Hanada, M. & Shimokochi, M. (1986). Multiple unit activities recorded from the medial preoptic area during copulatory behavior in freely moving male rats. *Neuroscience Research, 3*, 311–320.

Hull, E.M., Bazzett, T.J., Warner, R.K., Eaton, R.C. & Thompson, J.T. (1990). Dopamine receptors in the ventral tegmental area modulate male sexual behavior in rats. *Brain Research, 512*, 1–6.

Hull, E.M., Weber, M.S., Eaton, R.C., Dua, R., Markowski, V.P., Lumley, L. & Moses, J. (1991). Dopamine receptors in the ventral tegmental area affect motor, but not motivational or reflexive components of copulation in male rats. *Brain Research, 554*, 72–76.

Jovanović, K., Petrov, T. & Stein, R.B. (1999). Effects of inhibitory neurotransmitters on the mudpuppy (Necturusmaculatus) locomotor pattern in vitro. *Experimental Brain Research, 129* (2), 172–184.

Katz, P.S. & Frost, W.N. (1995). Intrinsic neuromodulation in the trifonia swim CPG: The serotonergic dorsal swim interneurons act presynaptically to enhance transmitter release from interneuron C2. *Journal of Neuroscience, 15* (9), 6035–6045.

Kolb, B. (1990). Prefrontal cortex. In B. Kolb and R.C. Tees (Eds.), *The Cerebral Cortex of the Rat* (pp. 437–58). Cambridge, EE.UU: The MIT Press.

Komisaruk, B.R. (1970). Synchrony between limbic system theta activity and rhythmical behaviour in rats. *Journal of Comparative and Physiological Psychology, 70,* 483–492.

Komisaruk, B.R. (1977). The role of rhythmical brain activity in sensorimotor integration. *Progress in Psychology and Physiological Psychology, 7,* 55–90.

Komisaruk, B.R. & Olds, J. (1968). Neuronal correlates of behavior in freely moving rats. *Science, 161,* 810–12.

Kurtz, R. & Adler N. (1973). Electrophysiological correlates of copulatory behavior in the male rat. *Journal of Comparative and Physiological Psychology, 84*, 225–239.

Larsson, K. (1979). Features of the neuroendocrine regulation of masculine sexual behavior. In C. Beyer (Ed.), *Endocrine Control of Sexual Behavior* (pp. 77–163). New York, EE.UU: Raven Press.

Lillvis, J.L. & Katz, P.S. (2013). Parallel evolution of serotonergic neuromodulation underlies independent evolution of rhythmic motor behavior. *Journal of Neuroscience, 33* (6), 2709–2717.

Lindvall, O. & Bjorklund, A. (1974). The organization of the ascending catecholamine neuron systems in the rat brain as revealed by the glyoxylic acid fluorescence method. *Acta Physiologica Scandinavica, 412*, 1–48.

Liu, Y., Sachs, B. & Salamone, J. (1998). Sexual behavior in male rats after radiofrequency or dopamine-depleting lesions in nucleus accumbens. *Pharmacology Biochemestry and Behavior, 60*, 585–592.

Maras, P.M. & Petrulis, A. (2006). Chemosensory and steroid-responsive regions of the medial amygdala regulate distinct aspects of opposite-sex odor preference in male Syrian hamsters. *European Journal of Neuroscience, 24*, 3541–3552.

Marder, E. & Bucher, D. (2001). Central pattern generators and the control of rhythmic movements. *Current Biology, 11*, 986–996.

McIntosh, R., Barfield, R. & Thomas, D. (1984). Electrophysiological and ultrasonic correlates of reproductive behavior in the male rat. *Behavioral Neuroscience, 98*, 1100–1103.

McLean, D.L. & Sillar, K.T. (2003). Spinal and supraspinal functions of noradrenaline in the frog embryo: consequences for motor behavior. *Journal of Physiology, 551* (2), 575–587.

Mogenson, G.J., Jones, D.L. & Yim, C.Y. (1980). From motivation to action: Functional interface between the limbic system and the motor system. *Progress in Neurobiology, 14*, 69–97.

Mogenson, G.J. & Yong, C.R. (1991). The contribution of basal forebrain to limbic-motor integration and the mediation of motivation to action. *Advances in Experimental Medicine and Biology, 295*, 267–290.

Moralí, G. & Beyer, C. (1992). Motor aspects of masculine sexual behavior in rats and rabbits. In P.J.R. Slater, J.S. Rosenblatt, C. Beer and M. Milinski (Eds.), *Advances in the Study of Behavior* (pp. 201–238). New York, EE.UU: Academic Press.
Moralí, G., Carrillo, L. & Beyer, C. (1985). Neonatal androgen influences sexual motivation but not the masculine copulatory pattern in the rat. *Physiology & Behavior, 34*, 265–275.
Moralí, G., Contreras, J.L. & Beyer, C. (1982). Effects of penile denervation and genital anaesthetization on the motor copulatory pattern of the rat and the rabbit. *Conf. Reprod. Behav. 14th* (p. 77). East Lansing, Michigan.
Morali, G., Hernández, G. & Beyer, C. (1986). Restoration of the copulatory pelvic thrusting pattern in castrated male rats by the intracerebral implantation of androgen. *Physiology & Behavioral, 36*, 495–499.
Moralí, G., Komisaruk, B.R. & Beyer, C. (1989). Copulatory pelvic thrusting in the male rat is insensitive to the perispinal administration of glycine and GABA antagonists. *Pharmacology, Biochemistry and Behavior, 32*, 169.
Morali, G. & Larsson, K. (1984). Differential effects of a new serotoninomimetic drug, 8-OH-DPAT, on copulatory behavior and pelvic thrusting pattern in the male rat. *Pharmacology, Biochemistry and Behavior, 20* (2), 185–187.
Olivier, B., Chan, J.S.W., Snoeren, E.M., Olivier, J.D.A., Veening, J.G., Vinkers, C.H., Waldinger, M.D. & Oosting, R.S. (2011). Differences in Sexual Behaviour in Male and Female Rodents: Role of Serotonin. In J.C. Neill and J. Kulkarni (Eds.), *Biological Basis of Sex Differences in Psychopharmacology.* Current Topics in Behavioral Neurosciences 8. Berlin, Germany: Springer-Verlag.
Oomura, Y., Yoshimatsu, H. & Aou, S. (1983). Medial preoptic and hypothalamic neuronal activity during sexual behavior of the male monkey. *Brain Research, 266*, 340–343.
Paredes, R.G., Highland, L. & Karam, P. (1993). Socio-sexual behaviour in male rats after lesions of the medial preoptic area: Evidence of reduced sexual motivation. *Brain Research, 618*, 271–276.
Paredes, R.G., Tzschentke, T. & Nakach, N. (1998). Lesions of the medial preoptic area/anterior hypothalamus (MPOA/AH) modify partner preference in male rats. *Brain Research, 813*, 1–8.
Pearson, K.G. & Gordon, J.E. (2013). Locomotion. In E.R. Kandel, J.H. Schwartz, T.M. Jessel, S.A. Siegelbaum and A.J. Hudspeth (Eds.), *Principles of Neural Sciences* (pp. 812–834). New York, EE.UU: Mc Graw Hill Companies Inc.
Pratt, C.A. & Jordan, L.M. (1987). Ia inhibitory interneurons and Renshaw cells as contributors to the spinal mechanisms of fictive locomotion. *Journal of Neurophysiology, 57*, 56–71.
Qureshi, G.A. & Södersten, P. (1986). Sexual activity alters the concentration of amino acids in the cerebralspinal fluid of male rats. *Neuroscience Letters, 70*, 374–378.
Rose, J.E. & Woolsey, C.N. (1948). The orbitofrontal cortex and its connections with the mediodorsal nucleus in rabbit, sheep and cat. *Association for Research in Nervous and Mental Disease, 27*, 210–232.
Sachs, B.D. & Barfield, R.J. (1976). Functional analysis of masculine copulatory behavior in the rat. In J.S. Rosenblatt, R.A. Hinde, E. Shaw and C. Beer (Eds.), *Advances in the Study of Behavior* (p. 91). New York, EE.UU: Academic Press.
Semba, K., Szechtman, H. & Komisaruk, B.R. (1980). Synchrony among rhythmical facial tremor, neocortical "alpha" waves, and thalamic non-sensory bursts in intact awake rats. *Brain Research, 195,* 281–298.
Shimura, T., Yamamoto, T. & Shimokochi, M. (1994). The medial preoptic area is involved in both sexual arousal and performance in male rats: Reevaluation of neuron activity in freely moving animals. *Brain Research, 640*, 215–222.
Skinner, R.D. & García-Rill, E. (1984). The mesencephalic locomotor region (MLR) in the rat. *Brain Research, 323*, 385–389.

Swanson, L.W. & Mogenson, G.J. (1981). Neural mechanisms for the functional coupling of autonomic, endocrine and somatomotor responses in adaptive behavior. *Brain Research Reviews*, *3*, 1–34.

Uylings, H.B.M., Groenewegen, H.J. & Kolb, B. (2003). Do rats have a prefrontal cortex? *Behavioural Brain Research, 146,* 3–17.

Uylings, H.B.M. & Van Eden, C.G. (1990). Qualitative and quantitative comparison of the prefrontal cortex in rat and in primates, including humans. In H.B.M. Uylings, C.G. Van Eden, J.P.C. de Bruin, M.A. Corner and M.P.G. Feenstra (Eds.), *The Prefrontal Cortex: Its Structure, Function and Pathology. Progress in Brain Research, 85* (31–62). Amsterdam, the Netherlands: Elsevier.

Wiener, S.I., Paul, C.A. & Eichenbaum, H. (1989). Spatial and behavioral correlates of hippocampal neuronal activity. *The Journal of Neuroscience, 9*, 2737–2763.

Wiesenfeld-Hallin, Z. (1987). Intrathecal noradrenaline has a dose-dependent inhibitory or facilitatory effect on the flexion reflex in the rat. *Acta Physiologica Scandinava, 130*, 507–511.

Wood, R.I. (1998). Integration of chemosensory and hormonal input in the male Syrian hamster brain. *Annals of the New York Academy of Sciences*, *855*, 362–372.

Yim, C.Y. & Mogenson, G.J. (1980). Electrophysiological studies of neurons in the ventral tegmental area of Tsai. *Brain Research, 181*, 301–313.

5 Male Sexual Satiety and the Coolidge Effect in Rats

Relation between Behavioral and Seminal Parameters

Rosa Angélica Lucio, Alonso Fernández-Guasti, and Knut Larsson

CONTENTS

5.1 Copulation ... 84
    5.1.1 Motor Patterns and Genital Responses ... 84
5.2 Ejaculation and Ejaculate ... 84
    5.2.1 Seminal Plug and Transcervical Sperm Transport ... 85
    5.2.2 Different Kinds of Ejaculate ... 86
5.3 Sexual Satiety ... 88
    5.3.1 Why Copulating So Much? ... 88
    5.3.2 Sexual Incentive and Copulation ... 89
    5.3.3 The Coolidge Effect ... 90
    5.3.4 Does Copulation after Satiety Include Seminal Expulsion? ... 91
    5.3.5 What is the Adaptive or Evolutionary Significance of the Coolidge Effect? ... 92
5.4 Recovery from Sexual Satiety ... 93
    5.4.1 Reestablishment of Copulatory Behavior ... 93
    5.4.2 Restoration of Ejaculation ... 95
5.5 Summary ... 97
5.6 Concluding Remarks ... 97
Epilogue ... 98
Acknowledgments ... 98
References ... 98

5.1 COPULATION

Copulation consists of behavioral and physiological events. For males, the behavioral events consist of the skeletal muscle movements that comprise the copulatory motor patterns (usually mounts, intromissions and ejaculation), and the physiological events of penile erection and seminal expulsion.

5.1.1 MOTOR PATTERNS AND GENITAL RESPONSES

Male rats display stereotypic motor patterns during copulation, identified as mounts, intromissions and ejaculation. During the mount, the male approaches the female rear, palpates her flanks with his forepaws and executes pelvic thrusts. Intromissions are characterized by deeper intravaginal thrusts followed by an abrupt dismount, indicating that penile insertion has occurred (Pollak and Sachs 1976; Sachs and Meisel 1988). Some treatments prevent the male from achieving penile insertion but do not prevent display of the motor pattern of intromission (Pollak and Sachs 1976). A series of mounts and intromissions leads to ejaculation, which is behaviorally characterized by a deeper long thrust and a slower dismount. Ejaculation is the forceful expulsion of the ejaculate from the distal urethra. Rhythmic contractions of skeletal muscles of the hips, hind limbs, forelimbs, and perineum (bulbospongiosus, ischiocavernosus, and external anal sphincter) characteristically accompany ejaculation. Seminal expulsion does not necessarily occur when the skeletal motor pattern of ejaculation is observed (Sachs and Meisel 1988; Hull et al. 2006). Furthermore, under certain experimental conditions, seminal expulsion may occur in the absence of the characteristic ejaculatory motor pattern (Larsson and Swedin 1971; McGlynn and Erpino 1974; Queen et al. 1981).

Thus, during copulation, ejaculation and other responses should be verified directly, rather than based solely on observation of the motor patterns. In rats, it is difficult to observe penile erections because they are brief; nevertheless, they can be observed directly by using a mirror placed beneath the observation cage at 45°. Ascertaining that ejaculation has occurred is more certain than penile erection, as it can be verified by the presence of a seminal plug in the female's vagina using a radiopaque contrast medium (Wallach and Hart 1983). If further analyses are required, the seminal plug can be obtained by dissection of the vaginal wall or by collecting the plugs dislodged after intromissions executed by a sexually experienced male (McClintock et al. 1982; Lucio et al. 1994; Tlachi-López et al. 2012; Lucio et al. 2014). Additional methods may be used, for example, registering the activity of the ischiocavernosus (during erection) or bulbospongiosus muscles (during ejaculation; Holmes et al. 1991).

5.2 EJACULATION AND EJACULATE

Ejaculation involves the coordinated succession of physiological events, including seminal emission and seminal expulsion. The emission phase corresponds to the secretion of the different components of the seminal fluid from the accessory sex glands and sperm transport from the epididymis caudae. During emission, the bladder neck sphincters close and the vas deferens, seminal vesicles, prostate and coagulating glands contract. Contractions result in the deposition of seminal secretions into the prostatic urethra. Once emission is completed, seminal expulsion then occurs. The semen is rapidly advanced through the urethra (membranous and penile) and spurts out the urethral meatus. Contractions of the perineal adjacent muscles to the base of the penis also contribute to the seminal expulsion (Shafik et al. 2005; Hull and Rodríguez-Manzo 2009; Clément and Giuliano 2011).

Seminal expulsion consists of the ejaculate, gametic cells, and fluids of the accessory sexual glands. The gametic cells, that is, spermatozoa produced by the testes, undergo a prior process of maturation and storage into the epididymis, which is a tubular coiled structure. The ejaculatory fluids, that is, seminal plasma, are produced by the accessory sexual glands: seminal vesicles, prostate, coagulating and bulbourethral glands. The seminal plasma provides a nutritive and protective medium for the spermatozoa during their transit along the female's reproductive tract. In species with multipartner mating, the typical seminal coagulation process is intensified, in which case a solid copulatory plug forms (Poiani 2006).

5.2.1 Seminal Plug and Transcervical Sperm Transport

The rat seminal plug principally consists of secretions of the accessory sex glands and some dead spermatozoa (Carballada and Esponda 1993). In this species, the weight, size, hardness, and adhesion to the cervix–vagina of the seminal plug depends on the seminal vesicles, the coagulating glands (Carballada and Esponda 1992) and the prostate (Tlachi-López et al. 2011).

Upon ejaculation, the sperm are expelled first, followed by the secretions of some sex glands. The secretions deposited into the vagina harden to form the seminal plug (Blandau 1945). In the rat, the rostral end of the plug forms a cup at the cervix–vagina junction holding the sperm under pressure until transcervical sperm transport is completed, approximately 6 minutes after ejaculation. Most of the transport occurs during the first 2 minutes after ejaculation (Matthews and Adler 1977).

The seminal plug can be dislodged by a subsequent few intromissions executed by the same or a different male (Wallach and Hart 1983; Lucio et al. 1994). The post-ejaculatory period that persists 5–7 minutes (after the first ejaculation) prevents the male from removing his own plug before sperm transport occurs. Removal of the plug within less than 6 minutes after ejaculation disrupts the transcervical sperm transport, resulting in a reduced number of sperm in the uterus (Matthews and Adler 1977). If the plug is removed later than 6 minutes after ejaculation, the sperm will have been transported through the cervix into the uterus. Under natural and seminatural conditions, sexually receptive female rats may copulate with different males (McClintock and Adler 1978; McClintock et al. 1982). Consequently, if two different males ejaculate into the same female, there will be sperm "competition," in which sperm of different males "compete" to fertilize the ova (Parker 1998). This situation may provide evolutionary advantage by increasing the genetic diversity of the offspring.

Dislodgement of the seminal plug may be facilitated by the existence of backward-facing penile spines, which can pull the plug back toward the vaginal orifice during the rapid dismount, which is characteristic of the motor pattern of intromission. Only 3–4 intromissions may be sufficient to remove the seminal plug from the vagina

(Wallach and Hart 1983; Lucio et al. 1994; Tlachi-López et al. 2012; Lucio et al. 2014). Once a male dislodges the seminal plug that was deposited by a prior male, he can deposit his own ejaculate. When two male rats mate with the same female, both males have offspring. However, the second male has the advantage in paternity only when ejaculation occurs very close to the first male's ejaculation. If more than 5 minutes elapse, the advantage in fathering pups is lost (Coria-Avila et al. 2004).

5.2.2 Different Kinds of Ejaculate

The ejaculate can be evaluated in samples obtained from the male or in samples taken from a recently mated female, that is, direct/indirect spermatobioscopy, respectively. Because electroejaculation is difficult to implement in male rats, the ejaculate samples had been primarily analyzed by indirect spermatobioscopy. To obtain the ejaculate some authors inject saline solution into each uterine horn of a recently mated female and then aspirate the content (Matthews and Adler 1977; Austin and Dewsbury 1986; Carballada and Esponda 1992). Others remove the uteri from the abdominal cavity and after the elimination of blood, fat tissue and external uterine vessels; each uterine horn is squeezed gently. Then the uterine fluid content is emptied. The parameters evaluated in the sample obtained include: sperm count, motility, viability, and morphology. The seminal plug may also be analyzed. The plug is obtained from the vagina by separating the pubic symphysis and making an incision in the vaginal wall (for details, see Lucio and Tlachi-López 2008).

Under typical laboratory conditions, the male rat is exposed to a sexually receptive female in a single-chambered arena. In this paradigm, a succession of 5–18 mounts and intromissions precedes ejaculation. Within each such ejaculatory series, the number of mounts and intromissions preceding ejaculation and the ejaculation latency are recorded. After ejaculation there is a postejaculatory interval, which increases exponentially with successive copulations (Larsson 1956). The ejaculate obtained after the first ejaculatory series contains 13–20 million spermatozoa, of which 99%–100% show normal morphology, 61%–67% are alive and 70%–87% show progressive motility, that is, the spermatozoa move forward. The plugs were found to be 104–126 mg in weight, 11–13 mm in length and 4.9–5.7 mm in width (Lucio and Tlachi-López 2008). The studies analyzing the ejaculate parameters primarily refer to the sperm count, weight, and size of the seminal plug (Austin and Dewsbury 1986; Carballada and Esponda 1992, 1993).

Austin and Dewsbury (1986) analyzed the variation in seminal parameters through seven consecutive ejaculatory series, each with a different female. The sperm count obtained in the first and second ejaculates were similar (approximately 20 million), those obtained after the third to fifth ejaculation showed a progressive decrease, and by the sixth and seventh ejaculation diminished sharply to approximately 2 and 1 million, respectively. The seminal plugs weighed 100 mg and 40 mg in the first and seventh ejaculates, respectively, and the plug length diminished from 13 to 7 mm, while the width of the plug persisted at 5 mm (Austin and Dewsbury 1986). In our laboratory, we evaluated these parameters after eight ejaculatory series and found no sperm in the uteri and a minuscule seminal plug (Tlachi-López et al. 2012; see Table 5.1).

TABLE 5.1
Behavioral and Seminal Parameters during Sexual Satiety Development and the Coolidge Effect

	Sexual Satiety Development Number of Ejaculations										
Parameters	**1**	**2**	**3**	**4**	**5**	**6**	**7**	**8**	**Sexual Satiation**	**Satiation with Coolidge Effect**	**References**
Behavior											
Males ejaculating (%)	100	100	100	100	100	*92*	*56*	*44*	0	*89*	Rodríguez-Manzo and
Number of intromissions	12	8	9	10	8	*6*	*4*	*4*	–	*4*	Fernández-Guasti
Ejaculation latency (min)	7 ± 0.7	5 ± 0.7	7 ± 0.8	9 ± 2	9 ± 3	*12 ± 2*	*8 ± 2*	*9 ± 4*	–	*2 ± 0.5*	1994; Rodriguez-
Postejaculatory interval (min)	6 ± 0.5	8 ± 0.4	9 ± 0.8	14 ± 3	24 ± 4	*47 ± 7*	*42 ± 7*	*55 ± 11*	–	–	Manzo et al. 2011
Ejaculate											
Sperm count in uterine horns (10^6)	19	12	12	8	11	2	1	0	–	0	Austin and Dewsbury
Seminal plug length (mm)	12	10	10	11	7	7	6	7	–	0	1986; Tlachi-López
Width (mm)	6	5	5	5	5	4	4	4	–	0	et al. 2012
Weight (mg)	100	80	90	80	50	40	30	35	–	0	
Sperm count in epididymis cauda (10^6)	550	–	–	–	–	–	–	–	–	309	
Plugs dislodgement											
Number of intromissions	3	–	–	–	–	–	–	–	–	2	Lucio et al. 2014
Intromission latency (sec)	13 ± 4	–	–	–	–	–	–	–	–	42 ± 10	
Time to remove plug (sec)	63 ± 25	–	–	–	–	–	–	–	–	97 ± 32	
Males interrupting other male's sperm transport (%)	17	–	–	–	–	–	–	–	–	33	
Males preventing other male's sperm transport (%)	83	–	–	–	–	–	–	–	–	67	
Pregnant females after exposed to control or satiated males (%)	0	–	–	–	–	–	–	–	–	0	

From the sixth behavioral ejaculation onward, the sperm count was drastically diminished (the minimum fertile sperm count is above 5-8 million, according to Toner and Adler [1985]). The behavioral ejaculation during the Coolidge Effect lacked sperm. Consequently, the mating during the Coolidge Effect was infertile, corresponding to the 6th to 8th behavioral ejaculations during sexual satiety development (represented by italic numbers). No differences were found in any plug dislodgement parameters between males ejaculating once and those exposed to the Coolidge Effect.

5.3 SEXUAL SATIETY

In males of many species, from worms to humans, there is a process of inhibition of copulatory behavior after repeated mating (for review in rats, see Phillips-Farfán and Fernández-Guasti 2009; for rabbits, see Jiménez et al. 2012). In the male rat, repeated mating results in changes in various parameters of the copulatory series, including a reduction in the number of intromissions preceding ejaculation and an increase in the ejaculatory latency (Table 5.1). However, the most drastic change is the exponential increase in the length of the refractory periods that follow ejaculation (Beach and Jordan 1956; Larsson 1956; Rodríguez-Manzo and Fernández-Guasti 1994). The absence of copulatory behavior during 30 minutes (Beach and Jordan 1956; Tlachi-López et al. 2012) or 90 minutes after the last ejaculation (Rodríguez-Manzo and Fernández-Guasti 1994) is considered to be "sexual satiety." In the male rat there are at least two methods to induce sexual satiety: copulatory testing during a fixed period or *ad libitum*. The fixed period method consists of providing an opportunity for 1 hour of copulation every 4 days (three repetitions) or every 3 days (four repetitions), or during 5 consecutive days (Larsson 1956). The *ad lib* paradigm provides an opportunity for repetitive ejaculatory series until the male stops copulating for a selected duration. In our laboratory conditions, the males may vary in the number of consecutive ejaculatory series before reaching sexual exhaustion, ranging from 5 (Phillips-Farfán and Fernández-Guasti 2009) to 18 (Tlachi-López et al. 2012) during a period of 4–6 hours with the same female. It is generally assumed that the male ceases to copulate because sexual motivation is diminished (Agmo 1999).

Along with the development of sexual satiety, there is a parallel reduction in multiple behavioral and physiological functions, which raises the question of its adaptive significance (see Table 5.1; Phillips-Farfán and Fernández-Guasti 2009). Biological evolution has taught us to consider the likely adaptive significance of specific behavioral patterns (Dewsbury 1981; Dimijian 2005), which leads us to ask what may be the adaptive significance of sexual satiety. One possibility is that it facilitates the replenishment of the male rat's energy stores (Phillips-Farfán and Fernández-Guasti 2009).

5.3.1 Why Copulating So Much?

Copulatory behavior is generally considered to be a series of stereotypic interactive patterns having functional or adaptive significance. Agmo (1999) concluded, after a detailed review, that a particular behavior pattern does not have any purpose or function; instead, it may have consequences that may be advantageous or disadvantageous for the individual or the species. What about sexual behavior? Promiscuous animals copulate when they have the opportunity. Many species considered to be monogamous frequently display extra-pair sexual behavior, that is, not always with the stable pair. When the animal engages in sexual behavior, it is not for the sake or "purpose" of reproduction as far as the individual is concerned (Agmo 1999). Humans use contraceptives precisely to differentiate copulation from reproduction. It is reported that there is just one single fertilization per 1,100 copulations among

Swedes, thus rendering fertile copulation a rare event (Zetterberg 1969; referred in Agmo 1999).

In many species, including humans, mating partners possibly never see each other again after copulation. Most likely, copulation has no functional significance for the individuals other than that it is perhaps reinforcing (Agmo 1999, 2010). The reinforcement intensity is determined by the cues present during copulation, such as those of the partner. If the sexual activity was strongly rewarding, the environmental stimuli, including the partner, will become incentives for an increased likelihood of approach in the future. However, if the sexual activity was only modestly rewarding, the incentive qualities of these same stimuli will be less (Agmo 1999).

In naïve male rats the occurrence of one ejaculation is a powerful reinforcing stimulus (Kippin and Pfaus 2001). It is unknown if multiple ejaculations are more rewarding than a single ejaculation (Crawford et al. 1993). Male rats engage in copulation for extended periods (Beach and Jordan 1956; Larsson 1956; Fisher 1962; Wilson et al. 1963; Hsiao 1965; Rodríguez-Manzo and Fernández-Guasti 1994; Tlachi-López et al. 2012). Thus, it may be concluded that repetitive mating occurs in response to its rewarding properties and that sexual satiety results from the male's losing sexual motivation for a particular female (Agmo 1999). In support of the latter, we and others have found that a novel receptive female induces sexual activity in sexually satiated males, which has been termed, the "Coolidge Effect" (see below, Wilson et al. 1963; Hsiao 1965; Tlachi-López et al. 2012; Jiménez et al. 2012 for rabbits).

5.3.2 Sexual Incentive and Copulation

The process that induces an animal to seek sexual contact is termed "sexual motivation." The earliest systematic analysis of sexual motivation was that of Frank Beach (1956). He first coined the term "sexual arousal mechanism" that was described as "the processes underlying the activation of sexual behavior, which increase the male's sexual excitement to such a pitch that the copulatory threshold is attained." Over the years the concept "sexual arousal mechanisms" was defined as sexual motivation, libido or sexual drive, as it is currently termed. "Drive" is a term for neural states that energize a specific behavior pattern and it has two components: the first is necessary for "energizing" behavior, that is, "arousal"; the second is the mechanism that responds to particular physiological signals arising from specific biological imperatives such as sexual behavior, essential to species survival. This second component gives "direction" to the motivated behavior (Pfaff 1999).

Events or stimuli may be distinguished as hedonically potent and hedonically neutral. The former are separated as positive or negative (Bindra 1976; referred to in Agmo 1999). Hedonically positive events/stimuli are related to approach behavior and are termed positive incentives. Therefore, the events/stimuli having rewarding properties are called incentives. The motivation aroused by such events/stimuli is known as incentive motivation. Sexual motivation in males is activated when an appropriate stimulus is perceived, for example, an estrous female. Stimuli with sexual significance, that is, sexual incentives, are able to activate approach behavior. Once approach behavior occurs it is also likely that copulation will be executed (Agmo 1999, 2010).

Because masculine copulatory patterns are highly stereotypic motor events (Moralí and Beyer 1992) they are considered somatic reflexes. Pelvic thrusting related to mount, intromission and ejaculation is remarkably stable in duration, amplitude and frequency, as indicated by accelerometric studies (Beyer et al. 1982). Penile events such as erection and activity of the perineal muscles constitute the viscerosomatic reflexes of copulation (Holmes et al. 1991). Male copulatory patterns are relatively insensitive to environmental disturbances, for example, intense flashing light or loud noise that do not modify mount-bout intervals (Paredes and Agmo, unpublished results; Agmo 1999). This observation suggests that, once initiated, the execution of copulatory patterns can continue to ejaculation relatively independent of surrounding events. However, it is also the case that reproductive behavior can be interrupted and even coexpressed with other behaviors by external stimuli and contexts that have biological relevance (Agrati et al. 2011).

5.3.3 The Coolidge Effect

The term, "Coolidge Effect" was first used in print by Wilson et al. (1963) and its history was detailed by Bermant (1976). The term has its origins in an anecdote about the President of the United States, Calvin Coolidge.

> One day President and Mrs. Coolidge were visiting a government farm. Soon after their arrival they were taken off on separate tours. When Mrs. Coolidge passed the chicken pens she paused to ask the man in charge if the rooster copulates more than once each day. "Dozens of times" was the reply. "Please tell that to the President," Mrs. Coolidge requested. When the President passed the pens and was told about the rooster, he asked, "Same hen every time?" "Oh no, Mr. President, a different one each time." The President nodded slowly, "Tell *that* to Mrs. Coolidge" (Bermant 1976, pp. 76–77).

As mentioned above, sexually exhausted male rats do not execute mounts, intromissions, or ejaculations for a period of 30–90 minutes after they have copulated *ad lib* with the same female (Phillips-Farfán and Fernández-Guasti 2009; Tlachi-López et al. 2012). A large body of evidence indicates that sexual motivation is diminished after copulation with the same female. However, if the female is replaced by a different sexually receptive female, the male will resume copulation. Restoration of mating behavior with a different female from that with which the male attained sexual satiety is termed the "Coolidge Effect" (Wilson et al. 1963; Hsiao 1969). It is also generally assumed that the new female increases sexual motivation in the male, reinforcing the notion that sexual satiety is due to a low level of motivation and not to an inability of the male to display sexual behavior after repeated mating. Copulation of previously satiated males, during the Coolidge Effect, includes not only mounts and intromissions but also ejaculation (Beach and Jordan 1956; Larsson 1956; Rodríguez-Manzo and Fernández-Guasti 1994; Phillips-Farfán and Fernández-Guasti 2009).

The Coolidge Effect is observed in approximately 90% of male rats. That is, in sexually satiated males, replacing the initial female with a different female resulted in 88% of the males recommencing sexual activity (Tlachi-López et al. 2012). It has been speculated that the Coolidge Effect may benefit the contribution of both females and males to reproductive success of the species (Adler 1978, 1979; Wedell et al. 2002; Steiger et al. 2008). That is, in the female it increases the probability of receiving enough sperm to ensure complete ova fertilization. Ejaculates of different males promote sperm competition (Parker 1998), increasing the genetic diversity of the pups. The advantage to the males includes their wider sperm distribution among different available females, thereby increasing his offspring (Wedell et al. 2002).

While these concepts and biological outcomes of the Coolidge Effect seem reasonable, they imply that the ejaculation during the Coolidge Effect is fertile. In a series of experiments, we tested this assumption (see Table 5.1). We questioned the fertile capacity of ejaculation during the Coolidge Effect primarily on the basis of previous findings that after several ejaculations the sperm count and the size of the seminal plug are drastically reduced (Austin and Dewsbury 1986; Tlachi-López et al. 2012).

5.3.4 Does Copulation after Satiety Include Seminal Expulsion?

Regarding the possible adaptive significance of the Coolidge Effect, we asked: do sexually satiated males in the presence of a new female expel semen when they perform the behavioral ejaculation pattern? Recall that the ejaculate of males at or near sexual satiety—after eight ejaculatory series—lacks spermatozoa and that the size of the seminal plugs is minuscule (Tlachi-López et al. 2012). Such lack of sperm was unrelated to the number of ejaculations required to reach sexual satiety or with the number of intromissions needed to ejaculate during the Coolidge Effect (Tlachi-López et al. 2012). Is sexual novelty sufficient to stimulate seminal expulsion? No, on the basis that we recently demonstrated that sexually satiated males during the Coolidge Effect failed to expel semen. In the reproductive tract of the females mated with sexually exhausted males exposed to the Coolidge Effect, no sperm or seminal plugs were found (Table 5.1; Tlachi-López et al. 2012). This finding led us to conclude that the behavioral motor pattern of ejaculation is independent of seminal expulsion. Then we may ask why sexually satiated males, exposed to the Coolidge Effect, are able to display the behavioral motor pattern of ejaculation. Do they have functional penile erections? The answer to these questions is related to the "ejaculatory threshold." Sexually satiated males are able to display the behavioral motor pattern of ejaculation, which requires penile "excitation" (Bermant 1967; Hull and Rodríguez-Manzo 2009). While these males displayed the overt behavioral pattern characteristic of intromission, the authors provided no direct evidence of their erectile capacity. Thus, we asked whether they removed the seminal plug deposited by other males in the female's vagina, which requires penile erection (Wallach and Hart 1983; Lucio et al. 1994; Tlachi-López et al. 2012; Lucio et al. 2014). We found that sexually satiated males exposed to the Coolidge Effect do indeed expel the seminal plugs deposited by other males. Thus, the mean (±SEM) number of intromissions needed to dislodge another male's seminal plug was 2.8 ± 0.5 for the (satiated) experimental group *vs.* 2.1 ± 0.4 for the (nonexhausted) controls. These data provide evidence that sexually satiated male rats retain penile erection sufficient for intromission (Table 5.1; Lucio et al. 2014). Consequently, the intromissions provide stimulation sufficient for the male to surpass the threshold of the behavioral pattern of ejaculation.

Recall that sexually satiated males exposed to the Coolidge Effect fail to expel semen or deposit a seminal plug (Tlachi-López et al. 2012; Lucio et al. 2014). To analyze the physiological basis, we measured the amount of sperm stored in the caudae epididymis that is ready to be expelled during seminal expulsion. We found that after sexual satiation there was a 44% decrease in the epididymal sperm count. Despite this drastic reduction, there were still approximately 300 million spermatozoa (Tlachi-López et al. 2012), that is, far more than the amount observed in a first ejaculation, which is approximately 20 million (Austin and Dewsbury 1986; Lucio y Tlachi-López 2008). Regarding the lack of production of a seminal plug after sexual exhaustion, seminal plasma contains the various secretions of the accessory sexual glands—seminal vesicles, coagulating glands and prostate. Their content gradually declines after repeated ejaculations (Pessah and Kochva 1975). Therefore, even though there are sufficient sperm in the caudae epididymis, there are insufficient sexual gland secretions.

During the Coolidge Effect, males preserve their behavioral motor patterns of intromission and ejaculation, maintain penile erection, but they do not ejaculate. Thus, the behavioral motor pattern of ejaculation is dissociable from semen expulsion (Tlachi-López et al. 2012).

5.3.5 What is the Adaptive or Evolutionary Significance of the Coolidge Effect?

Does the Coolidge Effect facilitate the diffusion of the male's genes among different females? (Adler 1978). If so, it would imply that the behavioral ejaculation would be accompanied by seminal expulsion (with a reasonable sperm count of approximately 20 million) and an optimal seminal plug deposition.

It is often implied that the energetic "expense" of sperm production is minimal, but that is actually not the case (Parker 1982; Olsson et al. 1997). Sperm production is energetically costly; males would be selected to maximize the number of females they are able to inseminate (Parker 1970; Wedell et al. 2002). On the other hand, male mammals are capable of delivering a limited number of ejaculations (Dewsbury 1981) as evidenced for rats (Austin and Deswbury 1986; Tlachi-López et al. 2012) and rabbits (Jiménez et al. 2012). Then, what could be the adaptative value of copulating with a novel female if no semen is expelled? Immediately after sexual exhaustion, most of the male rats exposed to the Coolidge Effect display 1–3 ejaculatory series (Hsiao 1965). Rats are gregarious and promiscuous, and under natural or seminatural conditions, females cohabitating with others, synchronize their cycles, and at certain phases, there are multiple females available to mate (Robitaille and Bovet 1976; McClintock and Adler 1978; McClintock et al. 1982). Males in this environment can copulate several times with different females during the same or consecutive days (Robitaille and Bovet 1976; Calhoun 1963). If males attain sexual satiety after repeated encounters, their ejaculate is absent, but when they copulate during the Coolidge Effect, they can still dislodge the seminal plug deposited by others (Tlachi-López et al. 2012). Thus, it seems plausible that males that copulate without expelling semen remain as behavioral competitors, alternating copulations with other nonsexually satiated males. Male rats do not fight each other, but instead copulate alternately (McClintock et al. 1982; Wallach and Hart 1983). It is very likely that males ignore the quality of their ejaculate. Nevertheless, they copulate repeatedly as a result of its rewarding property (Agmo 1999). During the Coolidge Effect, sexually satiated males display a similar number of intromissions, and a similar latency to dislodge the seminal plugs deposited by other males (Lucio et al. 2014). Moreover, sexually satiated males dislodged other male's plugs in less than 60 seconds, effectively interrupting or preventing the sperm transport from the vagina to the uteri, and consequently preventing pregnancy induced by others (Table 5.1; Lucio et al. 2014). Thus, copulation postsatiation, during the Coolidge Effect, may be adaptative due to the interference with rival males' impregnating the females. In fact, impregnation is avoided if satiated males copulate with recently mated females (Lucio et al. 2014). Consequently, the evolutionary significance of the Coolidge Effect can be viewed as a strategy that offers reproductive advantages.

5.4 RECOVERY FROM SEXUAL SATIETY

Repeated copulation has behavioral and physiological consequences: sex behavior is inhibited and the ejaculate is absent. After sexual satiety, a prolonged period of inactivity enables recovery of copulation and ejaculation (Table 5.2).

5.4.1 Reestablishment of Copulatory Behavior

The earlier studies of sexual satiation in male rats by Knut Larsson (1956) and Beach and Jordan (1956) also analyzed the recuperation of sexual behavior. Larsson indicated that the males are recovered from sexual satiety when their copulatory parameters are similar to those obtained by these same males before sexual satiety. Larsson reported that a 15-day postsatiation interval is required for the full recovery of sexual behavior. Beach and Jordan (1956) emphasized the number of ejaculatory series displayed by males as the most important parameter indicative of the level of recovery after satiety. They recognized that the execution of more than five consecutive copulatory series indicated that males have recovered from sexual satiety. They found that males displayed five series at day 15–16 postsatiety, and seven series at day 18 postsatiety. Normally, nonsatiated males perform a median of seven ejaculatory series fluctuating between 5 and 12 according to Rodríguez-Manzo and Fernández-Guasti (1994), and 6–18 according to Tlachi-López et al. (2012). Consequently, it is appropriate to wait at least 18 days after sexual satiation for the total recovery of male sexual behavior. The recovery from sexual satiety is also related to the protocol used to satiate the males and the satiety criteria. After 4 hours exposure to a sexually receptive female, male rats are sexually satiated. Most of the sexual activity is restricted to the first 2.5 hours and in the subsequent 90 minutes, they are mostly inactive and may be considered sexually satiated. If we test these males 24 hours later, 67% will not show sexual activity, even if provided with a sexually receptive female (i.e., the Coolidge Effect is restricted to the period of inactivity immediately after exhaustion),

TABLE 5.2
Behavioral and Seminal Parameters: Recovery after Sexual Satiation of Male Rats

Percentage of males showing	Standard values	Sexual Satiety Recovery Number of Postsatiety Days 1	2	3	4	5	7	10	15	References
Behavior										
One series with motor pattern of ejaculation		*30*	*30*	*70*	*95*		*100*	*92*	100	Rodríguez-Manzo and Fernández-Guasti 1994; Rodríguez-Manzo et al. 2011; Romano-Torres et al. 2007
Two series with motor pattern of ejaculation		*0*	*8*	*60*	*95*		*100*	*92*	100	
Three series with motor pattern of ejaculation		–	*8*	*40*	*86*		*95*	*88*	100	
Four series with motor pattern of ejaculation		–	–	–	*38*		*85*	*84*	100	
Five series with motor pattern of ejaculation		–	–	–	–		*60*	*76*	100	
Six series with motor pattern of ejaculation		–	–	–	–		*25*	*48*	*74*	
Seven series with motor pattern of ejaculation		–	–	–	–		*5*	*12*	*48*	
Eight series with motor pattern of ejaculation		–	–	–	–		–	–	*0*	
Parameters	**Standard values**									
Ejaculate										
Sperm count in uterine horns (10^6)	21 ± 3	–	–	–	–	0	–	0	15 ± 2	Lucio et al. 2014
Seminal plug length (mm)	12 ± 0.4	–	–	–	–	0	–	12 ± 0.1	12 ± 0.1	
Width (mm)	5 ± 0.1	–	–	–	–	0	–	5 ± 0.1	5 ± 0.1	
Weight (mg)	108 ± 7	–	–	–	–	0	–	116 ± 2	119 ± 0.3	

At postsatiety days 5 and 10 no sperm was found in the uterine horns of mated females. Then, all behavioral ejaculations executed during 1–10 days after satiation are infertile (represented by italic numbers). At 15 days, postsatiety males are similarly fertile as controls (for details, see Lucio et al. 2014).

while 33% will show a single ejaculatory series from which they do not recover to display a second series (Table 5.2; Rodríguez-Manzo and Fernández-Guasti 1994; Fernández-Guasti and Rodríguez-Manzo 2003; Rodríguez-Manzo et al. 2011). Two days after the sexual exhaustion test, 70% of the males showed a complete inhibition, while 30% displayed a single ejaculation. However, 3 days after satiety 70% of the animals displayed one ejaculation and 30% ejaculated 3 times. Four days later (one week after the sexual exhaustion test), all males ejaculated twice and a 30% showed six ejaculations. This latter percentage increased to 64% and 100% at 4 and 7 days after satiation, respectively (Table 5.2; Romano-Torres et al. 2007). These data indicate that the sexual inhibition occurs between 24 and 48 hours postsatiety (Fernández-Guasti and Rodríguez-Manzo 2003) and that the most dramatic shift occurred between 48 and 72 hours postsatiety. This effect may be related to androgen receptor expression in particular brain areas (Romano-Torres et al. 2007). It is noteworthy that all these percentages are exclusively based upon behavioral ejaculation without determining other parameters, for example, seminal expulsion.

It has been proposed that males are sexually active only when they are physiologically potent, that is, able to impregnate (Dewsbury 1982). However, this notion is invalidated by the finding that mated males produced no litter with their very late copulations (Toner and Adler 1985), and by the evidence that sexually satiated males, during the Coolidge Effect, executed the behavioral pattern of ejaculation without seminal expulsion. Therefore, the behavioral motor pattern of ejaculation is not necessarily correlated with semen expulsion (Tables 5.1 and 5.2; Tlachi-López et al. 2012).

Ejaculation is a sexual function that consists of two stages: emission and expulsion. The autonomic component of ejaculation involves the confluence of seminal secretions and sperm in the prostatic urethra, while the somatic component involves the control of the striated muscular contractions during ejaculation (Clément and Giuliano 2011). Abolition of seminal and prostatic fluids, produced by the bilateral section of the hypogastric nerves or of the vas deferens, does not alter any aspect of copulatory performance in male rats (Larsson and Swedin 1971; McGlynn and Erpino 1974). Furthermore, the administration of (antiadrenergic) guanethidine monosulfate prevents seminal emission without affecting motor copulatory parameters or bulbospongiosus muscles activity (Holmes and Sachs 1991). This indicates that the stimulation of the urethral mucosa (by the ejaculate flow) is not required for generating the rhythmic motor patterns associated with the ejaculatory response in copula (Holmes and Sachs 1991). Sexually satiated male rats can be considered a natural model demonstrating that seminal emission is not essential for the execution of the behavioral motor pattern of ejaculation.

5.4.2 Restoration of Ejaculation

The Coolidge Effect-related reduction in sperm and the seminal plug imply depletion of the content of the sexual accessory glands (Pessah and Kochva 1975; Purvis et al. 1986), and a drastic decrease in sperm concentration in the epididymis (Austin and Dewsbury 1986; Toner and Adler 1986; Tlachi-López et al. 2012). Secretions of the seminal vesicles are the substrate for the fluid synthesized by the coagulating glands (Beil and Hart 1973). To our knowledge, only we have studied the characteristics of the ejaculate recovery after sexual satiety (Lucio et al. 2014). We found that after 5 days postsatiety, males failed to expel semen when displaying the behavioral ejaculatory pattern, suggesting that the secretions of the accessory glands, and possibly also the epididymal sperm concentration, remain unrecovered. At 10 days postsatiety, the ejaculate consisted of seminal plug and sperm. The weight and size of the plug indicated that the seminal vesicle and coagulating gland secretions were reestablished, even though the sperm was located in the vagina instead of in the uteri, most likely due to the lack of vaginal adhesion of the coagulated plug. The seminal plug adhesion depends on prostate secretions (Tlachi-López et al. 2011) and is crucial for transcervical sperm transport, which depends on the hydrostatic pressure that pushes the spermatozoa from the vagina through the cervix to the uteri (Blandau 1945). These data, taken together, suggest that the recovery of prostatic secretion is slower than that of the seminal vesicles and coagulating glands, and still has not occurred by 10 days after sexual satiety. The complete recuperation of the secretions of the accessory glands seems to be completed only by 15 days after satiation, when coagulated seminal plugs adhere tightly to the vagina, thereby promoting transcervical sperm transport (see Table 5.2). The sperm count is still reduced at this time by 67% of the expected quantity—approximately 20 million. These data were obtained from males that successively ejaculated at 5, 10, and 15 days postsatiety. However, the sperm count was reduced only by 26% in males that did not copulate 15 days after sexual satiety. The sperm count was much higher in males that rested for 15 days postsatiety (15 million) than in animals also tested 15 days after satiety but that were sequentially tested three times with a 5-day interval (6 million; Lucio et al. 2014).

Thus, the ejaculate components recover at different intervals after sexual satiety: the seminal vesicles and coagulating gland secretions recover first (within 10 days), the prostate secretions recover later (within 10–15 days postsatiety). The testicular and epididymal functions are still incomplete even at 15 postsatiety days, suggesting that full sperm count restoration more time than that required for the seminal plasma.

We analyzed when, after sexual satiety, the ejaculate recovered the characteristics necessary to impregnate the female (Lucio et al. 2014). Sexually satiated males were tested immediately after satiety or after 15 days of sexual abstinence. Males of both conditions, as well as controls (nonsatiated males—those that displayed only one ejaculatory series) were placed in cages in which they were allowed to cohabit with intact females (1 male and 3 females per cage; in all: 6 males and 18 females per condition) during 15 days, corresponding to three estrous cycles. Females were isolated when pregnant and the date of parturition was recorded to determine the day of impregnation. Sexually satiated males were able to impregnate only five females during the second and third cycles, and those females had the lowest number of pups, while males that rested for 15 days after satiety impregnated almost all the females: 50%, 28%, and 11% during the first, second, and third cycles. These females produced a number of pups that did not differ from those produced by control males. Considering the aforementioned sperm count results, it seems that males that rested for 15 days after satiety were equally able, as controls, to impregnate most of the females (89% vs. 100% of the control males) and to produce the same number of pups per litter. These data reveal that a sperm count of around 15 million is as effective as 20 million to fertilize and that 15 days of sexual abstinence are enough to recover full fertility, although the sperm count seemed reduced (Lucio et al. 2014).

5.5 SUMMARY

This chapter summarizes recent contributions to our understanding of male sexual satiety and the Coolidge Effect, from a behavioral perspective and its ecophysiological implications. Copulation in male rats consists of behavioral patterns of mounting, intromitting, and ejaculatory thrusting; but they are not necessarily accompanied by penile erection and seminal expulsion. For example, sexually satiated males exposed to a novel female, that is, during the Coolidge Effect, showed behavioral ejaculatory movements but failed to expel semen. Such findings invite reconsideration of other putative biological implications of mating other than impregnation. We discuss the association among sexual behavior, parenting and competition in males, in relation to two different classes of ejaculate, the critical role of deposition and dislodgement of seminal plugs, transcervical sperm transport and competition.

5.6 CONCLUDING REMARKS

After repeated copulation there is a gradual decrease in the ejaculate primarily reflected in the sperm count and the size of the seminal plug. If copulation persists, males display the behavioral motor ejaculation pattern, but are unable to expel seminal fluid. This condition is also observed during the Coolidge Effect (Table 5.1). The observation that sexually satiated males exposed to the Coolidge Effect are as able as controls to remove the seminal plugs deposited by others, thus interfering with sperm transport, indicates that copulation postsatiety could be a reproductive strategy. This strategy is not a conventional spermatic competition because sperm and seminal plug of only one male is in the female genital tract. However, copulation postsatiety can be considered a "variant" of spermatic competition, as dislodging the seminal plug of another male interferes with sperm transport from the vagina to the uterus.

In general, copulation may have, at least, two main adaptive functions: fertilization and competition. The former is attained by males during the first ejaculatory series, while the latter occurs by males copulating through the last series in sequence or during the Coolidge Effect (italic numbers given in Tables 5.1 and 5.2). Such competitive advantage remains for days after sexual satiety, when rats recover their behavioral capacity to copulate, but before their fertilizing capacity is restored. The limits imposed on fertilization and competition primarily depend upon the functional capacity of the sexual accessory glands and the optimal amount of sperm available to be expelled from the epididymis caudae. The competitive component of copulation seems to prevail over its impregnation component (italic numbers given in Tables 5.1 and 5.2). Regardless of the "ultimate cause" of copulation, its "proximate cause" is its intrinsically rewarding properties.

EPILOGUE

We write this chapter in honor of our teacher and colleague Carlos Beyer, who was surprised that these findings challenge the apparently well established notion that genetic diversity is the biological function of the Coolidge Effect. Carlos Beyer was one of the most outstanding researchers in reproductive biology, who loved to question deep-rooted, inadequately studied concepts.

ACKNOWLEDGMENTS

The authors wish to thank M. Sc. Rebeca Reyes for careful editing and preparing the tables. R.A.L. is also thankful to Universidad Autónoma de Tlaxcala, Project CACyPI-UATX-2014.

REFERENCES

Adler, N. T. 1978. On the mechanisms of sexual behavior and their evolutionary constraints. In *Biological Determinants of Sexual Behavior,* ed. J. B. Hutchison, 655–95. New York: Wiley Press.

Adler, N. T. 1979. On the physiological organization of social behavior: Sex and aggression. In *Handbook of Behavioral Neurobiology. Vol. 3. Social Behavior and Communication,* ed. P. Marler and J. G. Vanderbergh, 29–71. New York: Plenum Press.

Agmo, A. 1999. Sexual motivation—an inquiry into events determining the occurrence of sexual behavior. *Behav Brain Res* 105:129–50.

Agmo, A. 2010. La conducta sexual desde el punto de vista epicúreo: Reforzamiento, recompensa e incentivos sexuales. In *Aproximaciones al estudio de la motivación y ejecución sexual,* 2nd Edition, ed. M. A. Guevara-Pérez, M. Hernández-González, L. Chacón-Gutiérrez, and J. A. Barradas-Bribiesca, 13–51. Guanajuato: Univ Guanajuato.

Agrati, D., A. Fernández-Guasti, M. Ferreño, and A. Ferreira. 2011. Coexpression of sexual behavior and maternal aggression: The ambivalence of sexually active mother rats toward male intruders. *Behav Neurosci* 125:446–51.

Austin, D. and D. A. Dewsbury. 1986. Reproductive capacity of male laboratory rats. *Physiol Behav* 37:627–32.

Beil, E. R. and G. R. Hart. 1973. Cowper's gland secretion in rat semen coagulation. II. Identification of the potentiating factor secreted by the coagulating glands. *Biol Reprod* 8:613–17.

Beach, F. A. 1956. Characteristics of masculine "sex drive." In *Nebraska Symposium on Motivation,* ed. M. R. Jones, 1–32. Lincoln, NE: University of Nebraska Press.

Beach, F. A. and L. Jordan. 1956. Sexual exhaustion and recovery in the male rat. *Q J Exp Psychol* 8:121–33.

Bermant, G. 1967. Copulation in rats. *Psychol Today* 1: 52–60.

Bermant, G. 1976. Sexual behavior: Hard times with the Coolidge effect. In *Psychological Research: The Inside Story,* ed. M. H. Siegel and H. P. Zeigler, 76–105. New York: Harper & Row.

Beyer, C., J. L. Contreras, K. Larsson, M. Olmedo, and G. Moralí. 1982. Patterns of motor and seminal vesicle activities during copulation in the male rat. *Physiol Behav* 29:495–500.

Bindra, D. A. 1976. *A Theory of Intelligent Behavior*. New York: Wiley.

Blandau, R. J. 1945. On the factors involved in sperm transport through the cervix uteri of the albino rat. *Am J Anat* 77:253–72.

Calhoun, J. B. 1963. *The Ecology and Sociology of the Norway Rat.* Public Health Service Publication No. 1008. Bethesda, MD: Govt. Print.

Carballada, R. and P. Esponda. 1992. Role of fluid from seminal vesicles and coagulating glands in sperm transport into the uterus and fertility in rats. *J Reprod Fertil* 95:639–48.

Carballada, R. and P. Esponda. 1993. Structure of the vaginal plugs generated by normal rats and by rats with partially removed seminal vesicles. *J Exp Zool* 265:61–8.

Clément, P. and F. Giuliano. 2011. Physiology of ejaculation. In *Cancer and Sexual Health, Current Clinical Urology,* ed. J. P. Mulhall, L. Incrocci, I. Goldstein, and R. Rosen, 77–89. New York: Humana Press.

Coria-Ávila, G. A., J. Pfaus, M. E. Hernández, J. Manzo, and P. Pacheco. 2004. Timing between ejaculations changes paternity success. *Physiol Behav* 80:733–7.

Crawford, L. L., K. S. Holloway, and M. Domjan. 1993. The nature of sexual reinforcement. *J Exp Anal Behav* 60:55–66.

Dewsbury, D. A. 1981. Effects of novelty on copulatory behavior: The Coolidge effect and related phenomena. *Psychol Bull* 89:464–82.

Dewsbury, D. A. 1982. Ejaculate costs and mate choice. *Am Nat* 119:601–10.

Dimijian, G. G. 2005. Evolution of sexuality: Biology and behavior. *Proc Bayl Univ Med Cent* 18:244–58.

Fernández-Guasti, A. and G. Rodríguez-Manzo. 2003. Pharmacological and physiological aspects of sexual exhaustion in male rats. *Scand J Psychol* 44:257–63.

Fisher, A. E. 1962. Effects of stimulus variation on sexual satiation in the male rat. *J Comp Physiol Psychol* 55:614–20.

Holmes, G. M., W. D. Chapple, R. E. Leipheimer, and B. D. Sachs. 1991. Electromyographic analysis of male rat perineal muscles during copulation and reflexive erections. *Physiol Behav* 49:1235–46.

Holmes, G. M. and B. D. Sachs. 1991. The ejaculatory reflex in copulating rats: Normal bulbospongiosus activity without apparent urethral stimulation. *Neurosci Lett* 125:195–7.

Hsiao, S. 1965. Effect of female variation on sexual satiation in the male rat. *J Comp Physiol Psychol* 60:467–9.

Hsiao, S. 1969. The Coolidge effect in male rat copulatory behaviour: Failure to replicate Fisher's results. *Psychon Sci* 14:1–2.

Hull, E. M. and G. Rodríguez-Manzo. 2009. Male sexual behavior. In *Hormones, Brain and Behavior,* ed. D. W. Pfaff, A. P. Arnold, A. M. Etgen, S. E. Fahrbach, and R. T. Rubin, 5–65. San Diego, CA: Academic Press.

Hull, E. M., R. I. Wood, and K. E. McKenna. 2006. Neurobiology of male sexual behavior. In *Physiology of Reproduction,* 3rd Edition, ed. J. D. Neill, 1729–924, St. Louis, MO: Elsevier.

Jiménez, P., M. A. Serrano-Meneses, E. Cuamatzi, and G. González-Mariscal. 2012. Analysis of copulatory parameters in male rabbits across successive tests leading to sexual exhaustion. *World Rabbit Sci* 20:13–23.

Kippin, T. E. and J. G. Pfaus. 2001. The development of olfactory conditioned ejaculatory preferences in the male rat: I. Nature of the unconditioned stimulus. *Physiol Behav* 73:457–69.

Larsson, K. 1956. *Conditioning and Sexual Behavior in the Male Albino Rat.* Stockholm, Sweden: Almqvist &Wiksell.

Larsson, K. and G. Swedin. 1971. The sexual behavior of male rats after bilateral section of the hypogastric nerve and removal of the accessory genital glands. *Physiol Behav* 6:251–3.

Lucio, R. A., J. Manzo, M. Martínez-Gómez, B. D. Sachs, and P. Pacheco. 1994. Participation of pelvic nerve branches in male rat copulatory behavior. *Physiol Behav* 55:241–6.

Lucio, R. A., V. Rodríguez-Piedracruz, J. L. Tlachi-López, M. García-Lorenzana, and A. Fernández-Guasti. 2014. Copulation without seminal expulsion: The consequence of sexual satiation and the Coolidge effect. *Andrology* 2:450–7.

Lucio, R. A. and J. L. Tlachi-López. 2008. *Análisis de la cópula y el eyaculado en la rata albina (Rattus norvegicus). Manual de Laboratorio.* Tlaxcala: Universidad Autónoma de Tlaxcala y Consejo Nacional de Ciencia y Tecnología.

Matthews, M. and N. T. Adler. 1977. Facilitative and inhibitory influences of reproductive behavior on sperm transport in rats. *J Comp Physiol Psychol* 91:727–41.

McClintock, M. K. and N. T. Adler. 1978. The role of the female during copulation in the wild and domestic Norway rat *(Rattusnorvegicus). Behavior* 67:67–95.

McClintock, M. K., J. J. Anisko, and N. T. Adler. 1982. Group mating among Norway rats II. The social dynamics of copulation: Competition, cooperation, and mating choice. *Anim Behav* 30:419–25.

McGlynn, J. M. and M. J. Erpino. 1974. Effects of vasectomy on the reproductive system and sexual behavior of rats. *J Reprod Fertil* 40:241–7.

Moralí, G. and C. Beyer. 1992. Motor aspects of masculine sexual behavior in rats and rabbits. *Adv Study Behav* 21:201–38.

Olsson, M., T. Madsen, and R. Shine. 1997. Is sperm really so cheap? Costs of reproduction in male adders, *Viperaberus. Proc R Soc Lond B* 264:455–9.

Parker, G. A. 1970. Sperm competition and its evolutionary consequences in the insects. *Biol Rev* 45:525–67.

Parker, G. A. 1982. Why are there so many tiny sperm? Sperm competition and the maintenance of two sexes. *J Theor Biol* 96:281–94.

Parker, G. A. 1998. Sperm competition and the evolution of ejaculates: Towards a theory base. In *Sperm Competition and Sexual Selection,* ed. T. R. Birkhead and A. P. Moller, 3–54. San Diego, CA: Academic Press.

Pessah, H. and E. Kochva. 1975. The secretory activity of the seminal vesicles in the rat after copulation. *Biol Reprod* 13:557–60.

Pfaff, D. W. 1999. *Drive. Neurobiological and Molecular Mechanisms of Sexual Motivation.* Cambridge, MA: The Massachusetts Institute of Technology Press.

Phillips-Farfán, B. V. and A. Fernández-Guasti. 2009. Endocrine, neural and pharmacological aspects of sexual satiety in male rats. *Neurosci Biobehav Rev* 33:442–55.

Poiani, A. 2006. Complexity of seminal fluid. A review. *Behav Ecol Sociobiol* 60:289–310.

Pollak, E. I. and B. D. Sachs. 1976. Penile movements and the sensory control of copulation in the rat. *Behav Biol* 117:177–86.

Purvis, K., E. Haug, Y. Thomassen, B. Mevag, and H. Rui. 1986. Short-term effects of mating on the accessory sex glands of the male rat. *J Reprod Fertil* 77:373–80.

Queen, K., C. B. Dhabuwala, and C. G. Pierrepoint. 1981. The effect of the removal of the various accessory sex glands on the fertility of male rats. *J Reprod Fertil* 62:423–6.

Robitaille, J. A. and J. Bovet. 1976. Field observations on the social behavior of the Norway rat. *Rattusnorvegicus* (Berkenhout). *Biol Behav* 1:289–308.

Rodríguez-Manzo, G. and A. Fernández-Guasti. 1994. Reversal of sexual exhaustion by serotonergic and noradrenergic agents. *Behav Brain Res* 62:127–34.

Rodríguez-Manzo, G., I. L. Guadarrama-Bazante, and A. Morales-Calderón. 2011. Recovery from sexual exhaustion-induced copulatory inhibition and drug hypersensitivity follow a same time course: Two expressions of the same process? *Behav Brain Res* 217:253–60.

Romano-Torres, M., B. V. Phillips-Farfán, R. Chavira, G. Rodríguez-Manzo, and A. Fernández-Guasti. 2007. Relationship between sexual satiety and brain androgen receptors. *Neuroendocrinology* 85:16–26.

Sachs, B. D. and R. L. Meisel. 1988. The physiology of male sexual behavior. In *The Physiology of Reproduction,* 1st Edition, ed. E. Knobil and J. D. Neill, 1393–485. New York: Raven Press Ltd.

Shafik, A., A. A. Shafik, I. Shafik, and O. el-Sibai. 2005. Urethral sphincters response to cavernosus muscles stimulation with identification of cavernoso-urethral reflex. *Arch Androl* 51:335–43.

Steiger, S., R. Franz, A. K. Eggert, and J. K. Müller. 2008. The Coolidge effect, individual recognition and selection for distinctive cuticular signatures in a burying beetle. *Proc R Soc* B 275:1831–8.

Tlachi-López, J. L., J. R. Eguibar, A. Fernández-Guasti, and R. A. Lucio. 2012. Copulation and ejaculation in male rats under sexual satiety and the Coolidge effect. *Physiol Behav* 106:626–30.

Tlachi-López, J. L., A. A. López, K. Hoffman, J. Velázquez-Moctezuma, M. García-Lorenzana, and R. A. Lucio. 2011. Rat dorsal prostate is necessary for vaginal adhesion of the seminal plug and sperm motility in the uterine horns. *Biol Res* 44:259–67.

Toner, J. P. and N. T. Adler. 1985. Potency of rats ejaculations varies with their order and with male age. *Physiol Behav* 35:113–15.

Toner, J. P. and N. T. Adler. 1986. The pre-ejaculatory behavior of male and female rats affects the number of sperm in the vagina and uterus. *Physiol Behav* 36:363–7.

Wallach, S. J. R. and B. L. Hart. 1983. The role of striated penile muscles of the male rat in seminal plug dislodgement and deposition. *Physiol Behav* 31:815–21.

Wedell, N., M. J. G. Gage, and G. A. Parker. 2002. Sperm competition, male prudent and sperm limited females. *Trends Ecol Evol* 17:313–20.

Wilson, J. R., R. E. Kuehn, and F. A. Beach. 1963. Modification in the sexual behavior of male rats produced by changing the stimulus female. *J Comp Physiol Psychol* 56:636–44.

Zetterberg, H.L. 1969. Sex i Sverige: Huvudavsnittet av Undersökningen om Sexuallivet i Sverige Företagen av SIFO. Stockholm: Aldus/Bonniers.

6 The Sexual Cerebellum

Jorge Manzo, Porfirio Carrillo,
Genaro A. Coria-Avila, and Luis I. Garcia

CONTENTS

6.1 Introduction 103
6.2 The Cerebellum 103
6.3 Non-Contact Stimulation 105
6.4 Sexual Behavior 106
6.5 Motor Learning 107
6.6 Neuroendocrinology 109
6.7 Conclusions 110
References 110

6.1 INTRODUCTION

Sexual reproduction is a successful strategy in the evolution of mammals. It is organized as a highly complex behavioral repertoire of each mating pair that relies on the function of multiple systems that involve an elaborate neural network. While multiple regions of the brain and spinal cord are recognized as main structures underlying sexual behavior, our laboratory has shown that the cerebellum in particular is a key structure in reproductive processes. The cerebellum controls complex movements during courtship, mating, and parental behavior. Furthermore, it seems that the cerebellum integrates signals coming from sense organs that trigger processes leading to copulation. In the present review, we compile the current knowledge of the role of the cerebellum in the execution of sexual behavior.

6.2 THE CEREBELLUM

The cerebellum (literally, "little brain") is a central nervous system component covering the dorsal part of the brainstem caudal to the brain hemispheres. Although it is smaller than the whole brain, in humans it consists of about 50 billion neurons, representing almost half the total number of neurons in the brain (Ramnani 2006). The mature cerebellum is in the shape of a hemispherical ellipse with the major axis in a coronal position and the minor in a sagittal position, with three well recognized regions: the central vermis, a hemisphere at each side of the vermis, and the flocculus and paraflocculus complex at the caudal border of the cerebellar hemispheres (Figure 6.1).

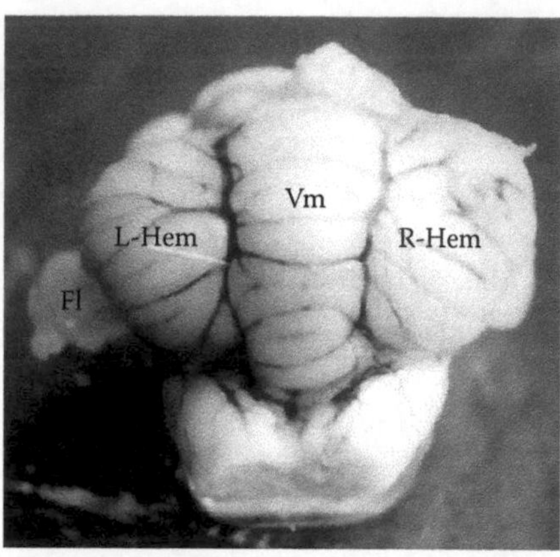

FIGURE 6.1 Dorsal view of an adult male rat cerebellum. The picture shows the central vermis (Vm) and the left (L-Hem) and right (R-Hem) hemispheres. Lateral to L-Hem it is possible to see the protuberance of the flocculus and paraflocculus region (Fl).

The sagittal view at mid-cerebellum reveals a foliated ("leaf-like") structure with a gross division of its cortex into 10 lobules. A noteworthy characteristic of the cerebellar cortex is its three distinctive layers: the molecular, Purkinje and granular (Figure 6.2). In these layers, the main groups of neurons are orderly distributed and consist of several main types of cells: the granular, Purkinje, basket, and stellate neurons (Voogd and Glickstein 1998). Embedded in the cerebellar body, there is a complex of three deep nuclei that have a key function, along with Purkinje neurons, in generating efferent activity from the cerebellum

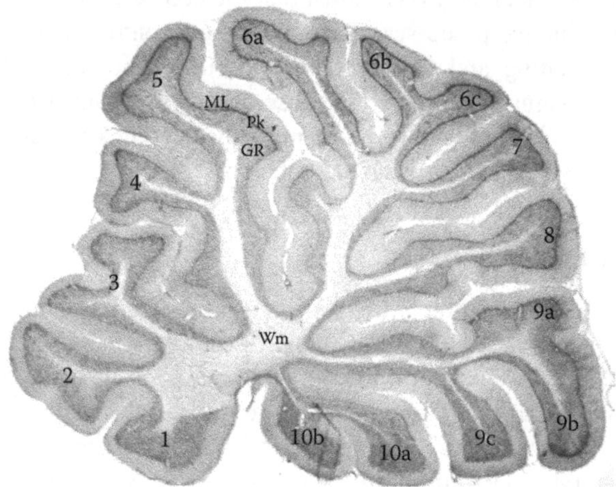

FIGURE 6.2 Sagittal view of the cerebellum at the mid-vermis area. The picture shows the 10 lobules and sublobules (1 to 10) and the clear distinction of the molecular (ML), Purkinje (Pk), and granule layers. Also there is a clear distinction of the white matter (Wm).

(Habas 2010). The biochemical diversity of the cerebellar neurons is complex and its functional significance is not well understood. Thus, in recent years there is increasing evidence that, in addition to its widely known role in the control of movement, the cerebellum plays a key role in cognitive function, learning, language, and emotional expression, and as we review herein, a significant role in reproduction.

6.3 NON-CONTACT STIMULATION

Courtship and its accompanying sexual arousal may include absence of body contact, with vision, olfaction and audition each playing a greater or lesser role, depending on the species. These non-contact stimuli can trigger neural responses that can lead to physical contact, mating, and reproduction (Sachs 1995). The cerebellum receives information from these three non-contact sensory systems.

Thus, the visual pathway projects to specific regions of the cerebellum in cats (Blanks 1990), rats (Blanks and Precht 1983; Blanks et al. 1983), pigeons (Pakan et al. 2010), non human primates (Langer et al. 1985), and humans (Voogd et al. 2012). Although the functions of these projections are largely unknown, recently it has been observed that cerebellar neurons are strongly activated when the visual stimulation is in a sexual context. Visual stimulation evoked by body ornaments plays a role in reproduction in birds and mammals, and has received attention from ethology and neuroscience. Nonhuman primates depend on visual cues for sexual attraction (Dubuc et al. 2014). Stimulation of the human visual field with erotic movies evokes intense activation of the cerebellum, not only in heterosexual, but also in homosexual and male-to-female transsexual subjects (Hu et al. 2008; Oh et al. 2012). Furthermore, while the pattern of activation of brain regions in heterosexual men differs from that in homosexual men, both groups experience activation of the cerebellum (Hu et al. 2008).

The cerebellum also receives auditory input (Baumann and Mattingley 2010). While there are few studies on sexual arousal evoked by erotic auditory stimuli (Hawk et al. 2007), audiovisual arousing erotic stimulation activates the cerebellar vermis in men (Tsujimura et al. 2006).

Olfaction plays a significant role in the social organization in mammals, including sexual arousal (Doty 1986). In studies in which a male rat was placed with a receptive female in an arena that prevented physical contact, the males displayed penile erections and hormonal responses (Sachs et al. 1994; Cruz et al. 1999; Manzo et al. 1999), all mediated by olfactory cues (Sachs 1997). In humans, odorants activate the cerebellum (Sobel et al. 1998), and cerebellar lesions alter olfaction (Mainland et al. 2005). In our laboratory, we found that airborne sexual odorants from receptive females or from the wood chip bedding from estrous females cages activated granule cells in the mid-cerebellar vermis of male rats (Manzo et al. 2008; García et al. 2014; Figure 6.3), that is, in lobule VII involved in sniffing behavior displayed during detection of sexual airborne cues (Ortiz-Pulido et al. 2011).

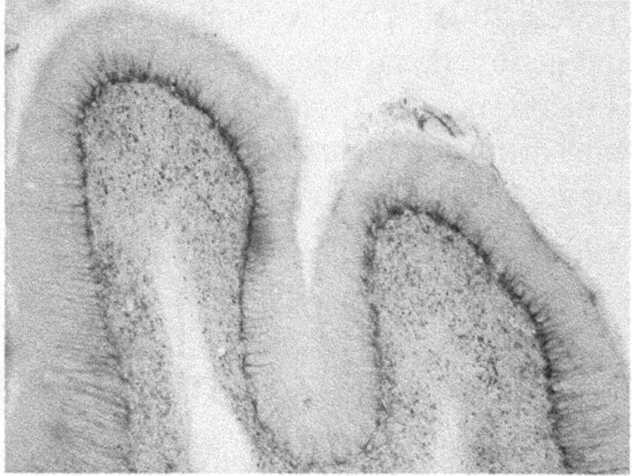

FIGURE 6.3 Non-contact stimulation arena (left panel) that allows the distant stimulation of the male (in the right side of the arena) by a receptive female. The wire mesh in the middle of the arena prevents the physical contact of the couple. The right panel is a micrograph of the lobule 6 of the cerebellar cortex showing expression of Fos-IR evoked in the male after 30 minutes of being stimulated in the noncontact arena.

6.4 SEXUAL BEHAVIOR

There is extensive evidence that the cerebellum is activated in relation to sexual behavior. Manual stimulation of the penis activates the cerebellum in humans (Holstege et al. 2003) and rats (Ortiz-Pulido et al. 2011). Holstege et al. (2003) reported activation of the cerebellum in the vermis, hemispheres and deep nuclei in relation to penile stimulation, perhaps due to bodily movement during stimulation and ejaculation. The authors commented that the response of the cerebellum to the manipulation of the penis is similar to that during heroin rush (Sell et al. 1999), listening to music (Blood and Zatorre 2001), or receiving monetary reward (Martin-Sölch et al. 2001); hence the response could be related to the emotional component. In rats, Ortiz-Pulido et al. (2011) lesioned lobule VI of the vermis and in sexual behavior tests, observed an increased latency to the first intromission and increased intervals between consecutive intromissions, suggesting reduced processing of penile afferent stimuli. The application of an anesthetic to the rat penis has a similar effect of reducing intromissions, indicating reduced ability to detect the vaginal orifice (Hart 1972; Stefanick et al. 1983). A similar response was reported after lesion of the dorsal penile nerve (Dahlöf et al. 1988; Sachs and Liu 1992). These findings suggest the existence of a penis–cerebellum neural pathway, with a proprioception-like process for a fine spatial orientation of the erected penis toward the vaginal orifice during thrusting for successful intromission.

In humans, the cerebellar vermis becomes activated during sexual arousal leading to erection of the penis (Tsujimura et al. 2006). Similarly in rats, the mid-vermis becomes active in relation to both sexual sensory stimuli and motor performance of sexual behavior (Manzo et al. 2008). In the rat, Manzo et al. (2008) using Fos immuno reactivity (Fos-IR) reported that the granule cells of the cerebellar cortex became highly active during the first sexual bout. In the second and third ejaculations the number of activated granule cells was reduced. Thus, they proposed a dual

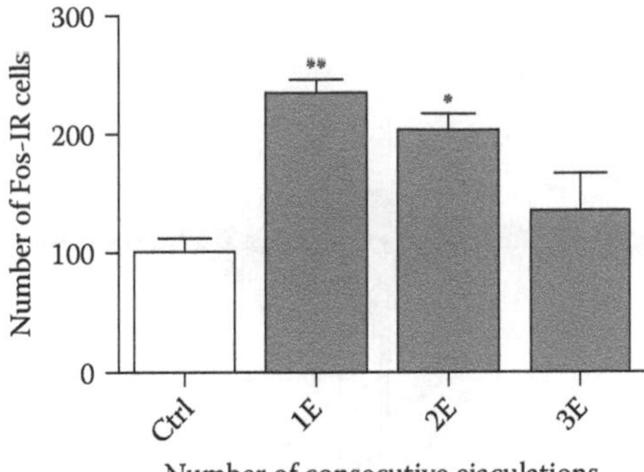

FIGURE 6.4 Graph showing the number of cells expressing Fos-IR before male copulation (Ctrl) or after the male executes one (1E), two (2E), or three (3E) consecutive ejaculations. Bars represent the mean ± SEM. ** = $p < 0.01$; * = $p < 0.05$.

mechanism, in which sexual sensory stimulation activates the granule cells, while the performance of sexual behavior is inhibitory (Figure 6.4). The cerebellum is also activated in female rats during sexual behavior. Females that could pace their copulations in a sexual reward paradigm showed a significant increase in Fos-IR activation of granule and Purkinje neurons in the cerebellar cortex, with lobule 3 being the most sensitive to sexual reward (Paredes-Ramos et al. 2011). In women, Komisaruk et al. (2004) using functional magnetic resonance imaging (MRI) during vaginal self-stimulation-produced orgasm, reported an increase of activity in deep cerebellar nuclei, and a subsequent study (Georgiadis et al. 2006) using positron emission tomography (PET) found activation of deep cerebellar nuclei during clitorally induced orgasm.

6.5 MOTOR LEARNING

The proper execution of sexual behavior involves a complex display of patterns that males develop as they mature. In the male rat there are three well-characterized patterns: mount, intromission, and ejaculation (Meisel and Sachs 1994). In order to execute the characteristic display of the motor patterns of each behavior, males gain developmental experience. Typically, male rats display multiple mounts and intromissions prior to ejaculation; following a refractory period, they can display another similar series toward a second ejaculation and refractory period, repetitively until reaching a state of exhaustion after several consecutive ejaculations. The cerebellum, with its long-known function in the control of acquisition and control of movement, was considered a central structure underlying these developmental experience processes. Garcia-Martinez et al. (2010) obtained multiunit recordings of the cerebellar cortex in naïve male rats during the course of their development of sexual experience. They reported that the first time that a naïve male was with a receptive female, his copulatory movements were disoriented. At that phase, lobules 6 and 7 of the cerebellar vermis and the inferior olive, which projects directly to the Purkinje cell dendritic arbor, showed the highest recorded activity, while the activity of the

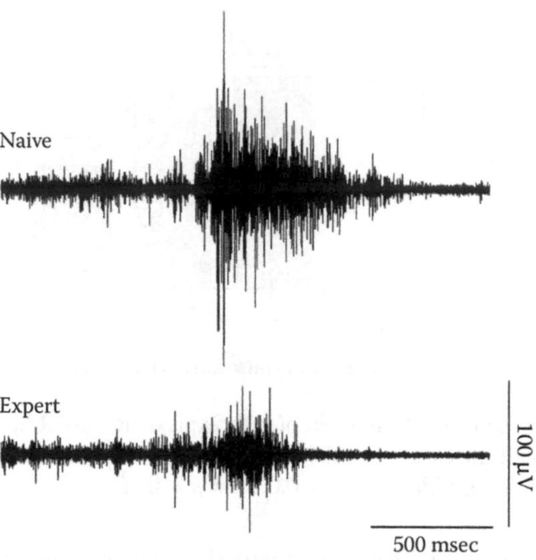

FIGURE 6.5 Multiunit recording of lobule 6a during mounting behavior. Traces are from a naïve and a sexually expert male rat. The duration of the parameter is about 500 ms, and this duration was not modified by experience. As shown, the trace is increased above the baseline during mounting behavior and the maximum amplitude is always lower in expert subjects (from Garcia-Martinez et al. 2010, *Cerebellum* 9:96–102).

deep fastigial nucleus was lowest. As the male became sexually expert, the multiunit activity reversed, that is, in the expert male the lobules exhibit a lowest activity, while the deep nuclei the highest activity. Thus, in the cerebellar cortex in naïve males there is a high level of neuronal activity, which becomes reduced as the male develop copulatory efficacy (Figure 6.5). This observation could be related to synaptic efficiency, in which improvement of a movement sequence is correlated with activation of fewer neurons (Zhang and Poo 2001).

This process of experiential development of efficient movement seems to depend on a process termed "Long-Term Depression or LTD," that is, a reduction of the synaptic potency between the parallel fiber from the granule cell and the dendrite of the Purkinje neuron; it is the basis of the learning process (Ito 2002). In this mechanism, endocannabinoids are released retrogradely by the Purkinje dendrites, which block the release of glutamate by the parallel fiber. Based on this concept, we proposed that the development of the sexual motor patterns could also be reflected in the levels of cannabinoid 1 (CB1) receptors (Manzo et al. 2010). By using sexually naïve male rats in a behavioral paradigm to develop effective sexual performance, we observed that the cerebellar vermis showed a transient reduction in the level of CB1 receptors in lobules 1, 6, 7, and 10. The level of the receptors returned to basal levels after the male became sexually "expert" (Figure 6.6). Thus, considering the reduction in the multiunit activity and the reduction in the CB1 receptors during sexual experience, it seems that the activity of the cerebellum during sexual behavior is an essential mechanism that could be reduced and increased in an experiential sequence that enables the development of the motor skills of copulation.

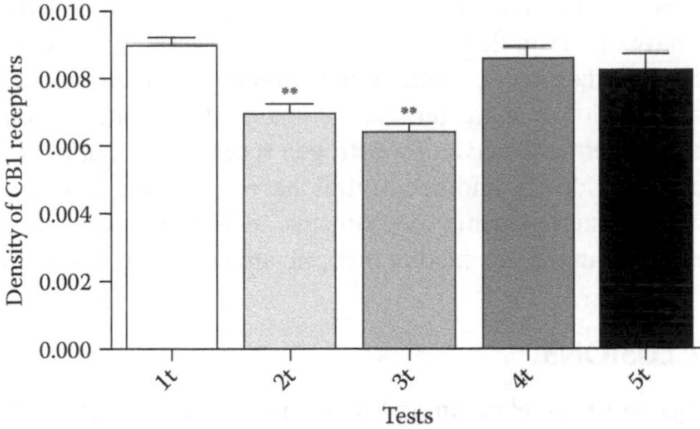

FIGURE 6.6 Density of cannabinoid 1 (CB1) receptors in lobules 1, 6, 7, and 10 at the vermis area in male rats acquiring sexual experience. Males were analyzed as naïve males after their first encounter with a receptive female (1t), and during each test executed every other day (2t to 5t). In this period, males changed from naïve (white bar) to sexually expert (black bar), and CB1 receptors were significantly reduced during the two tests. Bars represent the mean ± SEM. ** = $p < 0.01$ (modified from Manzo et al. 2010, *eNeurobiologia* 1:280610).

6.6 NEUROENDOCRINOLOGY

Reproduction is evoked by an orchestrated mechanism depending on neural and hormonal regulation, and the cerebellum is a part of this system. In a book published in 1838 by Franz Joseph Gall et al. (Gall et al. 1838), they review their prior work stating that the cerebellum "is the organ of the instinct of reproduction." In that book, they refer to the influence of castration on the cerebellum, indicating that castration in youth arrests the development of the cerebellum, which fails to reach full size, a situation that does not occur if castration is performed during adulthood. Also, they indicated that unilateral castration of animals attenuates the size of the contralateral lobe of the cerebellum. Furthermore, they describe that after 6 to 8 months of being unilaterally castrated on the left or right side, male rabbits, with no exception, show a smaller contralateral lobe of the cerebellum. Carlos Beyer and Barry Komisaruk repeated the rabbit experiment, but did not see any effect on the cerebellum (personal communication). However, although these discrepant observations are unreconciled, certainly the cerebellum has a highly complex neuroendocrine system.

It is known that the cerebellum depends on steroids for its normal development. Starting as early as the second postnatal day, the cerebellum contains estrogen, progesterone, and androgen receptors that exhibit different peaks leading up to adulthood, and there is evidence that the cerebellum contains the mechanism to synthesize these neurosteroids during development (Dean and McCarthy 2008; Tsutsui 2012). Although we are not aware of evidence regarding the role of these "cerebellar" neurosteroids in sexual behavior, it is known that the Purkinje neurons of adult rats express androgen receptors (Simerly et al. 1990), and synthesize neurosteroids (Ukena et al. 1999). These reports led us to obtain preliminary data toward ascertaining whether cerebellar androgen in adult male rats is modified during sexual behavior. We have observed that Purkinje neurons in adult rats have a basal level of androgen receptors; they are reduced after castration and recovered after androgen replacement. They show a characteristic fluctuation during the execution of sexual behavior. A similar process occurs for estrogen receptors (unpublished data). Thus, the cerebellum is evidently another central nervous system structure underlying sexual behavior via steroidogenic mechanisms, as in the case of other intensively studied nuclei, for example, the medial preoptic area (Murphy and Hoffman 2001).

6.7 CONCLUSIONS

Recent findings on the cerebellum and its relation to sexual behavior indicate that there are specific regions in the vermis that are involved during the experiential development and execution of copulation in male rats. Despite the scant literature on the role of the cerebellum in sexual behavior in males, and even less in females, clearly there is a cerebellar "zone" involved in sexual behavior that requires more research as to its role in all aspects of reproduction.

REFERENCES

Baumann, O. and J. B. Mattingley. 2010. Scaling of neural responses to visual and auditory motion in the human cerebellum. *J Neurosci* 30:4489–4495.

Blanks, R. H. 1990. Afferents to the cerebellar flocculus in cat with special reference to pathways conveying vestibular, visual (optokinetic) and oculomotor signals. *J Neurocytol* 19:628–642.

Blanks, R. H. and W. Precht. 1983. Responses of units in the rat cerebellar flocculus during optokinetic and vestibular stimulation. *Exp Brain Res* 53:1–15.

Blanks R. H., W. Precht, and Y. Torigoe. 1983. Afferent projections to the cerebellar flocculus in the pigmented rat demonstrated by retrograde transport of horseradish peroxidase. *Exp Brain Res* 52:293–306.

Blood, A. J. and R. J. Zatorre. 2001. Intensely pleasurable responses to music correlate with activity in brain regions implicated in reward and emotion. *Proc Natl Acad Sci U S A* 98:11818–11823.

Cruz, M. R., Y. C. Liu, J. Manzo, P. Pacheco, and B. D. Sachs. 1999. Peripheral nerves mediating penile erection in the rat. *J Auton Nerv Syst* 76:15–27.

Dahlöf, L. G., S. Ahlenius, and K. Larsson. 1988. Copulatory performance of penile desensitized male rats following the administration of 8-OH-DPAT. *Physiol Behav* 43:841–843.

Dean, S. L. and M. M. McCarthy. 2008. Steroids, sex and the cerebellar cortex: Implications for human disease. *Cerebellum* 7:38–47.

Doty, R. L. 1986. Odor-guided behavior in mammals. *Experientia* 42:257–271.

Dubuc, C., W. L. Allen, D. Maestripieri, and J. P. Higham. 2014. Is male rhesus macaque red color ornamentation attractive to females? *Behav Ecol Sociobiol* 68:1215–1224.

Gall, F. J., J. Vimont, and F. J. V. Broussais. 1838. *On the Functions of the Cerebellum*. London: Maclachlan & Stewart, Longman & Company.

García, L. I., P. García-Bañuelos, G. E. Aranda-Abreu, G. Herrera-Meza, G. A. Coria-Avila, and J. Manzo. 2014. Activation of the cerebellum by olfactory stimulation in sexually naive male rats. *Neurologia*. doi:10.1016/j.nrl.2014.02.002.

Garcia-Martinez, R., M. Miquel, L. I. Garcia, G. A. Coria-Avila, C. A. Perez, G. E. Aranda-Abreu, R. Toledo, M. E. Hernandez, and J. Manzo. 2010. Multiunit recording of the cerebellar cortex, inferior olive, and fastigial nucleus during copulation in naive and sexually experienced male rats. *Cerebellum* 9:96–102.

Georgiadis, J. R., R. Kortekaas, R. Kuipers, A. Nieuwenburg, J. Pruim, A. A. Reinders, and G. Holstege. 2006. Regional cerebral blood flow changes associated with clitorally induced orgasm in healthy women. *Eur J Neurosci* 24:3305–3316.

Habas, C. 2010. Functional imaging of the deep cerebellar nuclei: a review. *Cerebellum* 9:22–28.

Hart, B. L. 1972. Sexual reflexes in the male rat after anesthetization of the glans penis. *Behav Biol* 7:127–130.

Hawk, S. T., R. Tolman, and C. W. Mueller. 2007. The effects of target attractiveness on men's sexual arousal in response to erotic auditory stimuli. *J Sex Res* 44:96–103.

Holstege, G., J. R. Georgiadis, A. M. Paans, L. C. Meiners, F. H. van der Graaf, and A. A. Reinders. 2003. Brain activation during human male ejaculation. *J Neurosci* 23:9185–9193.

Hu, S. H., N. Wei, Q. D. Wang, L. Q. Yan, E. Q. Wei, M. M. Zhang, J. B. Hu, M. L. Huang, W. H. Zhou, and Y. Xu. 2008. Patterns of brain activation during visually evoked sexual arousal differ between homosexual and heterosexual men. *AJNR Am J Neuroradiol* 29:1890–1896.

Ito, M. 2002. The molecular organization of cerebellar long-term depression. *Nat Rev Neurosci* 3:896–902.

Komisaruk, B. R., B. Whipple, A. Crawford, S. Grimes, W-C. Liu, A. Kalnin, and K. Mosier. 2004. Brain activation during vaginocervical self-stimulation and orgasm in women with complete spinal cord injury: fMRI evidence of mediation by the Vagus nerves. *Brain Res* 1024:77–88.

Langer, T., A. F. Fuchs, C. A. Scudder, and M. C. Chubb. 1985. Afferents to the flocculus of the cerebellum in the rhesus macaque as revealed by retrograde transport of horseradish peroxidase. *J Comp Neurol* 235:1–25.

Mainland, J. D., B. N. Johnson, R. Khan, R. B. Ivry, and N. Sobel. 2005. Olfactory impairments in patients with unilateral cerebellar lesions are selective to inputs from the contralesional nostril. *J Neurosci* 25:6362–6371.

Manzo, J., M. R. Cruz, M. E. Hernández, P. Pacheco, and B. D. Sachs. 1999. Regulation of noncontact erection in rats by gonadal steroids. *Horm Behav* 35:264–270.

Manzo, J., M. Miquel, M. Perez-Pouchoulen, G. A. Coria-Avila, L. I. Garcia, R. Toledo, and M. E. Hernandez. 2010. Aprendizaje motor y receptores a canabinoidesen la corteza del cerebelo *eNeurobiologia* 1:280610.

Manzo, J., M. Miquel, R. Toledo, J. A. Mayor-Mar, L. I. Garcia, G. E. Aranda-Abreu, M. Caba, and M. E. Hernandez. 2008. Fos expression at the cerebellum following non-contact arousal and mating behavior in male rats. *Physiol Behav* 93:357–363.

Martin-Sölch, C., S. Magyar, G. Künig, J. Missimer, W. Schultz, and K. L. Leenders. 2001. Changes in brain activation associated with reward processing in smokers and nonsmokers. A positron emission tomography study. *Exp Brain Res* 139:278–286.

Meisel, R. L. and B. D. Sachs. 1994. The physiology of male sexual behavior. In *The Physiology of Reproduction,* ed. E. Knobil and J. D. Neill, 3–105. New York: Raven Press.

Murphy, A. Z. and G. E. Hoffman. 2001. Distribution of gonadal steroid receptor-containing neurons in the preoptic-periaqueductal gray-brainstem pathway: A potential circuit for the initiation of male sexual behavior. *J Comp Neurol* 438:191–212.

Oh, S. K., G. W. Kim, J. C. Yang, S. K. Kim, H. K. Kang, and G. W. Jeong. 2012. Brain activation in response to visually evoked sexual arousal in male-to-female transsexuals: 3.0 tesla functional magnetic resonance imaging. *Korean J Radiol* 13:257–264.

Ortiz-Pulido, R., M. Miquel, L. I. Garcia, C. A. Perez, G. E. Aranda-Abreu, R. Toledo, M. E. Hernandez, and J. Manzo. 2011. Sexual behavior and locomotion induced by sexual cues in male rats following lesion of Lobules VIa and VII of the cerebellar vermis. *Physiol Behav* 103:330–335.

Pakan, J. M., D. J. Graham, and D. R. Wylie. 2010. Organization of visual mossy fiber projections and zebrin expression in the pigeon vestibulo cerebellum. *J Comp Neurol* 518:175–198.

Paredes-Ramos, P., J. G. Pfaus, M. Miquel, J. Manzo, and G. A. Coria-Avila. 2011. Sexual reward induces fos in the cerebellum of female rats. *Physiol Behav* 102:143–8.

Ramnani, N. 2006. The primate cortico-cerebellar system: anatomy and function. *Nat Rev Neurosci* 7:511–522.

Sachs, B. D. 1995. Placing erection in context: the reflexogenic-psychogenic dichotomy reconsidered. *Neurosci Biobehav Rev* 19:211–224.

Sachs, B. D. 1997. Erection evoked in male rats by airborne scent from estrous females. *Physiol Behav* 62:921–924.

Sachs, B. D., K. Akasofu, J. H. Citron, S. B. Daniels, and J. H. Natoli. 1994. Noncontact stimulation from estrous females evokes penile erection in rats. *Physiol Behav* 55:1073–1079.

Sachs, B. D. and Y. C. Liu. 1992. Copulatory behavior and reflexive penile erection in rats after section of the pudendal and genitofemoral nerves. *Physiol Behav* 51:673–680.

Sell, L. A., J. Morris, J. Bearn, R. S. Frackowiak, K. J. Friston, and R. J. Dolan. 1999. Activation of reward circuitry in human opiate addicts. *Eur J Neurosci* 11:1042–1048.

Simerly, R. B., C. Chang, M. Muramatsu, and L. W. Swanson. 1990. Distribution of androgen and estrogen receptor mRNA-containing cells in the rat brain: an in situ hybridization study. *J Comp Neurol* 294:76–95.

Sobel, N., V. Prabhakaran, C. A. Hartley, J. E. Desmond, Z. Zhao, G. H. Glover, J. D. Gabrieli, and E. V. Sullivan. 1998. Odorant-induced and sniff-induced activation in the cerebellum of the human. *J Neurosci* 18:8990–9001.

Stefanick, M. L., E. R. Smith, and J. M. Davidson. 1983. Penile reflexes in intact rats following anesthetization of the penis and ejaculation. *Physiol Behav* 31:63–65.

Tsujimura, A., Y. Miyagawa, K. Fujita, Y. Matsuoka, T. Takahashi, T. Takao, K. Matsumiya, Y. Osaki, M. Takasawa et al. 2006. Brain processing of audiovisual sexual stimuli inducing penile erection: a positron emission tomography study. *J Urol* 176:679–683.

Tsutsui, K. 2012. Neurosteroid biosynthesis and action during cerebellar development. *Cerebellum* 11:414–415.

Ukena, K., C. Kohchi, and K. Tsutsui. 1999. Expression and activity of 3b-hydroxysteroid dehydrogenase/D5-D4-isomerase in the rat Purkinje neuron during neonatal life. *Endocrinology* 140:805–813.

Voogd, J. and M. Glickstein. 1998. The anatomy of the cerebellum *Trends Cogn Sci* 2:307–313.

Voogd, J., C. K. Schraa-Tam, J. N. van der Geest, and C. I. De Zeeuw. 2012. Visuomotor cerebellum in human and nonhuman primates. *Cerebellum* 11:392–410.

Zhang, L. I. and M. M. Poo. 2001. Electrical activity and development of neural circuits. *Nat Neurosci* (4 Suppl):1207–1214.

7 Intracellular Signaling Involved in Progestin Regulation of Female Sexual Behavior in Rodents

Oscar González-Flores, Ignacio Camacho-Arroyo, and Anne M. Etgen

CONTENTS

7.1 Introduction 113
7.2 Background 114
7.3 Progestin Receptors 115
7.3.1 PR Expression in the Brain 116
7.3.2 Mechanisms of P Action 117
7.3.2.1 Regulation of Gene Expression 117
7.3.2.2 Modulation of Neurotransmitter Receptors 117
7.3.2.3 Activation of Signaling Cascades 117
7.3.3 PR Isoforms and the Display of Female Sexual Behavior in Rodents 118
7.3.4 Cellular Mechanisms Proposed to Explain the Regulation of Estrous Behavior by Different Agents 118
7.3.4.1 Protein Kinase A 119
7.3.4.2 The Nitric Oxide (NO) Pathway 119
7.3.4.3 Leptin and Female Sexual Behavior 120
7.3.4.4 Vaginocervical Stimulation and Female Sexual Behavior 122
7.3.4.5 Src Tyrosine Kinase, MAPK Pathway, and Lordosis Behavior 122
References 125

7.1 INTRODUCTION

Female sexual behavior has been studied in different fields of science and from distinct points of view. Some researchers have studied this topic from a primarily behavioral or comparative angle or from an evolutionary perspective. Dr. Carlos Beyer, throughout his brilliant career, integrated neuroendocrine evidence to elucidate the physiological basis of female sexual behavior. He used this approach as a model to explore the cellular and molecular mechanisms by which various hormones, for example, those produced in the ovary (estradiol [E_2] and progesterone [P]), neuropeptides, prostaglandins, and other chemical messengers, exert their effects on different brain areas involved in the regulation of specific and physiologically relevant behavior. Thus, in the 1980s Dr. Beyer proposed a model in which some agents, due to their lipophobicity, do not penetrate into the cell, but exert their effects on receptors located in the cell membrane. Here they would interact with molecules that, in turn, activate intracellular receptors, a phenomenon known as cross-talk, that occurs in a variety of cells including neurons and glial cells, ultimately leading to female sexual behavior. As detailed later in this chapter, it is important to mention that Dr. Beyer pioneered pharmacological strategies to investigate the cellular mechanisms underlying the expression of female sexual behavior and was the first to propose a cross-talk model in which progestin receptors (PRs) act as a common effector for lordosis facilitation by steroidal and nonsteroidal agents in rodents (Beyer et al. 1981; Beyer and González-Mariscal, 1986). In this chapter we focus on our findings in collaboration with Dr. Beyer over more than a decade. These results have contributed to the understanding of the cellular mechanisms involved in the regulation of female sexual behavior in rats and represent a significant aspect of Dr. Beyer's scientific legacy.

7.2 BACKGROUND

Estrous behavior occurs when a female is receptive to, and mates with, a male. In many mammals, receptivity involves the adoption of the lordosis posture, which facilitates penile insertion and ejaculation. In rats, lordosis comprises an arching of the back and elevation of the pelvis that is frequently accompanied by tail deviation. Another behavioral aspect of estrus is proceptivity, the repertoire of female behavior directed toward the male to initiate, establish, and maintain sexual interaction. These behavior patterns include a zigzag locomotion termed "darting" and a presentation posture that may be accompanied by ear-wiggling produced by rapid head shaking (for review, see Pfaff and Sakuma 1980).

Estrous behavior (lordosis and proceptivity) is induced by the combined effects of E_2 and P; gonadectomy performed in rats and hamsters shortly before the preovulatory P peak interferes with the subsequent expression of lordosis, whereas this response is consistently displayed when animals are ovariectomized (ovx) *after* the preovulatory P peak (Ciaccio and Lisk 1971; Powers 1970). P only facilitates lordosis behavior if females have been primed with E_2 (12–18 hours in rats and guinea pigs). Several species of rodents respond to treatment with E_2-only under some conditions. For example, Beyer and colleagues reported that estrogens activate lordosis in ovx rats, with 17β–E_2 being the most potent estrogen, followed in decreasing order by estrone and estriol (Beyer et al. 1971).

In several female mammals, P also exerts an inhibitory effect on sexual behavior; this effect is observed during the formation of the corpus luteum after ovulation occurs, or during pregnancy. This may serve to terminate the period of behavioral estrus and to prevent mating when females are pregnant. In addition, P depresses estrous behavior in ovx rats or prevents its induction by E_2 when administered concurrently with, or prior to, estrogen, a phenomenon termed concurrent inhibition (Powers and Moreines 1976). After an initial period of behavioral facilitation, P also induces sexual inhibition; if a second injection of P is administered 24 hours after the first, it fails to induce lordosis in the estrogen-primed rodents. This is referred to as sequential inhibition (Feder and Marrone 1977; Morin 1977; González-Mariscal et al. 1993).

It has been suggested that both the facilitation and sequential inhibition of female sexual behavior are due to the interaction of P with its intracellular receptor, PR. Thus, abundant literature shows a close correlation between estrogen induction of hypothalamic PR and the onset and decay of estrous behavior (Moguilewsky and Raynaud 1979; Blaustein et al. 1980; Sá et al. 2013). For example, implantation and removal of E_2-filled Silastic capsules was highly correlated with the appearance and disappearance of estrogen-inducible PR and the expression or decay of estrous behavior. Some researchers have proposed that a declining level of estrogen-inducible PRs leads to sequential inhibition of lordosis (Feder and Marrone 1977; Morin, 1977). Thus, in rats and guinea pigs, this behavioral refractoriness to P is correlated with a decrease in the concentration of estrogen-inducible PRs in the hypothalamus (Blaustein and Feder 1979; Moguileswsky and Raynaud 1979; Parsons et al. 1981; Etgen and Shamamian 1986; Pleim et al. 1989). Indeed, maintaining PR levels by administration of agents that inhibit the action of 26S proteasome, which would otherwise degrade the PRs, blocked sequential inhibition of lordosis (González-Flores et al. 2004a).

7.3 PROGESTIN RECEPTORS

The intracellular PR is a member of the superfamily of ligand-dependent transcription factors. PRs in many species are mainly expressed as two isoforms, PR-A and PR-B, which are encoded from the same gene but are regulated by different promoters and exhibit distinct regulation and function (Kastner et al. 1990; Cabrera-Muñoz et al. 2011). The PR-B isoform is the longest (110–120 kDa), because it contains a unique N-terminal domain, termed the B upstream segment (BUS), that is not present in the truncated isoform PR-A (72–94 kDa). Both PR-A and PR-B have a DNA-binding domain that includes approximately 70 amino acids that make up two zinc fingers that interact with hormone response elements (HREs). These isoforms also have the ligand-binding domain (LBD), which consists of approximately 250 amino acids at the C-terminal of the receptor. The LBD has a specific three-dimensional structure responsible not only for hormone binding, but also for other functions, including the interaction with heat shock proteins. PRs also contain a flexible hinge region that functions, in part, to aid DNA binding. Additionally, steroid receptors contain at least two activation functions (AFs) necessary to mediate transcriptional activation, designated AF-1 and AF-2. AF-1 is located within the N-terminal domain, whereas AF-2 is in the C-terminal domain. Although early immunohistochemistry studies suggested that receptors are mainly found in the cell nucleus, loosely bound to chromatin, more recent evidence indicates that the unliganded PR isoforms rapidly shuttle between the cytoplasm and the nucleus as part of a heteroprotein complex in association with heat-shock protein chaperone molecules (Lange 2004; Lange et al. 2007). Consequently, as described below, PRs are positioned to participate in cytoplasmic or membrane-associated signaling cascades.

Recently, a membrane-associated PR (mPR) was cloned from spotted sea trout oocytes and was subsequently determined to be a member of a new family of membrane steroid hormone receptors structurally distinct from the intracellular steroid receptors. The mPR has homologues in diverse species, including mouse and human (Karteris et al. 2006; Mani and Oyola 2012). The mPR and other members of this steroid receptor family have seven transmembrane spanning domains, typical of G-protein coupled receptors. The mPR is localized in the plasma membrane, binds P with high affinity, and in sea trout oocytes is involved in P-induced meiotic maturation (Thomas et al. 2007). Human homologues of mPR (alpha, beta, gamma, delta, and epsilon subtypes) have been identified. These human mPR subtypes are expressed in a tissue- and cell-type-specific manner and belong to the family of progestin and Adipo Q receptors in eukaryotes and eubacteria (Pang and Thomas 2011). The mPRs expressed in human breast cancer cells (MDAMB-231) that lack classical nuclear PR, mediate a rapid and transient progestin-mediated activation of mitogen-activated protein kinase (MAPK) that modifies the production of cyclic adenosine monophosphate (cAMP; Mani and Oyola 2012).

7.3.1 PR Expression in the Brain

The classical intracellular PR is expressed in several brain regions associated with reproductive behavior. In the intact ovarian-cycling rat, PR is present in the anteroventral periventricular nucleus, medial preoptic area, arcuate nucleus, and ventromedial nucleus (VMN) of the hypothalamus during diestrus and proestrus-estrus (Numan et al. 1999; Blaustein and Mani 2007). During mid-pregnancy, the expression of PR in the arcuate and VMN markedly decreases, most likely due to the high levels of circulating P. PR-A and PR-B isoforms are regulated in a brain region-specific manner by ovarian steroids throughout the ovarian cycle. In ovx rats, via the western blot technique, E_2 administration increased the content of both PR-A and PR-B proteins in the hypothalamus and preoptic area, whereas the subsequent administration of P decreased the content of both PR isoforms in the hypothalamus but not in the preoptic area. During diestrus, intact females exhibited high PR-A and PR-B expression in the preoptic area, but low expression in the hypothalamus, compared with the other days of the estrous cycle (Guerra-Araiza et al. 2003). The cerebral expression of PR in other species is generally consistent with that observed in rats. In estrous ferrets, which are induced ovulators, PR was detected in the medial and lateral preoptic area, lateral hypothalamus, arcuate nuclei, and VMN of the hypothalamus (Tobet et al. 1986). A similar distribution of PR was observed in guinea pigs, rhesus macaques (Garris et al. 1981) and cynomolgus monkeys (Rees et al. 1985), all of which are spontaneous ovulators. In ovx, E_2-treated female rabbits (induced ovulators), PR detected by means of immunocytochemistry were highly expressed in the hypothalamus and preoptic area; administration of P dramatically reduced PR protein in these brain regions (Caba et al. 2003; Beyer et al. 2007).

7.3.2 Mechanisms of P Action

P exerts its effects on the brain and other target tissues through three main mechanisms: (a) regulation of gene expression, (b) modulation of neurotransmitter systems, and (c) activation of protein kinase-associated signaling cascades.

7.3.2.1 Regulation of Gene Expression

P is a lipophilic molecule that readily passes from the circulation across the cell membrane into the cells. Once inside a target cell, it binds to, and converts, its intracellular PR from an inactive to an active complex. Hormone–receptor complexes then dimerize and bind to HREs within the regulatory promoter region of P-responsive genes. The interaction between the steroid receptor and general transcription machinery involves the recruitment of coactivator proteins (steroid receptor coactivators, SRC-1 and SRC-3). Some coactivators possess intrinsic enzyme activity for acetylation (histone acetyltransferase activity) or methylation of core histones. Thus, through its binding to PR, P has a major role in regulating the expression of specific gene networks in the female reproductive tract and other target tissues (Beyer et al. 2003; Blaustein and Olster 1989; Blaustein and Erskine 2002).

7.3.2.2 Modulation of Neurotransmitter Receptors

In 1986, it was first shown that the 3alpha-reduced metabolites of P, 5beta-pregnan-3alpha-ol-20 one (5beta,3alpha-pregnanolone), and 5alpha-pregnan-3alpha-ol-20-one,-tetrahydroprogesterone (5alpha,3alpha-pregnanolone or allopregnanolone) are potent positive allosteric modulators of γ-aminobutyric acid-A (GABA-A) receptors. These neurosteroids displace the GABA-derived ligand t-butyl bicyclophosphorothionate (TBPS) from the chloride channel and enhance the binding of muscimol and benzodiazepines (Majewska et al. 1986).

The expression of recombinant GABA-A receptors and subsequent electrophysiological recording has enabled the molecular characterization of neurosteroid effects at the subunit level (Paul and Purdy 1992). Although it has been postulated that neurosteroids act through unique recognition sites on the GABA-A receptor, a specific neurosteroid binding site has not yet been determined. However, the subunit composition of the GABA-A receptor (i.e., $alpha_{1-6}$, $beta_{1-3}$, $gamma_{1-3}$, delta, epsilon, theta, pi, rho_{1-3}) seems to play a pivotal role in the sensitivity of the receptor to neurosteroids. Thus, several steroids, including the endogenous P metabolite, 5α-pregnan-3α-ol-20-one, potentiate the function of GABA-A receptors preferably through the subunits $alpha_4beta_3$delta (Brown et al. 2002). Moreover, glycine (Wu et al. 1990) and *N*-methyl-D-aspartate (NMDA) receptors (Wu et al. 1991) are sensitive to modulation by specific neurosteroids. For a more detailed view of neurotransmitter–receptor mediated effects of neurosteroids, see Paul and Purdy (1992).

7.3.2.3 Activation of Signaling Cascades

In addition to the rapid membrane-initiated effects exerted by the interaction of 3alpha-reduced metabolites of P with GABA-A receptors and other ligand-gated ion channels, there is increasing evidence that P has rapid effects that alter the activity of intracellular signal transduction pathways. Some of the rapid effects of P are inhibited by the PR antagonist, RU486, and are mediated through the same intracellular PR that regulates gene transcription, although PR activation of transcription is not required (Edwards et al. 2002; Boonyaratanakornkit et al. 2007; Mani and Oyola 2012). Other rapid effects of P appear to be mediated by a cell membrane-associated PR distinct from the intracellular PR (Zhu et al. 2003), the mPR. Edwards and colleagues have proposed that intracellular PRs both activate cytoplasmic signaling pathways and act as a ligand-activated transcription factor (Edwards et al. 2002; Leonhardt et al. 2003; Boonyaratanakornkit et al. 2007). PRs may also act as an anchoring protein in the membrane where, in concert with other proteins, they produce a cascade of events that result in the activation of MAPK (see below) which is capable of activating nuclear transcription factors independent of nuclear steroid receptors. This raises the possibility that the activation of such pathway by ligand-bound PR could represent another means by which P modulates gene expression, independent of its classical direct interaction with HREs.

7.3.3 PR Isoforms and the Display of Female Sexual Behavior in Rodents

In collaboration with Carlos Beyer, we showed that PR isoforms play a differential role in the display of female sexual behavior. By inhibiting PR isoform expression through the intracerebroventricular (icv) administration of PR-B and total PR (PR-A + PR-B) antisense oligonucleotides, we demonstrated that the lordosis induced by P, 5alpha-DHP, and 5beta,3beta-pregnanolone in ovx rats is mainly due to PR-B activation (Guerra-Araiza et al. 2009). The role of specific PR isoforms in female sexual behavior seems to be species-specific, because PR-A plays a key role in sexual receptivity in mice (Blaustein and Mani 2007; Mani and Portillo 2010). Using the same experimental strategy with antisense oligonucleotides we also showed that P induced the expression of tryptophan hydroxylase and glutamic acid decarboxylase in the hypothalamus of ovx rats via PR-B, whereas P reduced the expression of tyrosine hydroxylase through both PR isoforms (González-Flores et al. 2011). Together, these results suggest that the PR-B isoform is essential for the display of the lordosis behavior in rats.

7.3.4 Cellular Mechanisms Proposed to Explain the Regulation of Estrous Behavior by Different Agents

Recent results obtained in our collaborative work with Carlos Beyer and by other laboratories have enabled us to extend and propose new mechanisms that may be involved in the regulation of estrous behavior induced by various agents and compounds with various chemical structures. The mechanisms that we explored include diverse intracellular pathways, for example, those that involve protein kinase A (PKA), nitric oxide (NO)/protein kinase G (PKG), and the Src/PR/MAPK system. As described below, these pathways participate in the estrous behavior induced by multiple progestins, peptides including gonadotropin-releasing hormone (GnRH) and leptin, and the effects produced by vaginocervical stimulation (VCS). This led us, with Dr. Beyer, to propose that the PR could be the common effector for lordosis facilitation induced by these diverse agents (Beyer et al. 1997).

7.3.4.1 Protein Kinase A

Regulation of cellular function often requires that proteins undergo changes in conformation (allosteric processes) or chemical composition (e.g., methylation, acetylation, glycosylation, and phosphorylation). Protein phosphorylation is the major mechanism through which a variety of hormones regulate intracellular events in mammalian cells (Nestler and Greengard 1994). Thus, protein kinases, the enzymes that phosphorylate proteins, are specifically activated by one or several second messengers, among them cyclic adenosine monophosphate (cAMP), cyclic guanosine monophosphate (cGMP), peptides, calcium, and phospholipids (Nestler and Greengard 1994; Beyer et al. 2003).

Among the agents that regulate protein kinases, cAMP is a major mediator for a large variety of hormones through stimulation of PKA. Beyer, Canchola, and Larsson were the first to explore the role of this intracellular signaling pathway in the display of estrous behavior of ovx, E_2-primed rats. They reported that intracerebral or subcutaneous administration of dibutyryl cAMP as well as various drugs that activate cAMP synthesis facilitated lordosis behavior in ovx, E_2-primed rats (Beyer et al. 1981). From these early results they proposed a unitary model to explain the facilitation of lordosis by P and other hormones with different chemical structures in estrogen-primed rats. Thus, estrogens bind to an intracellular receptor and the ligand-receptor complex translocates to the nucleus to stimulate the synthesis of an estrogen-induced protein, most of which is in an inactive condition (iEIP). Activation of that iEIP could occur through phosphorylation by PKA. The cAMP is produced when adenylyl cyclase is activated by the binding of various triggering hormones to their cognate membrane receptors. A likely possibility, at least for P and GnRH, is that cAMP is produced by the activation of noradrenaline receptors linked to adenylyl cyclase. This was the first proposed model of cross-talk between membrane and intracellular receptors to explain the activation of lordosis behavior, and recent evidence from our laboratories and others suggests that the iEIP could be the PR. This extension of the model accommodates findings, summarized below, that kinases activated by the agents that induce lordosis (i.e., PKA) do not directly phosphorylate PR, but work indirectly by activating MAPK, which phosphorylates serine residues on PR. However, it is important to note that diverse agents that induce lordosis may act through different protein kinase systems. For example, the PKC inhibitor, H7, prevented lordosis facilitation in ovx, E_2-primed rats by several ring A-reduced progestins, but not by P (González-Flores et al. 2006) whereas the lordosis induced by P, GnRH, prostaglandin (PG), E_2 and db-cAMP were blocked by RpAMPC, an inhibitor of PKA (Ramírez-Orduña et al. 2007).

7.3.4.2 The Nitric Oxide (NO) Pathway

NO is a peculiar chemical transmitter that freely diffuses through aqueous and lipid environments and plays a role in major aspects of brain function related to reproduction (Dawson et al. 1994; García-Juárez et al. 2012). NO is produced in the brain from L-arginine by activation of NO synthase (NOS), which presents three isoforms produced by separate genes: neuronal NOS (nNOS), endothelial NOS (eNOS), and inducible NOS (iNOS). The nNOS and eNOS require calcium/calmodulin for activation and are responsible for most release of NO, whereas iNOS is only expressed in response to inflammatory cytokines and lipopolysaccharides (Yun et al. 1996).

Hormones or neurotransmitters can elevate intracellular calcium, and this process has the potential to activate NOS, producing NO, which stimulates soluble guanylyl cyclase which, in turn, induces the synthesis of the second messenger, cGMP. The hypothalamic content of this messenger doubles between the afternoon and evening of proestrous, when the surge and the period of behavioral receptivity occur (Calka 2006).

There are reports that NO/cGMP facilitates lordosis as well as the preovulatory LH surge. Studies by Carlos Beyer and his group in the early 1980s evaluated the effect of systemic administration of different analogs of cGMP in ovx-adrenalectomized, E_2-primed rats and found that cGMP, dpc-GMP, and 5′GMP induced significant lordosis behavior (Beyer et al. 1982). Later, Mani et al. (1994) showed that intracerebral injection of a NOS inhibitor reduced lordosis in rats pretreated with E_2 plus P, and that a NO donor facilitated the behavior in E_2-primed rats in the absence of P.

Chu and Etgen (1999) then proposed a model in which the coupling of hypothalamic alpha$_1$-adrenoreceptors to the NO-cGMP pathway is steroid hormone-dependent and that this pathway mediates noradrenaline (i.e., alpha$_1$-adrenergic), E_2 and P-dependent facilitation of lordosis behavior in rats. They used the selective inhibitor of NO-stimulated cGMP, 1H-[1,2,4]oxadiazolo[4,3-a]quinoxalin-1-one (ODQ), and found that this agent suppressed lordosis in E_2 and P-treated rats when infused icv or when concurrently infused with P (Chu and Etgen 1999). Therefore, they also evaluated whether cGMP, through the activation of PKG, is involved in facilitating lordosis behavior by using KT5823, a cell-permeable, highly specific inhibitor of PKG. Icv infusion of this inhibitor decreased lordosis behavior in E_2 + P-primed female rats. In addition, PR also was involved in the NO/cGMP/PKG pathway, because the administration of the PR antagonist RU486 blocked the lordosis behavior induced by icv infusion of 8-br-cGMP.

Recently, we collaborated with Dr. Beyer to explore whether the induction of estrous behavior by diverse agents involves the NO/cGMP/PKG pathway. We found that lordosis induced by P, two of its ring A-reduced metabolites, 5alpha-DHP and 5alpha,3alpha-Pgl, or by VCS was transiently reduced by the previous injection of either a NOS inhibitor or by ODQ. Lordosis behavior was suppressed at 2 hours but returned to control values by 4 hours. Icv infusion of the PKG inhibitor, KT5823, significantly inhibited the lordosis behavior induced by all three progestins at 2 hours. In addition, we tested the hypothesis that GnRH, PGE2, and db-cAMP act via the NO-cGMP pathway to facilitate estrous behavior in E_2-primed female rats. NOS inhibitors and ODQ blocked the estrous behavior induced by GnRH, PGE2, and db-cAMP (González-Flores et al. 2009). The PKG inhibitor, KT5823, reduced PGE2, and db-cAMP facilitation of estrous behavior but did not affect the behavioral response to GnRH. Thus, the NO-cGMP pathway, via PKG, represents an important mechanism by which progestins and several neurotransmitters/neuromodulators, including GnRH and PGE2, can facilitate estrous behavior in estrogen-primed rodents.

7.3.4.3 Leptin and Female Sexual Behavior

Leptin is a versatile 16 kDa peptide hormone synthesized and secreted by adipocytes; it informs the brain as to the status of energy stores present in adipose tissue.

Leptin binds to multiple receptor isoforms termed OB-Rs, all of which present an extracellular domain of over 800 amino acids, a transmembrane domain, of 34 amino acids and a variable intracellular domain, characteristic for each of the isoforms. OB-Rs belong to the class I cytokine receptor family, which is known to act through Janus kinases (JAK) and signal transducers and activators of transcription (STATs). Leptin, through OB-Rb, is able to trigger the Ras/Raf/MAPK pathway, leading to the expression of specific target genes.

It has been established that leptin also stimulates other signaling cascades; for example, it induces NO production in white adipocytes through PKA and MAPK activation (Mehebik et al. 2005). Leptin appears to have both stimulatory and inhibitory effects on PKC (Sweeney 2002) and it also regulates the 5′-AMP-activated protein kinase (AMPK) activity in a tissue-specific manner. For instance, leptin activates AMPK in muscle and hepatocytes but inhibits AMPK activity in multiple hypothalamic regions, including the arcuate and paraventricular nuclei, thereby inhibiting food intake (Park and Ahima 2014). Dominant negative AMPK expression in the hypothalamus is sufficient to reduce food intake and body weight (Park and Ahima 2014).

Humoral stimuli originating in the periphery can modulate estrous behavior, and leptin has been reported to influence estrous behavior in rodents (Wade et al. 1997; Fox and Olster 2000; García-Juárez et al. 2012). For example, Wade et al. (1997) showed that leptin facilitated sexual behavior in ad libitum fed, but not in food-deprived, female hamsters. The possibility of a link between leptin and reproduction became evident when it was determined that a homozygous mutation in the leptin (ob) gene was responsible for the obesity syndrome and impaired reproductive function in genetically obese (ob/ob) mice (Zhang et al. 1994) or in genetically obese Zucker rats (fa/fa), which have an altered leptin receptor protein (Truett et al. 1995). Intracerebral administration of leptin failed to augment the lordosis response induced by E2 and P, and inhibited proceptivity in Zucker rats (Fox and Olster 2000). These results suggest that leptin has complex effects on the sexual behavior of rodents.

In an initial experiment, we found that various dosages of leptin, infused icv, significantly facilitated lordosis behavior in ovx, E_2-primed rats. These findings led us to explore the participation of receptors, such as PRs and GnRH-1 receptors, on lordosis behavior induced by leptin. We found that the facilitation of lordosis behavior by leptin is mediated through these receptors, because the administration of RU486 or antide (an antagonist of the GnRH-1 receptor) prior to leptin inhibited the ability of leptin to facilitate lordosis (García-Juarez et al. 2011).

We recently tested the hypothesis that the NO-cGMP-PKG pathway is involved in the facilitation of lordosis behavior induced by icv administration of leptin (García-Juárez et al. 2012). Administration of L-NAME, ODQ, and KT5823 significantly reduced the lordosis behavior induced by this peptide. We also explored the participation of several other protein kinases including JAK2, Src tyrosine kinase and MAPK in lordosis behavior produced by icv infusion of leptin in ad libitum-fed, ovx, E_2-primed rats. Leptin-facilitated lordosis behavior was significantly decreased by AG490 (JAK2 inhibitor), PP2 (Src tyrosine kinase inhibitor), and PD98059 (MAPK inhibitor). These findings provide evidence that the stimulation of lordosis behavior in estrogen-primed rats by leptin requires the simultaneous or sequential activation of several signaling pathways, most likely converging on hypothalamic neurons expressing PR (García-Juárez et al. 2013). The mechanism through which PR participates in the facilitation of lordosis by leptin is unclear but it may involve PR phosphorylation.

7.3.4.4 Vaginocervical Stimulation and Female Sexual Behavior

Komisaruk and Diakow (1973) showed that vaginal stimulation produces analgesia and facilitates lordosis in rats. When VCS is applied to ovx rats, with or without E_2 priming, using a glass rod combined with manual flank-perineal stimulation, the lordosis reflex is elicited (Komisaruk and Diakow, 1973). Natural or artificial stimulation releases neuromodulators, such as GnRH and several neurotransmitters (noradrenaline, dopamine, NO, etc.). As described above, many of these agents activate intracellular signaling mechanisms through G protein linked membrane receptors, inducing the formation of second messengers and the facilitation of lordosis behavior in rodents (González-Flores et al. 2007). Thus, we worked with Carlos Beyer to explore the role of such signaling pathways in VCS facilitation of lordosis. We showed that lordosis behavior induced by VCS was prevented by intracerebral injection of the GnRH-1 receptor antagonist, antide. In addition, blockers of PKA, PKG, and MAPK interfere with the stimulatory effect of VCS on receptive behavior in E_2-primed rats (González-Flores et al. 2007). Additional results supported our hypothesis that alpha$_1$-adrenergic receptors mediate VCS facilitation of estrous behavior via the NO-cGMP-PKG pathway. Thus, administration of prazosin and phenoxybenzamine (alpha$_1$-adrenergic receptor antagonists) as well as inhibitors of NOS (L-NAME), NO-stimulated guanylyl cyclase (ODQ), and PKG (KT5823) all attenuated the VCS-induced increase in lordosis and proceptive behaviors observed in E_2-primed female rats (González-Flores et al. 2007). Other studies provided evidence that PR is a key mediator of the facilitation of receptive behavior because the antiprogestin RU486 also blocked the estrous behavior induced by VCS (González-Flores et al. 2008).

7.3.4.5 Src Tyrosine Kinase, MAPK Pathway, and Lordosis Behavior

One of the most fundamental players for the integration of steroid and growth factor signaling is the cellular tyrosine kinase molecule c-Src. This kinase is present in essentially all eukaryotic cells, where its activation is regulated by diverse growth factors, cytokines, adhesion molecules, and antigen receptors critical for generating an appropriate cellular response to external stimuli (Brown and Cooper 1996; Thomas and Brugge 1997). Src kinases share a conserved domain structure consisting of consecutive SH3, SH2, and tyrosine kinase (SH1) domains, and all family members contain an SH4 membrane-targeting region at their N-terminus. MAPK is activated by Src kinase and is one of the most important signaling pathways studied in a multitude of organisms, ranging from yeast to humans. It is at the heart of a molecular-signaling network that governs the growth, proliferation, differentiation and survival of many, if not all, cell types.

PRs actively shuttle between the cytoplasm and nucleus by active import and export mechanisms and, therefore, they have the potential to interact with cytoplasmic signaling molecules. In cell culture systems, the ligand-bound PR interacts with the Src kinases, thereby increasing their activity and stimulating the Src/ras/raf/MAPK signaling pathway (Migliaccio et al. 1996, 1998). Migliaccio et al. (1998) first demonstrated that P induced the rapid activation of MAPK through a Src-dependent mechanism. Subsequent protein–protein interaction experiments using endogenous and/or exogenous proteins in different cellular systems showed that there was an interaction of Src-PR with the estrogen receptor, leading the authors to suggest that the estrogen receptor appears to be required for the interaction of PR with Src, as well as the subsequent activation of MAPK (Migliaccio et al. 1996, 1998). Distinct work by Boonyaratanakornkit et al. (2008) showed that there are also ER-independent actions of PR in the activation of Src/MAPK following treatment with the synthetic progestin R5020.

Acosta-Martínez et al. (2006) were the first to show that the activation of MAPK by E_2 and insulin-like growth factor-I (IGF-I) is critical for hormonal facilitation of lordosis behavior. They infused MAPK inhibitors (PD98059 and U0126) during E2 priming and found that lordosis behavior was abolished. We found that estrous behavior (lordosis and proceptivity) induced in E_2-primed rats by several agents, such as ring A-reduced progestins, the delta opioid agonist DPDPE, GnRH, PGE2, cAMP, leptin, or by VCS was mediated through the MAPK pathway, on the basis that intracerebral infusion of PD98059 significantly reduced lordosis behavior induced by all these agents (González-Flores et al. 2008, 2009; García-Juárez et al. 2013). The inhibition of MAPK activity, or PR with RU486, decreased lordosis induced by 8-br-cGMP, a cell-permeable cGMP analog, suggesting that cGMP enhancement of lordosis involves ligand independent activation of PR in the brain by MAPK phosphorylation (González-Flores et al. 2004b).

Because MAPK phosphorylation of PRs eventually leads to receptor down-regulation via proteasomal degradation, we also studied MAPK participation in sequential inhibition induced by P. Thus, in E_2-primed rats, the MAPK inhibitor PD98059 was injected icv 30 minutes before the first injection of P, and these animals did not show lordosis behavior 4 hours later. The same animals received a second injection of P 24 hours later and were tested again for lordosis. Females showed high levels of lordosis in this test, indicating that sequential inhibition was blocked (González-Flores et al. 2004b). Down-regulation of hypothalamic PRs was also inhibited by PD98059 (González-Flores et al. 2004b). From these findings, we suggest that MAPK-dependent phosphorylation of PRs contributes to the facilitatory actions of P and then targets PR for degradation by the 26S proteasome pathway, leading to the sequential inhibition by P. This hypothesis was confirmed by the observation that a proteasome inhibitor blocked sequential inhibition induced by P (González-Flores et al. 2004a; Etgen et al. 2006).

In summary, based on collaborations with Carlos Beyer, we proposed the novel idea that the Src/PR/MAPK pathway is an essential component in the facilitation of estrous behavior (Figure 7.1).

This model has received support from our finding that the specific Src kinase inhibitor PP2 reduced lordosis behavior induced by progestins (P and its ring A-reduced metabolites), peptides, and VCS (González-Flores et al. 2010; Lima-Hernández et al. 2012). In addition, our model suggests that several agents (GnRH and PGE2) as well as VCS act on membrane G protein coupled receptors to activate second messenger-regulated serine/threonine kinases such as PKA, PKC, and PKG. Stimulation of these kinases may also activate Src kinase through several mechanisms. For example, ring A-reduced progestins may stimulate Src activity either

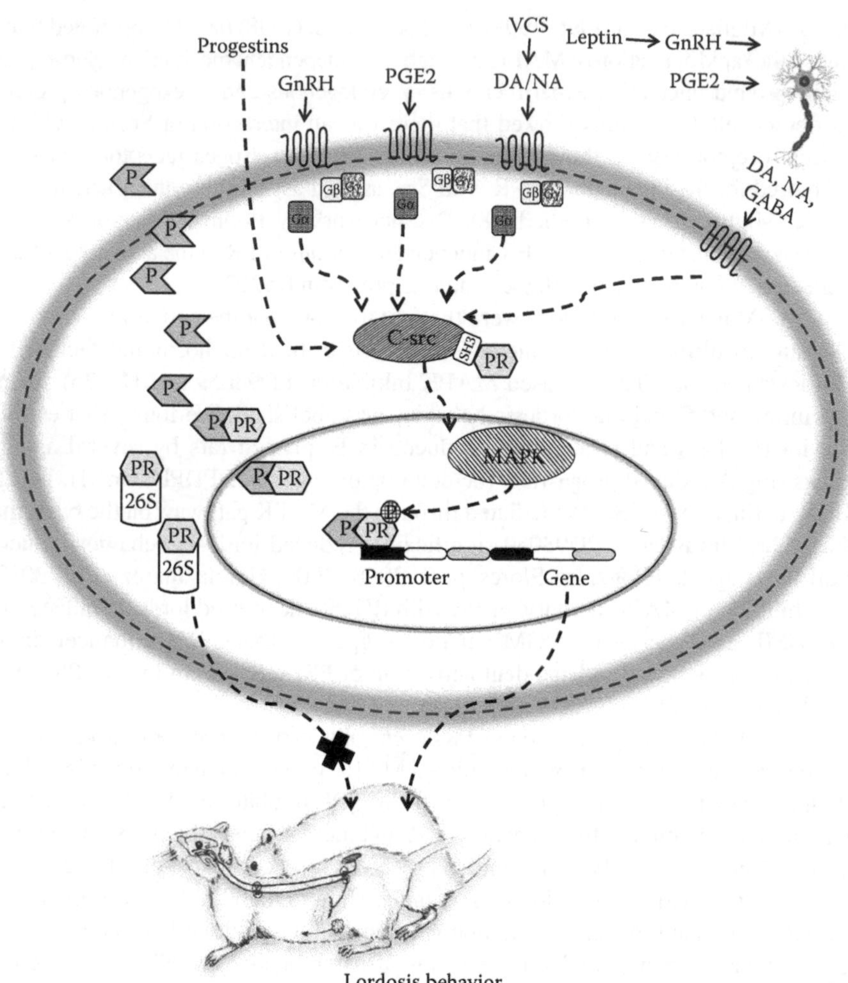

FIGURE 7.1 Activation of the Progesterone receptor-Src kinase (PR-Src) system and estrous behavior in the rat. Progestins, peptides, prostaglandin E2 (PGE2), and vaginocervical stimulation (VCS) can activate the PR–Src system to elicit estrous behavior in ovx, E_2B primed rats. As shown in the upper right section of the diagram, leptin releasing GnRH may act on interneurons projecting to neurons containing the PR–Src complex, resulting in release of neurotransmitters (dopamine [DA], norepinephrine [NA], GABA, etc.) capable of stimulating estrous behavior. Similarly, PGE2 or VCS stimulate the release of NA and DA, which act on G protein-coupled membrane receptors (GPCR). A direct action on the PR–Src neurons is shown on the left side of the diagram. GnRH, PGE2 act on membrane GPCRs activating Src kinase through several mechanisms (dashed line). Progestins can directly stimulate the Src–MAPK system. This signaling system can activate classical pathways of PR action by phosphorylating ser-294. The phosphorylation of this amino acid is targeted for degradation by the 26S proteasome.

by acting at a mPR or by releasing GnRH. Src signaling through the MAPK system occurs in the context of a multiprotein complex in which PR acts as a scaffold to recruit an assortment of enzymatic effectors. Src–MAPK signaling also interacts with the classical pathways of PR action by phosphorylating ser-294 in this protein, a reaction enhancing its transcriptional activity (Migliaccio et al. 1996, 1998; Lange et al. 2000).

In summary, one of the legacies of Dr. Beyer was to provide evidence that the stimulation of female sexual behavior in rats involved a nongenomic mechanism that could be triggered by a variety of steroidal and nonsteroidal agents. These findings led to the proposal that the stimulation of estrous behavior by several compounds requires an increase in hypothalamic second messengers and the consequent activation of protein phosphorylation of key regulators such as PR.

REFERENCES

Acosta-Martínez, M., O. González-Flores, and A. M. Etgen. 2006. The role of progestin receptors and the mitogen-activated protein kinase pathway in delta opioid receptor facilitation of female reproductive behaviors. *Horm Behav* 49; 4:458–462.

Beyer, C. 1980. A model for explaining estrogen progesterone interactions in induction of lordosis behavior. In: *Endocrinology*, ed. I. A. Camnusigs, J. W. Fander, and F. A. D. Mendelshon, 101–104. Australia: Canberra Press.

Beyer, C., A. Fernandez-Guasti, and G. Rodriguez-Manzo. 1982. Induction of female sexual behavior by GTP in ovariectomized estrogen primed rats. *Physiol Behav* 28; 6:1073–1076.

Beyer, C., E. Canchola, and K. Larsson. 1981. Facilitation of lordosis behavior in the ovariectomized estrogen primed rat by dibutyrylc AMP. *Physiol Behav* 26; 2:249–251.

Beyer, C. and G. González-Mariscal. 1986. Elevation in hypothalamic cyclic AMP as a common factor in the facilitation of lordosis in rodents: a working hypothesis. *Ann N Y Acad Sci* 474:270–281.

Beyer, C., G. Moralí, and R. Vargas. 1971. Effect of diverse estrogen on estrous behavior and genital tract development in ovariectomized rats. *Horm Behav* 191; 2:273–277.

Beyer, C., K. L. Hoffman, and O. González-Flores. 2007. Neuroendocrine regulation of estrous behavior in the rabbit: similarities and differences with the rat. *Horm Behav* 52; 1:2–11.

Beyer, C., O. González-Flores, and G. González-Mariscal. 1997. Progesterone receptor participates in the stimulatory effect of LHRH, prostaglandin E2, and cyclic AMP on lordosis and proceptive behaviors in rats. *J Neuroendocrinol.* 9; 8:609–614.

Beyer, C., O. González-Flores, M. García-Juárez, and G. González-Mariscal. 2003. Non-ligand activation of estrous behavior in rodents: cross-talk at the progesterone receptor. *Scand J Psychol.* 44; 3:221–229.

Blaustein, J. D. and D. H. Olster. 1989. Gonadal steroid hormone receptors and social behaviors. In *Molecular and Cellular Basis of Social Behavior in Vertebrates*, ed. J. Balthazart, 31–104. Berlin, Germany: Springer-Verlag.

Blaustein, J. D. and H. H. Feder. 1979. Cytoplasmic progestin-receptors in guinea pig brain: characteristics and relationship to the induction of sexual behavior. *Brain Res* 169; 3:481–497.

Blaustein, J. D., H. I. Ryer, and H. H. Feder. 1980. A sex difference in the progestin receptor system of guinea pig brain. *Neuroendocrinology* 31; 6:403–409.

Blaustein, J. D. and M. S. Erskine. 2002. Feminine sexual behavior: Cellular integration of hormonal and afferent information in the rodent forebrain. In *Hormones, Brain and Behavior*, ed. D. W. Pfaff, A. P. Arnold, A. M. Etgen, S. E. Fahrbach, and R. T. Rubin, 139–214. New York: Academic Press.

Blaustein, J. D. and S. K. Mani. 2007. Feminine sexual behavior from neuroendocrine and molecular neurobiological perspectives. In *Handbook of Neurochemistry and Molecular Neurobiology*, ed. A. Lajtha and J. D. Blaustein, 95–149. New York: Springer.

Boonyaratanakornkit, V., E. Mc Gowan, L. Sherman, M. A. Mancini, B. J. Cheskis, and D. P. Edwards. 2007. The role of extranuclear signaling actions of progesterone receptor in mediating progesterone regulation of gene expression and the cell cycle. *Mol Endocrinol* 21; 2:359–375.

Boonyaratanakornkit, V., Y. Bi, M. Rudd, and D. P. Edwards. 2008. The role and mechanism of progesterone receptor activation of extra-nuclear signaling pathways in regulating gene transcription and cell cycle progression. *Steroids* 73; 9/10:922–928.

Brown, M. T. and J. A. Cooper. 1996. Regulation, substrates and functions of src. *Biochem Biophys Acta* 1287; 2/3:121–149.

Brown, N., J. Kerby, T. P. Bonnert, P. J. Whiting, and Wafford. 2002. Pharmacological characterization of a novel cell line expressing human alpha(4)beta(3)delta GABA(A) receptors. *Br J Pharmacol* 136; 7:965–974.

Caba, M., M. J. Rovirosa, C. Beyer, and G. González-Mariscal. 2003. Immunocytochemical detection of progesterone receptor in the female rabbit forebrain: distribution and regulation by oestradiol and progesterone. *J Neuroendocrinol* 15; 9:855–864.

Cabrera-Muñoz, E., O. T. Hernández-Hernández, and I. Camacho-Arroyo. 2011. Role of progesterone in human astrocytomas growth. Curr Top Med Chem. 11; 13:1663–1667.

Calka, J. 2006. The role of nitric oxide in the hypothalamic control of LHRH and oxytocin release, sexual behavior and aging of the LHRH and oxytocin neurons. *Folia Histochem Cytobiol* 44; 1:3–12.

Chu, H. P. and A. M. Etgen. 1999. Ovarian hormone dependence of alpha(1)-adrenoceptor activation of the nitric oxide-cGMP pathway: relevance for hormonal facilitation of lordosis behavior. *J Neurosci* 19; 16:7191–7197.

Ciaccio, L. A. and R. D. Lisk. 1971. Hormonal control of cyclic estrus in the female hamster. *Am J Physiol* 221; 3:936–942.

Dawson, T.M., J. Zhang, V.L. Dawson, and S.H. Snyder. 1994. Nitric oxide: cellular regulation and neuronal injury. *Prog Brain Res*. 103; 365–369.

Edwards, D. P., S. E. Wardell, and V. Boonyaratanakornkit. 2002. Progesterone receptor interacting coregulatory proteins and cross talk with cell signaling pathways. *J Steroid Biochem Mol Biol* 83; 1–5:173–186.

Etgen, A. M., O. González-Flores, and B. J. Todd. 2006. The role of insulin-like growth factor-I and growth factor-associated signal transduction pathways in estradiol and progesterone facilitation of female reproductive behaviors. *Front Neuroendocrinol* 27; 4:363–375.

Etgen, A. M. and P. Shamamian. 1986. Regulation of estrogen-stimulated lordosis behavior and hypothalamic progestin receptor induction by antiestrogens in female rats. *Horm Behav* 20; 2:166–180.

Feder, H. H. and B. L. Marrone. 1977. Progesterone: its role in the central nervous system as a facilitator and inhibitor of sexual behavior and gonadotropin release. *Ann N Y Acad Sci* 11; 286:331–354.

Fox, A. S. and D. H. Olster. 2000. Effects of intracerebroventricular leptin administration on feeding and sexual behaviors in lean and obese female Zucker rats. *Horm Behav* 37; 4:377–387.

García-Juárez, M., C. Beyer, A. Soto-Sánchez, et al. 2011. Leptin facilitates lordosis behavior through GnRH-1 and progestin receptors in estrogen-primed rats. *Neuropeptides* 45; 1:63–67.

García-Juárez, M., C. Beyer, P. Gómora-Arrati, et al. 2012. The nitric oxide pathway participates in lordosis behavior induced by central administration of leptin. *Neuropeptides* 46; 1:49–53.

García-Juárez, M., C. Beyer, P. Gómora-Arrati, et al. 2013. Lordosis facilitation by leptin in ovariectomized, estrogen-primed rats requires simultaneous or sequential activation of several protein kinase pathways. *Pharmacol Biochem Behav* 110:13–18.

Garris, D. R., R. B. Billiar, Y. Takaoka, R. J. White, and B. Little. 1981. In situ estradiol and progestin (R5020) localization in the vascularly separated and isolated hypothalamus of the rhesus monkey. *Neuroendocrinology* 32; 4:202–208.

González-Flores, O., A. M. Etgen, B. K. Komisaruk, et al. 2008. Antagonists of the protein kinase A and mitogen-activated protein kinase systems and of the progestin receptor block the ability of vaginocervical/flank-perineal stimulation to induce female rat sexual behaviour. *J Neuroendocrinol* 20; 12:1361–1367.

González-Flores, O., C. Beyer, F. J. Lima-Hernández, et al. 2007. Facilitation of estrous behavior by vaginal cervical stimulation in female rats involves alpha1-adrenergic receptor activation of the nitric oxide pathway. *Behav Brain Res* 176; 2:237–243.

González-Flores, O., C. Beyer, P. Gómora-Arrati, et al. 2010. A role for Src kinase in progestin facilitation of estrous behavior in estradiol-primed female rats. *Horm Behav* 58; 2:223–229.

González-Flores, O., C. Guerra-Araiza, M. Cerbón, I. Camacho-Arroyo, and A. M. Etgen. 2004a. The 26S proteasome participates in the sequential inhibition of estrous behavior induced by progesterone in rats. *Endocrinology* 145; 5:2328–2336.

González-Flores, O., J. Shu, I. Camacho-Arroyo, and A. M. Etgen. 2004b. Regulation of lordosis by cyclic 3′,5′-guanosine monophosphate, progesterone, and its 5alpha-reduced metabolites involves mitogen-activated protein kinase. *Endocrinology* 145; 12:5560–5567.

González-Flores, O., J. M. Ramírez-Orduña, F. J. Lima-Hernández, M. García-Juárez, and C. Beyer. 2006. Differential effect of kinase A and C blockers on lordosis facilitation by progesterone and its metabolites in ovariectomized estrogen-primed rats. *Horm Behav* 49; 3:398–404.

González-Flores, O., P. Gómora-Arrati, M. Garcia-Juárez, et al. 2009. Nitric oxide and ERK/MAPK mediation of estrous behavior induced by GnRH, PGE2 and db-cAMP in rats. *Physiol Behav* 96; 4/5:606–612.

González-Flores O., P. Gómora-Arrati, M. García-Juárez, et al. 2011. Progesterone receptor isoforms differentially regulate the expression of tryptophan and tyrosine hydroxylase and glutamic acid decarboxylase in the rat hypothalamus. *Neurochem Int* 59; 5:671–677.

González-Mariscal G., A. I. Melo, and C. Beyer. 1993. Progesterone, but not LHRH or prostaglandin E2, induces sequential inhibition of lordosis to various lordogenic agents. *Neuroendocrinology* 57; 5:940–945.

Guerra-Araiza, C., O. Villamar-Cruz, A. González-Arenas, R. Chavira, and I. Camacho-Arroyo. 2003. Changes in progesterone receptor isoforms content in the rat brain during the oestrous cycle and after oestradiol and progesterone treatments. *J Neuroendocrinol* 15; 10:984–990.

Guerra-Araiza, C., P. Gómora-Arrati, M. García-Juárez, et al. 2009. Role of progesterone receptor isoforms in female sexual behavior induced by progestins in rats. *Neuroendocrinology* 90; 1:73–81.

Karteris, E., S. Zervou, Y. Pang, J. Dong, E. W. Hillhouse, H. S. Randeva, and P. Thomas. 2006. Progesterone signaling in human myometrium through two novel membrane G protein-coupled receptors: potential role in functional progesterone withdrawal at term. *Mol Endocrinol* 20; 7:1519–1534.

Kastner, P., M. T. Bocquel, B. Turcotte, et al. 1990. Transient expression of human and chicken progesterone receptors does not support alternative translational initiation from a single mRNA as the mechanism generating two receptor isoforms. *J Biol Chem* 265; 21:2163–2167.

Komisaruk, B. R. and C. Diakow. 1973. Lordosis reflex intensity in rats in relation to the estrous cycle, ovariectomy, estrogen administration and mating behavior. *Endocrinology* 93; 3:548–557.

Lange, C. A. 2004. Making sense of cross-talk between steroid hormone receptors and intracellular signaling pathways: who will have the last word? *Mol Endocrinol* 18; 2:269–278.

Lange, C. A., D. Gioeli, S. R. Hammes, and P. C. Marker. 2007. Integration of rapid signaling events with steroid hormone receptor action in breast and prostate cancer. *Annu Rev Physiol* 69:171–199.

Lange, C. A., T. Shen, and K. B. Horwitz. 2000. Phosphorylation of human progesterone receptors at serine-294 by mitogen-activated protein kinase signals their degradation by the 26S proteasome. *Proc Natl Acad Sci USA* 97; 3:1032–1037.

Leonhardt, S. A., V. Boonyaratanakornkit, and D. P. Edwards. 2003. Progesterone receptor transcription and non-transcription signaling mechanisms. *Steroids* 68; 10–13:761–770.

Lima-Hernández, F. J., C. Beyer, P. Gómora-Arrati, et al. 2012. Src kinase signaling mediates estrous behavior induced by 5β-reduced progestins, GnRH, prostaglandin E2 and vaginocervical stimulation in estrogen-primed rats. *Horm Behav* 62; 5:579–584.

Majewska, M. D., N. L. Harrison, R. D. Schwartz, J. L. Barker, and S. M. Paul. 1986. Steroid hormone metabolites are barbiturate-like modulators of the GABA receptor. *Science* 232; 4753:1004–1007.

Mani, S. K., J. M. Allen, V. Rettori, S. M. McCann, B. W. O'Malley, and J. H. Clark. 1994. Nitric oxide mediates sexual behavior in female rats. *Proc Natl Acad Sci USA* 91; 14:6468–6472.

Mani, S. K. and M. G. Oyola. 2012. Progesterone signaling mechanisms in brain and behavior. *Front Endocrinol (Lausanne)* 3; 7:1–8.

Mani, S. and W. Portillo. 2010. Activation of progestin receptors in female reproductive behavior: interactions with neurotransmitters. *Front Neuroendocrinol* 31; 2:157–171.

Mehebik, N., A. M. Jaubert, D. Sabourault, Y. Giudicelli, and C. Ribière. 2005. Leptin-induced nitric oxide production in white adipocytes is mediated through PKA and MAP kinase activation. *Am J Physiol Cell Physiol* 289; 2:C379–C387.

Migliaccio, A., D. Piccolo, G. Castoria, et al. 1998. Activation of the Src/p21ras/Erk pathway by progesterone receptor via cross-talk with estrogen receptor. *EMBO J* 17; 7:2008–2018.

Migliaccio, A., M. Di Domenico, G. Castoria, et al. 1996. Tyrosine kinase/p21ras/MAP-kinase pathway activation by estradiol-receptor complex in MCF-7 cells. *EMBO J* 15; 6:1292–1300.

Moguilewsky, M. and J. P. Raynaud. 1979. The relevance of hypothalamic and hyphophyseal progestin receptor regulation in the induction and inhibition of sexual behavior in the female rat. *Endocrinology* 105; 516–522.

Morin, L. P. 1977. Theoretical review. Progesterone: inhibition of rodent sexual behavior. *Physiol Behav* 18; 4:701–715.

Nestler, E. J. and P. Greengard. 1994. Protein phosphorylation and the regulation of neuronal function. In *Basic Neurochemistry*, ed. G. J. Siegel, B. W. Agranoff, R. A. Albers, and P. B. Molinoff, 449–474. New York: Raven Press.

Nestler, E. J. and P. Greengard. 1999. Serine and threonine phosphorylation. In *Basic Neurochemistry, Cellular and Medical Aspects*, ed. G. J. Siegel, B. W. Agranoff, W. R. Albers, and P. B. Molinoff, 471–495. Philadelphia, PA: Lippincott-Raven Publishers.

Numan, M., J. K. Roach, M. C. del Cerro, et al. 1999. Expression of intracellular progesterone receptors in rat brain during different reproductive states, and involvement in maternal behavior. *Brain Res* 830; 2:358–371.

Pang, Y. and P. Thomas. 2011. Progesterone signals through membrane progesterone receptors (mPRs) in MDA-MB-468 and mPR-transfected MDA-MB-231 breast cancer cells which lack full-length and N-terminally truncated isoforms of the nuclear progesterone receptor. *Steroids* 76; 9:921–928.

Park, H. K. and R. S. Ahima. 2014. Leptin signaling. F1000 *Prime Rep* 4; 6:1–8.

Parsons, B., M. Y. McGinnis, and B. S. McEwen. 1981. Sequential inhibition of progesterone: effects on sexual receptivity and associated changes in brain cytosol progestin binding in the female rat. *Brain Res* 221; 1:149–160.

Paul, S. M. and R. H. Purdy. 1992. Neuroactive steroids. *FASEB J* 6; 6:2311–2322.

Pfaff, D. W. and Y. Sakuma. 1980. *Estrogens and Brain Function*. New York: Springer-Verlag.

Pleim, E. T., T. J. Brown, N. J. MacLusky, A. M. Etgen, and R. J. Barfield. 1989. Dilute estradiol implants and progestin receptor induction in the ventromedial nucleus of the hypothalamus: Correlation with receptive behavior in female rats. *Endocrinology* 124; 4:1807–1812.

Powers, J. B. 1970. Hormonal control of sexual receptivity during the estrous cycle of the rat. *Physiol Behav* 5; 8:831–835.

Powers, J. B. 1975. Anti-estrogenic suppression of the lordosis response in female rats. *Horm Behav* 6; 4:379–392.

Powers, J. B. and J. Moreines. 1976. Progesterone: examination of its postulated inhibitory actions on lordosis during the rat estrous cycle. *Physiol Behav* 17; 3:493–498.

Ramírez-Orduña, J. M., F. J. Lima-Hernández, M. García-Juárez, O. González-Flores, and C. Beyer. 2007. Lordosis facilitation by LHRH, PGE2 or db-cAMP requires activation of the kinase A signaling pathway in estrogen primed rats. *Pharmacol Biochem Behav* 86; 1:169–175.

Rees, H. D., R. W. Bonsall, and R. P. Michael. 1985. Localization of the synthetic progestin 3H-ORG 2058 in neurons of the primate brain: evidence for the site of action of progestins on behavior. *J Comp Neurol* 235; 3:336–342.

Sá, S. I., P. A. Pereira, V. Malikov, and M. D. Madeira. 2013. Role of estrogen receptor α and β in the induction of progesterone receptors in hypothalamic ventromedial neurons. *Neuroscience* 15; 238:159–167.

Sweeney, G. 2002. Leptin signalling. *Cell Signal* 14; 8:655–663.

Thomas, P., Y. Pang, J. Dong, et al. 2007. Steroid and G protein binding characteristics of the sea trout and human progestin membrane receptor alpha subtypes and their evolutionary origins. *Endocrinology* 148; 2:705–718.

Thomas, S. M. and J. S. Brugge. 1997. Cellular functions regulated by Src family kinases. *Annu Rev Cell Dev Biol* 13:513–609.

Tobet, S. A., D. J. Zahniser, and M. J. Baum. 1986. Differentiation in male ferrets of a sexually dimorphic nucleus of the preoptic/anterior hypothalamic area requires prenatal estrogen. *Neuroendocrinology* 44; 3:299–308.

Truett, G. E., H. J. Jacob, J. Miller, et al. 1995. Genetic map of rat chromosome 5 including the fatty (fa) locus. *Mamm Genome* 6; 1:25–30.

Wade, G. N., R. L. Lempicki, A. K. Panicker, R. M. Frisbee, and J. D. Blaustein. 1997. Leptin facilitates and inhibits sexual behavior in female hamsters. *Am J Physiol* 272; 4 Pt 2:R1354–R1358.

Wu, F. S., T. T. Gibbs, and D. H. Farb. 1990. Inverse modulation of gamma-aminobutyric acid- and glycine-induced currents by progesterone. *Mol Pharmacol* 37; 5:597–602.

Wu, F. S., T. T. Gibbs, and D. H. Farb. 1991. Pregnenolone sulfate: a positive allosteric modulator at the N-methyl-D-aspartate receptor. *Mol Pharmacol* 40; 3:333–336.

Yun, H. Y., V. L. Dawson, and T. M. Dawson. 1996. Neurobiology of nitric oxide. *Crit Rev Neurobiol* 10; 3/4:291–316.

Zhang, Y., R. Proenca, M. Maffei, M. Barone, L. Leopold, and J. M. Friedman. 1994. Positional cloning of the mouse obese gene and its human homologue. *Nature* 372; 6505:425–432.

Zhu, Y., C. D. Rice, Y. Pang, M. Pace, and P. Thomas. 2003. Cloning, expression, and characterization of a membrane progestin receptor and evidence it is an intermediary in meiotic maturation of fish oocytes. *Proc Natl Acad Sci USA* 100; 5:2231–2236.

8 The Delicate Line between "Wanting" (Desire) and "Liking" (Reward) Sexual Behavior

Elisa Ventura-Aquino, Jorge Baños-Araujo, Alonso Fernández-Guasti, and Raúl G. Paredes

CONTENTS

8.1 Why Copulate? ... 131
8.1.1 Hormonal Influence ... 132
8.1.2 *Wanting* Sexual Activity ... 133
8.2 How Reward is Measured ... 134
8.2.1 *Liking* Sexual Behavior ... 134
8.3 Factors Altering Sexual Reward and Partner Selection ... 135
8.3.1 Sexual Behavior in Males and Females under Sexual Satiety and the "Coolidge Effect" ... 136
8.4 Biological Basis of Sexual Reward ... 138
8.5 Concluding Remarks ... 139
Epilogue ... 139
References ... 140

8.1 WHY COPULATE?

In strictly biological terms, sexual behavior is commonly related to reproduction, which favors evolution (Ågmo 1999). This concept, however, is questionable, as evolution favors the persistence of the most convenient characteristics at a certain moment and in a particular context. Undoubtedly in species with sexual reproduction, evolution is not possible without sexual behavior, but the notion that the ultimate purpose of this behavior is to achieve fertilization is a teleological rather than biological perspective, because there is no linear relationship between sexual behavior and fertilization (Møller and Birkhead 1991; Manson, Perry, and Parish 1997; Ågmo 1999). Thus, reproduction is one of the consequences of mating, so mating cannot be considered exclusively a reproductive behavior (Ågmo 1999; Meisel and Mullins 2006; Ågmo 2007a). An additional explanation is that animals mate because it is rewarding, which in turn assures that the animal will seek its repetition, as occurs with other behaviors such as feeding or social behavior (Berridge and Robinson 1998). Furthermore, sexual intercourse has other implications, depending on the species; for example, it was recently proposed that copulation in rats may prevent other males from impregnating the female (Lucio, Fernández-Guasti, and Larsson, see Chapter 5). In primates, including bonobos and chimpanzees, sexual behavior has different nonreproductive connotations (Wrangham 1993; Hrdy 1995), that is, (1) paternity confusion, in which the female copulates with different males, which may inhibit later potential harm toward her offspring; (2) practice interactions, where adolescent females or males interact sexually, gaining experience for their future sexual performance; (3) sex in exchange for food, when provisions are scarce, or for protection from a male, which provides an opportunity for mating; and (4) sex communication: in this case, sexual behavior helps to develop social relationships (e.g., homosexual interactions between a recent immigrant female with an older local one, thereby gaining social integration and protection), that prevents aggressions or repairs social relationships after an aggressive outburst (Wrangham 1993; Hrdy 1995; Manson, Perry, and Parish 1997). In humans, the reasons to have sex also go beyond fecundation, and are even more complex, and there are many interpretations to account for sexual encounters (for a discussion, see Pfaus et al. 2012).

8.1.1 Hormonal Influence

The levels of gonadal hormones influence the display of the mating pattern in many mammals, and for both sexes. In females of several species, the performance of sexual behavior requires adequate steroid hormone levels, usually characterized by a surge of estradiol, followed by a peak of progesterone (Feder 1981; Blaustein 2009). In males, adequate testosterone levels are a key factor for the display of sexual behavior (Balthazar and Ball 2010; Golinski et al. 2014). Also, sexually experienced males that are castrated lose their ability to copulate, and testosterone replacement restores their sexual behavior (Beyer et al. 1981; Merkx 1984). Furthermore, with aging, testosterone levels decline, followed by a reduction in sexual behavior (Södersten et al. 1976; Phoenix 1978; Chambers and Phoenix 1983; Wee, Weaver, and Clemens 1988).

In both sexes, an adequate hormonal environment promotes the initiation of a sexual encounter, and also increases the sensitivity to genital stimulation. For example, in females, estradiol enlarges the sensory field of pudendal nerves that provide innervation surrounding the clitoris, favoring coital stimulation (Komisaruk, Adler, and Hutchison 1972; Erskine 1992). In males, androgens increase penile sensitivity, facilitating ejaculation (Balthazar and Ball 2010). Eventually, sexual stimulation will induce a reward state that increases the probability that the behavior is repeated, when the conditions are adequate (Paredes 2009). In addition to the internal hormonal state, there are other factors involved in two different processes: wanting and liking sexual behavior, which are orchestrated by other neuromodulators that make mating a natural rewarding behavior.

8.1.2 *Wanting* Sexual Activity

"Wanting" implies desire and motivation to approach an incentive, predicting reward. As a consequence of reward, the individual searches for its repetition by associative learning (Berridge and Robinson 1998). A fundamental question is that, if previous sexual experience modifies sexual reward, how does a sexually naive animal become motivated to engage in sexual behavior? Although precise information is limited, we know that female rats with no previous sexual experience prefer odors from intact males over those of castrated ones. For male rats, the acquisition of incentive properties to females occurs with sexual experience. It is possible that the male represents an unconditioned incentive that induces the female's approach, which in turn can activate proceptive behavior in the females and unconditioned reflexes in males such as mounting, and eventually mating (Ågmo 1999). Classical studies have measured the propensity of the animal to obtain sexual contact. Researchers usually measure how much an animal is willing to work for the opportunity to mate with a conspecific. Thus, males cross electrified grids, poke their nose in holes, or press levers to have access to a receptive female (McDonald and Meyerson 1973; Paredes and Vazquez 1999), which is evidently a strong incentive.

The neural substrates of decision making have begun to be described. Studies in mammals reveal that two major interconnected neural circuits regulate this process: the mesolimbic system and the social behavior network (O'Connell and Hofmann 2011). Comparative analyses have shown that both circuits are present inclusively from low vertebrates to mammals. The brain areas involved in wanting an unconditioned incentive are still poorly understood, due to the lack of methodological tools and studies addressing this issue. Aside from hormones, it is clear that several neuromodulators are implicated in the wanting of sexual behavior, including serotonin, prolactin, and oxytocin. Dopamine (DA) may be the neuromodulator most studied to date, especially its projection from the ventral tegmental area to the nucleus accumbens and neostriatum (Berridge and Robinson 1998; Paredes and Ågmo 2004). Once an animal has evaluated the environmental and sensory cues, it performs an appropriate behavioral response that might be modulated by prior experience (O'Connell and Hofmann 2011). In a sexual context, DA seems to participate in the activation of appetitive and approaching (i.e., precopulatory) behavior (Wenkstern, Pfaus, and Fibiger 1993; Pfaus et al. 1995; Fiorino, Coury, and Phillips 1997). Subsequently, goal-directed behavior is activated. Such behavioral patterns are influenced by past sexual experiences that increase the expectation of a rewarding event (Berridge and Robinson 1998). The studies relating DA and sexual behavior, especially those using microdialysis, have helped to understand the role of this neurotransmitter. In males, there is a consensus that DA is released in the nucleus accumbens, the caudate nucleus, and the medial preoptic area (MPOA), even in the first sexual experience with a receptive female. Such DA release occurs before the male begins to copulate; that is, just the exposure to a receptive female is sufficient. DA is also released during the execution of mating (Wenkstern, Pfaus, and Fibiger 1993; Fiorino, Coury, and Phillips 1997). The DA discharge occurs only with a receptive female as stimulus, whereas if the incentive is a nonreceptive female, DA is not released, even if the male performs several mounts (Wenkstern, Pfaus, and Fibiger 1993).

In females, exposure to a male, as well as copulation, also induces DA release in the nucleus accumbens and the dorsal striatum, although the release from the latter seems to be related to locomotion rather than to sexual behavior (Pfaus et al. 1995). However, the role of the dopaminergic system in the control of female sexual behavior is less clear, due to several misconceptions and contrasting results (Paredes and Ågmo 2004). This lack of clarity as to the role of DA in female sexual behavior may have several explanations: (1) females are tested under different hormonal combinations or replacement treatments, (2) different testing conditions, and (3) the use of lordosis as the main behavioral pattern measured without consideration of other motivational measurements, for example, proceptive behavior. However, the most important reason for uncertainty is the selection of nonrelevant brain areas for sexual behavior to analyze changes in DA concentration. Most studies have selected the nucleus accumbens and striatum for measuring DA, but it has been demonstrated that these are not key regulatory sites for female sexual behavior. For example, lesions of these areas fail to alter lordosis (receptivity), or proceptivity (Rivas and Mir 1990, 1991). A critical review of the lack of effect of dopaminergic regulation on sexual behavior has been presented before (Paredes and Ågmo 2004). In humans, DA seems to be involved in sexual response and orgasm, as the ventral tegmentum and nucleus accumbens become activated maximally at orgasm (Komisaruk, Beyer, and Whipple, 2009). This view is consistent with studies in animals suggesting that DA promotes generalized behavioral arousal (Alcaro, Huber, and Panksepp 2007) since it is released in situations associated with reward but also in response to stress and aversive stimuli (Abercrombie et al. 1989; Puglisi-Allegra et al. 1991; Rougé-Pont et al. 1993; Pruessner et al. 2004). Moreover, detailed behavioral analysis has demonstrated that DA is involved in wanting but not in liking (Berridge and Robinson 1998).

In both sexes, wanting sexual behavior is mediated by DA and can be explained as the desire for a sexual interaction, which augments the willingness and facilitates its occurrence when the environmental and internal hormonal conditions are optimal. As a result of adequate sexual stimulation, we propose that mating induces a positive affective state making sex a rewarding experience, that is, "liking."

8.2 HOW REWARD IS MEASURED

8.2.1 *Liking* Sexual Behavior

Measuring a rewarding event, that is, how pleasurable an action is (including mating), is complicated. For example, the reward produced by orgasm includes subjective aspects that are difficult to express and measure (Komisaruk, Beyer, and Whipple 2006). Of course, while we can ask humans whether the sexual experience was rewarding, measuring reward in animals is more complicated (Paredes 2009). In animals, "reward" can be indirectly inferred with paradigms such as the conditioned place preference (CPP) (Tzschentke 1998; Nasr et al. 2013). In this test, an animal is placed alone in a three-compartment chamber. Each lateral compartment has a different color, texture, and odor. Initially, the preferred compartment is determined based on the time the animal spends in each lateral compartment (Mucha et al. 1982;

Tzschentke 1998). Subsequently, a positive stimulus (drug, mating, etc.) is associated with the nonpreferred compartment. In this way the state induced by the stimulus is associated with the environmental cues of the originally nonpreferred compartment. If the stimulus produces a positive affective state of sufficient duration and intensity, the subject will change its original preference. The preference change is assessed by a significant increase in the time spent in the originally nonpreferred compartment and, therefore, in the preference index. CPP has been useful in studying the neuromodulators and/or conditions implicated in the establishment of the positive affective state that develops after a reinforcing stimulus such as sexual behavior (Paredes and Vazquez 1999; Martínez and Paredes 2001) in various species including rats, hamsters, and birds. The rewarding properties of mating can also be evaluated by the CPP paradigm, and previous reports have revealed that only when the female or the male has control of the timing of mating, sexual behavior induces a change in the original preference, indicative of a reward state (Martínez and Paredes 2001). This observation suggests that the timing, and thereby the quality, of coital stimulation affects the establishment of sexual reward.

8.3 FACTORS ALTERING SEXUAL REWARD AND PARTNER SELECTION

Sexual behavior is not essential to maintain vital processes in the individual, and its absence is not life threatening, as occurs with lack of water or food (Le Boeuf 1974; Crawford, Holloway, and Domjan 1993). In addition, sexual behavior varies among species, among individuals of the same species, and within the same individual at different times. Such differences depend upon hormonal, situational, nutritional, environmental, and seasonal factors (Bronson 1985; Anderson, Nordheim, and Boesch 2006; Blaustein 2009). Mating may also involve risk in natural environments. For example, it is almost impossible for subordinate individuals of hierarchical species to mate, and if they do, they may be excluded from the colony by higher-rank individuals (Crawford, Holloway, and Domjan 1993). Additionally, precopulatory and copulatory behavior patterns may attract predators' attention. Thus, there are situations that can increase, reduce, or even abolish the rewarding properties of copulation. To enhance the rewarding value of sexual behavior under laboratory conditions, most studies (the vast majority performed in rats) evaluate sexual behavior under the paced mating condition. In this context the animal has the opportunity to regulate the sexual interaction and, therefore, to time the stimulation received (Erskine and Baum 1982; Coopersmith and Erskine 1994; Paredes and Vazquez 1999; Zipse, Brandling-Bennett, and Clark 2000). This is achieved by designing appropriate test conditions, for example, a bilevel chamber or a chamber divided in two compartments interconnected by a small hole through which only the female can pass from the male's chamber to the other. With this particular cage, the female controls the timing of interactions with the male. This enables one to obtain measures such as the time spent in the female's compartment after receiving sexual stimulation (as a mount or an intromission) and the percentage of exits after receiving a mount or an intromission (Brandling-Bennett, Blasberg, and Clark 1999; Zipse, Brandling-Bennett, and Clark 2000). Use of this paradigm has found that the female rat can dissociate sexual contacts of different intensities (Erskine, Kornberg, and Cherry 1989), allowing inferences about the female's willingness to initiate, continue, or end the sexual behavior interaction.

Under more natural conditions, mating is regulated by both sexes (McClintock, Anisko, and Adler 1982). In natural environments, female rats live in groups and their estrous cycles become synchronized. Mating occurs in a group, and pacing—and reward—would be possible for both sexes. For females, paced mating has reproductive advantages because fewer intromissions are required to produce the progestational state necessary for fertilization than the number of intromissions required in nonpaced mating. Also, females that pace the sexual interaction have more pups per litter (Coopersmith and Erskine 1994).

While rats normally have a promiscuous mating system, there are some studies suggesting that the females can discriminate and show a clear preference for a particular male over other potential sexual partners, at least under laboratory conditions (Ferreira-Nuño et al. 2005; Winland et al. 2012). Female rats consistently prefer a particular male, even when the mating test is repeated two or more times (Ferreira-Nuño et al. 2005; Lovell et al. 2007; Winland et al. 2012).

In Soay sheep some characteristics of dominance and health may be related to better genes and breeding success. In many other species, including the rat, the fruit fly *Drosophila melanogaster*, and the yellow-toothed cavy, the features that influence male attractiveness remain unclear because no hormonal or behavioral differences have been found between preferred and nonpreferred males (Winland et al. 2012). In primates, it is not clear whether females prefer one male over another, but in some cases, they prefer males with higher dominance; however, they also mate and reproduce with males that are lower in the hierarchy (Marvan et al. 2006). For humans, the selection of a sexual partner is complex; while there are some detectable physical factors, for example, individual health, hormonal stage, and genetic information, such biological features are not sufficient to account for choice of a sexual partner (Pfaus et al. 2012). It is also well documented that pleasurable sexual behavior has a positive impact on health, and positive correlations have been found between sexual pleasure and mental and physical health (Brody 2010).

8.3.1 Sexual Behavior in Males and Females under Sexual Satiety and the "Coolidge Effect"

Males of several species including rats are able to ejaculate many times in a relatively short period. We do not know, however, whether two, three, or several ejaculations are as rewarding as just one (Phillips-Farfán and Fernández-Guasti 2009). In rats it is well documented that after several ejaculations in one mating session there is a drastic decline in sexual behavior. This phenomenon is termed sexual satiety or exhaustion, at which point male sexual motivation is reduced (Ågmo 1999). Sexual exhaustion is thus defined as the inhibition of sexual behavior, and is present in many species (Phoenix and Chambers 1988; Estrada-Reyes et al. 2009; Phillips-Farfán and Fernández-Guasti 2009; Jimenez et al. 2012). Such inhibition of sexual behavior is not due to physical weakness (Rodríguez-Manzo and Fernández-Guasti 1994), because if a new potential sexual partner is presented immediately after sexual exhaustion, the male restarts mating. This mating resumption is known as the Coolidge Effect, and has been extensively observed in males of many species. Recently, we reported that the behavioral pattern of ejaculation observed during the Coolidge Effect lacks seminal emission, and that the penile intromissions were sufficient to detach the vaginal seminal plug deposited by another male. On these bases, we suggested that the sexual behavior observed during the Coolidge Effect is displayed to maintain the male as a sexual competitor (see Lucio, Fernández-Guasti, and Larsson, present book, Chapter 5). A possible interpretation considers that the sexual inhibition that precedes the Coolidge Effect involves habituation, defined as a systematic decrement in the stimulus magnitude after repeated exposure to a particular female with whom the male has been repeatedly copulating for a relatively long time. This habituation cannot be explained by fatigue (O'Donohue and Geer 1985). Evidence in favor of the motivational role of DA mediating male sexual satiety and the Coolidge Effect is provided by an in vivo microdialysis study showing that DA levels in the nucleus accumbens return to basal levels at the time of sexual satiety, but when a new sexual partner is presented behind a screen, DA levels increase in this brain area (Fiorino, Coury, and Phillips 1997). Such DA increase would elevate the incentive value of the new female, thus restoring mating and inducing the Coolidge Effect.

There is an increased DA efflux in the nucleus accumbens also in female rats when a male was presented behind a screen (Pfaus et al. 1995). However, a main methodological difference between these two studies is that, for females, the DA measurement was done before mating and was not tested after the exposure to the sexual partner. By contrast with the numerous reports in males regarding sexual satiety and the Coolidge Effect, these processes have been only minimally explored in females. Regarding the former, most researchers assume that in females sexual behavior declines primarily due to the extinction of the endocrine milieu necessary for its display (Carter and Porges 1974; Blaustein 2009). Findings in hamsters and rats suggest the existence of a Coolidge Effect in females (Krames 1971; Lisk and Baron 1982). However, there are important methodological and interpretive concerns, primarily due to attempts to equate and translate sexual satiety and the Coolidge Effect from males to females, disregarding their fundamental behavioral differences. Another important factor that obscures the mechanism of sexual satiety in females is its strong association with male sexual behavior. That is, if a male is sexually inactive—as a result of sexual satiety—it is difficult to dissociate the extent to which the female's sexual behavior is inhibited. We hypothesized that as in males, sexual exhaustion in females occurs due to a reduced sexual motivation. In the female rat, proceptivity is considered an appetitive behavior and has been used as an indicator of sexual motivation, as it precedes most intromissions, and is modified by drugs with known effects on sexual motivation in humans (Rössler et al. 2006; Pfaus, Giuliano, and Gelez 2007; Ventura-Aquino and Fernández-Guasti 2013b). Beach defined proceptivity as species-specific behavior displayed by females that incites males to mate (Beach 1976). In rats these behaviors include ear wiggling, hopping, and darting. Another means of inferring the female's sexual motivation is by counting the number of female entrances into the male's compartment in a paced-mating test (Ventura-Aquino and Fernández-Guasti 2013a). In this condition, a receptive female rat continuously copulated with a male, but when the female stayed at least 20 min in the compartment inaccessible to the male, it was considered to be sexually satiated (Ågmo 2007b). Subsequently, a new male was presented, and the female not only resumed mating but also increased proceptive behaviors. This result was interpreted as a female Coolidge Effect (Ågmo 2007b). Recently we provided an opportunity for female rats to copulate with a male for 90 min under paced-mating and nonpaced-mating conditions. The male was replaced by another male and they mated for another 90 min. The rationale for using 90 min of continuous mating was that during this period, the males copulated ceaselessly without reaching sexual satiety (Phillips-Farfán and Fernández-Guasti 2009). Under these conditions, proceptivity ceased at the end of the 90-min test, regardless of whether the females were, or were not, pacing the copulation. Only under the paced-mating condition, did the females show an increase in proceptive behavior when a new sexual partner was presented; and that also declined after the last 90 min of mating (Ventura-Aquino and Fernández-Guasti 2013a). Receptivity did not change over the course of either test, possibly because of its reflexive nature. These findings reinforce the concept that sexual satiety in females is primary a function of motivational components.

Habituation of sexual arousal is also observed in humans, but with differences between the sexes. In men, it appears readily as habituation of sexual arousal, measured as a decrease in penile tumescence, an eye blink startle response, and subjective sexual arousal after repetitive exposure to sexually explicit stimuli conveyed in texts, audiotapes, or films (O'Donohue and Geer 1985; O'Donohue and Plaud 1991). The decrease in these responses varies; for example, self-identified heterosexual men present greater genital responses to women than to men (Suschinsky and Lalumière 2011). In women, habituation appears after a long latency (at least for the same kind of stimuli as men), measured by vaginal plethysmography and self-reported sexual arousal (Laan and Everaerd 1995).

8.4 BIOLOGICAL BASIS OF SEXUAL REWARD

Several lines of evidence indicate that the reward state induced by copulation in males and females is mediated mainly by the opioidergic system (Paredes 2009, 2014). Although there is no direct method to measure opioid release in vivo, several indirect lines of evidence suggest that this mechanism underlies sexual reward (see Paredes, 2014, for a discussion). Furthermore, orgasm-like sensations in humans are related to acute administration of heroin or morphine, mainly in the first experiences, and their chronic use is associated with problems experiencing orgasm (Pfaus and Gorzalka 1987). Opiates can also participate in the orgasmic response since the administration of naltrexone in healthy males increased the number of orgasms and their intensity, particularly of the first ones, during a masturbation session (Sathe et al. 2001). It is possible that opioids released during sexual behavior modulate sensory thresholds, so that the opiate-antagonist administration facilitates penile sensitivity. For women, pleasurable vaginal stimulation produces an increase in pain tolerance, but this effect is weaker if the stimulation is just pressure, or absent if it is uncomfortable (Whipple and Komisaruk 1985). In rats, vaginocervical stimulation produces analgesia (Komisaruk and Wallman 1977; Gintzler and Komisaruk 1991; Gómora et al. 1994; Johnson and Komisaruk 1996), an effect that may be mediated, at least in part by opioid release. This phenomenon has also been observed in women during parturition and can decrease pain and aversion, perhaps enhancing the mother-child bond (Komisaruk, Beyer, and Whipple 2009). In male rats, mating reduces vocalizations, evoked by a painful electric shock to the skin, an effect prevented by the administration of an opioid antagonist, naloxone (Szechtman, Hershkowitz, and Simantov 1981).

The two major neural systems crucial for rewarding processes are the mesolimbic and the social behavior networks. The first is related to the reward from drugs of abuse while the latter is related to naturally occurring rewards such as sexual behavior (O'Connell and Hofmann 2011). The social behavior network is formed by brain areas that include the MPOA, the ventromedial nucleus of the hypothalamus (VMH), and the medial amygdala. These systems are homologous among species from reptiles to mammals (O'Connell and Hofmann 2011). There is strong evidence that the MPOA is crucial for the expression of male sexual behavior, as its lesion or inactivation by lidocaine eliminates mating in all species studied to date (Hurtazo, Paredes, and Ågmo 2008; Paredes 2014). In addition, it is proposed that opioids released from the MPOA mediate the rewarding properties of mating, because naloxone infusion into this brain area blocks the development of CPP after mating (García-Horsman, Ågmo, and Paredes 2008). Copulation also activates μ opioid receptors in the MPOA and induces their internalization (Coolen et al. 2004).

The social behavior network is also involved in female sexual reward mediated by opioids because naloxone infusion into the MPOA, the amygdala, or the VMH blocks CPP after paced mating (García-Horsman, Ågmo, and Paredes 2008). It is proposed that opioids acting through the social behavior network are important neuromodulators implicated in naturally occurring rewarding behaviors.

8.5 CONCLUDING REMARKS

Mating is a naturally occurring rewarding behavior pattern that ensures its repetition and, thereby, species survival. "Wanting" sexual behavior refers to the willingness to approach an incentive. There are several neuromodulators implicated, especially DA through its projection from the ventral–tegmental area to the nucleus accumbens and neostriatum. "Liking" sexual behavior refers to the rewarding nature of a sexual interaction, and it can be inferred by CPP in laboratory animals. The social decision network involves a brain circuit that includes the MPOA, the amygdala, and the VMH, which participate in sexual reward through endogenous opioid release that, in turn, induces a positive affective state after mating in both sexes.

EPILOGUE

We write this chapter in memory of Carlos Beyer, who was a pioneer in neuroendocrinology and passionate about science. His revolutionary ideas transformed the current notion of sexual behavior.

Carlos Beyer taught us that there are many ways to study sexual behavior, ranging from an ethological to a molecular approach. The present chapter discusses the biological bases underlying the subtle differences between "wanting" and "liking" sexual behavior. That is, we analyze the motivational aspects of sexual behavior as well as its rewarding properties. These two conceptually independent phenomena influence one another and may synergize or antagonize, producing a large range of sexual responses. Most of these data refer to experimental animals and—needless to say—they must be translated to humans only with caution.

REFERENCES

Abercrombie, Elizabeth D., Kristen A. Keefe, Daniel S. Di Frischia, and Michael J. Zigmond. 1989. "Differential Effect of Stress on in Vivo Dopamine Release in Striatum, Nucleus Accumbens, and Medial Frontal Cortex." *Journal of Neurochemistry* 52 (5): 1655–58. doi:10.1111/j.1471-4159.1989.tb09224.x.

Ågmo, Anders. 1999. "Sexual Motivation—An Inquiry into Events Determining the Occurrence of Sexual Behavior." *Behavioural Brain Research* 105 (1): 129–50. doi:10.1016/s0166-4328(99)00088-1.

Ågmo, Anders. 2007a. "On the Purpose of Sex and Some Notes on Scientific Explanations." In *Functional and Dysfunctional Sexual Behavior*, edited by Anders Ågmo, 1–29. Oxford: Academic Press.

Ågmo, Anders. 2007b. "An Incentive Motivational Framework and the Description of Sexual Behaviors." In *Functional and Dysfunctional Sexual Behavior*, edited by Anders Ågmo, 30–76. Oxford: Academic Press.

Alcaro, Antonio, Robert Huber, and Jaak Panksepp. 2007. "Behavioral Functions of the Mesolimbic Dopaminergic System: An Affective Neuroethological Perspective." *Brain Research Reviews* 56 (2): 283–321. doi:10.1016/j.brainresrev.2007.07.014.

Anderson, Dean P., Erik V. Nordheim, and Christophe Boesch. 2006. "Environmental Factors Influencing the Seasonality of Estrus in Chimpanzees." *Primates; Journal of Primatology* 47 (1): 43–50. doi:10.1007/s10329-005-0143-y.

Balthazart, Jacques, and Gregory F. Ball. 2010. "Male Sexual Behavior and Hormones in Non-Mammalian Vertebrates." In *Encyclopedia of Animal Behavior*, edited by Michael D. Breed and Janice Moore, 340–54. Oxford: Academic Press. http://www.sciencedirect.com/science/article/pii/B9780080453378002424.

Beach, Frank A. 1976. "Sexual Attractivity, Proceptivity, and Receptivity in Female Mammals." *Hormones and Behavior* 7 (1): 105–38. doi:10.1016/0018-506X(76)90008-8.

Berridge, Kent C., and Terry E. Robinson. 1998. "What Is the Role of Dopamine in Reward: Hedonic Impact, Reward Learning, or Incentive Salience?" *Brain Research. Brain Research Reviews* 28 (3): 309–69. doi:10.1016/s0165-0173(98)00019-8.

Beyer, Carlos, José Luis Contreras, Gabriela Moralí, and Knut Larsson. 1981. "Effects of Castration and Sex Steroid Treatment on the Motor Copulatory Pattern of the Rat." *Physiology & Behavior* 27 (4): 727–30. doi:10.1016/0031-9384(81)90247-X.

Blaustein, Jeffrey D. 2009. "Feminine Reproductive Behavior and Physiology in Rodents: Integration of Hormonal, Behavioral, and Environmental Influences." In *Hormones, Brain and Behavior (2nd Edition)*, edited by Donald W. Pfaff, Arthur P. Arnold, Anne M. Etgen, Susan E. Fahrbach, and Robert T. Rubin, 67–108. San Diego: Academic Press. http://www.sciencedirect.com/science/article/pii/B9780080887838000024.

Brandling-Bennett, Erica M., Meg E. Blasberg, and Ann S. Clark. 1999. "Paced Mating Behavior in Female Rats in Response to Different Hormone Priming Regimens." *Hormones and Behavior* 35 (2): 144–54. doi:10.1006/hbeh.1998.1507.

Brody, Stuart. 2010. "The Relative Health Benefits of Different Sexual Activities." *The Journal of Sexual Medicine* 7 (4 Pt 1): 1336–61. doi:10.1111/j.1743-6109.2009.01677.x.

Bronson, Franklin H. 1985. "Mammalian Reproduction: An Ecological Perspective." *Biology of Reproduction* 32 (1): 1–26. doi:10.1095/biolreprod32.1.1.

Carter, Carol Sue, and Stephen W. Porges. 1974. "Ovarian Hormones and the Duration of Sexual Receptivity in the Female Golden Hamster." *Hormones and Behavior* 5 (4): 303–15. doi:10.1016/0018-506X(74)90017-8.

Coolen, Lique M., Maureen E. Fitzgerald, L. Yu, and Michael Lehman. 2004. "Activation of µ Opioid Receptors in the Medial Preoptic Area Following Copulation in Male Rats." *Neuroscience* 124 (1): 11–21. doi:10.1016/j.neuroscience.2003.10.045.

Coopersmith, Carol B., and Mary S. Erskine. 1994. "Influence of Paced Mating and Number of Intromissions on Fertility in the Laboratory Rat." *Journal of Reproduction and Fertility* 102 (2): 451–58. doi:10.1530/jrf.0.1020451.

Crawford, Lawrence L., Kevin S. Holloway, and Michael Domjan. 1993. "The Nature of Sexual Reinforcement." *Journal of the Experimental Analysis of Behavior* 60 (1): 55–66. doi:10.1901/jeab.1993.60-55.

Erskine, Mary S. 1992. "Pelvic and Pudendal Nerves Influence the Display of Paced Mating Behavior in Response to Estrogen and Progesterone in the Female Rat." *Behavioral Neuroscience* 106 (4): 690–97. doi:10.1037/0735-7044.106.4.690.

Erskine, Mary S., Eva Kornberg, and James A. Cherry. 1989. "Paced Copulation in Rats: Effects of Intromission Frequency and Duration on Luteal Activation and Estrus Length." *Physiology & Behavior* 45 (1): 33–39. doi:10.1016/0031-9384(89)90163-7.

Erskine, Mary S., and Michael Baum. 1982. "Effects of Paced Coital Stimulation on Termination of Estrus and Brain Indoleamine Levels in Female Rats." *Pharmacology, Biochemistry, and Behavior* 17 (4): 857–61. doi:10.1016/0091-3057(82)90373-2.

Estrada-Reyes, Rosa, P. Ortiz-López, J. Gutiérrez-Ortíz, and Lucía Martínez-Mota. 2009. "Turnera Diffusa Wild (Turneraceae) Recovers Sexual Behavior in Sexually Exhausted Males." *Journal of Ethnopharmacology* 123 (3): 423–29. doi:10.1016/j.jep.2009.03.032.

Feder, Harvey H. 1981. "Estrous Cyclicity in Mammals." In *Neuroendocrinology of Reproduction*, edited by Norman T. Adler, 279–348. New York: Springer US. http://link.springer.com/chapter/10.1007/978-1-4684-3881-9_10.

Ferreira-Nuño, Armando, Adriana Morales-Otal, Raúl G. Paredes, and Javier Velázquez-Moctezuma. 2005. "Sexual Behavior of Female Rats in a Multiple-Partner Preference Test." *Hormones and Behavior* 47 (3): 290–96. doi:10.1016/j.yhbeh.2004.11.012.

Fiorino, Dennis F., Ariane Coury, and Anthony G. Phillips. 1997. "Dynamic Changes in Nucleus Accumbens Dopamine Efflux during the Coolidge Effect in Male Rats." *The Journal of Neuroscience: The Official Journal of the Society for Neuroscience* 17 (12): 4849–55. http://www.jneurosci.org/content/17/12/4849.long.

García-Horsman, S. Patricia, Anders Ågmo, and Raúl G. Paredes. 2008. "Infusions of Naloxone into the Medial Preoptic Area, Ventromedial Nucleus of the Hypothalamus, and Amygdala Block Conditioned Place Preference Induced by Paced Mating Behavior." *Hormones and Behavior* 54 (5): 709–16. doi:10.1016/j.yhbeh.2008.07.011.

Gintzler, Alan R., and Barry R. Komisaruk. 1991. "Analgesia Is Produced by Uterocervical Mechanostimulation in Rats: Roles of Afferent Nerves and Implications for Analgesia of Pregnancy and Parturition." *Brain Research* 566 (1/2): 299–302. doi:http://dx.doi.org/10.1016/0006-8993(91)91713-B.

Golinski, Alison, Lukáš Kubička, Henry John-Alder, and Lukáš Kratochvíl. 2014. "Elevated Testosterone Is Required for Male Copulatory Behavior and Aggression in Madagascar Ground Gecko (Paroedura Picta)." *General and Comparative Endocrinology* 205: 133–41. doi:10.1016/j.ygcen.2014.05.012.

Gómora, Porfirio, Carlos Beyer, Gabriela González-Mariscal, and Barry R. Komisaruk. 1994. "Momentary Analgesia Produced by Copulation in Female Rats." *Brain Research* 656 (1): 52–58. doi:http://dx.doi.org/10.1016/0006-8993(94)91365-X.

Hrdy, S. Blaffer. 1995. "The Primate Origins of Female Sexuality, and Their Implications for the Role of Nonconceptive Sex in the Reproductive Strategies of Women." *Human Evolution* 10 (2): 131–44. doi:10.1007/BF02437536.

Hurtazo, H. A., Raúl G. Paredes, and Anders Ågmo. 2008. "Inactivation of the Medial Preoptic Area/Anterior Hypothalamus by Lidocaine Reduces Male Sexual Behavior and Sexual Incentive Motivation in Male Rats." *Neuroscience* 152 (2): 331–37. doi:10.1016/j.neuroscience.2007.10.063.

Jimenez, Pedro, Martín Alejandro Serrano-Meneses, Evelia Cuamatzi, and Gabriela González-Mariscal. 2012. "Analysis of Sexual Behaviour in Male Rabbits across Successive Tests Leading to Sexual Exhaustion." *World Rabbit Science* 20 (1): 13–23. doi:10.4995/wrs.2012.1034.

Johnson, Byron M., and Barry R. Komisaruk. 1996. "Antinociceptive Action of Vaginocervical Stimulation in Rat Spinal Cord: 2-DG Analysis." *Physiology & Behavior* 60 (3): 979–83. doi:10.1016/00319384(96)00146-1.

Komisaruk, Barry R., Carlos Beyer, and Beverly Whipple. 2006. "Definitions of Orgasm." In *The Science of Orgasm*, 1–6. Baltimore, MD: Johns Hopkins University Press.

Komisaruk, Barry R., and Joshua Wallman. 1977. "Antinociceptive Effects of Vaginal Stimulation in Rats: Neurophysiological and Behavioral Studies." *Brain Research* 137 (1): 85–107. doi:10.1016/0006-8993(77)91014-9.

Komisaruk, Barry R., Norman T. Adler, and John B. Hutchison. 1972. "Genital Sensory Field: Enlargement by Estrogen Treatment in Female Rats." *Science* 178 (4067): 1295–98. doi:10.1126/science.178.4067.1295.

Komisaruk, Barry R., Beverly Whipple, and Carlos Beyer. 2009. "Sexual Pleasure." In *Pleasures of the Brain: Neural Bases of Sensory Pleasure*, edited by Kent C. Berridge and Morten Kringelbach, 169–177. Oxford: Oxford University Press.

Krames, Lester. 1971. "Sexual Responses of Polygamous Female and Monogamous Male Rats to Novel Partners after Sexual Cessation." *Journal of Comparative and Physiological Psychology* 77 (2): 294–301. doi:10.1037/h0031669.

Laan, Ellen, and Walter Everaerd. 1995. "Habituation of Female Sexual Arousal to Slides and Film." *Archives of Sexual Behavior* 24 (5): 517–41. doi:10.1007/bf01541832.

Le Boeuf, Burney J. 1974. "Male-Male Competition and Reproductive Success in Elephant Seals." *Integrative and Comparative Biology* 14 (1): 163–76. doi:10.1093/icb/14.1.163.

Lisk, Robert D., and Gregory Baron. 1982. "Female Regulation of Mating Location and Acceptance of New Mating Partners Following Mating to Sexual Satiety: The Coolidge Effect Demonstrated in the Female Golden Hamster." *Behavioral and Neural Biology* 36 (4): 416–21. doi:10.1016/S0163-1047(82)90822-6.

Lovell, Jennifer L., Abby Diehl, Elizabeth Joyce, Jenifer Cohn, Jose Lopez, and Fay A. Guarraci. 2007. "'Some Guys Have All the Luck': Mate Preference Influences Paced-Mating Behavior in Female Rats." *Physiology & Behavior* 90 (4): 537–44. doi:10.1016/j.physbeh.2006.11.002.

Manson, Joseph H., Susan Perry, and Amy R. Parish. 1997. "Nonconceptive Sexual Behavior in Bonobos and Capuchins." *International Journal of Primatology* 18 (5): 767–86. doi:10.1023/A:1026395829818.

Martínez, Isadora, and Raúl G. Paredes. 2001. "Only Self-Paced Mating Is Rewarding in Rats of Both Sexes." *Hormones and Behavior* 40 (4): 510–17. doi:10.1006/hbeh.2001.1712.

Marvan, Richard, Jeroen M. G. Stevens, Amy D. Roeder, I. Mazura, Michael W. Bruford, and J. R. de Ruiter. 2006. "Male Dominance Rank, Mating and Reproductive Success in Captive Bonobos (Pan Paniscus)." *Folia Primatologica; International Journal of Primatology* 77 (5): 364–76. doi:10.1159/000093702.

McClintock, Martha K., Joseph J. Anisko, and Norman T. Adler. 1982. "Group Mating among Norway Rats II. The Social Dynamics of Copulation: Competition, Cooperation, and Mate Choice." *Animal Behaviour* 30 (2): 410–25. doi:10.1016/S0003-3472(82)80052-3.

McDonald, Peter G., and Bengt J. Meyerson. 1973. "The Effect of Oestradiol, Testosterone and Dihydrotestosterone on Sexual Motivation in the Ovariectomized Female Rat." *Physiology & Behavior* 11 (4): 515–20. doi:10.1016/0031-9384(73)90038-3.

Meisel, Robert L., and Amanda J. Mullins. 2006. "Sexual Experience in Female Rodents: Cellular Mechanisms and Functional Consequences." *Brain Research* 1126 (1): 56–65. doi:10.1016/j.brainres.2006.08.050.

Merkx, Jeroen. 1984. "Effect of Castration and Subsequent Substitution with Testosterone, Dihydrotestosterone and Oestradiol on Sexual Preference Behaviour in the Male Rat." *Behavioural Brain Research* 11 (1): 59–65. doi:10.1016/0166-4328(84)90008-1.

Møller, Anders P., and Tim R. Birkhead. 1991. "Frequent Copulations and Mate Guarding as Alternative Paternity Guards in Birds: A Comparative Study." *Behaviour* 118 (3/4): 170–86. doi:10.2307/4534963.

Mucha, Ronald F., Derek van der Kooy, Martha O'Shaughnessy, and Peter Bucenieks. 1982. "Drug Reinforcement Studied by the Use of Place Conditioning in Rat." *Brain Research* 243 (1): 91–105. doi:10.1016/0006-8993(82)91123-4.

Nasr, Mohammed A. F., William J. Browne, Gina Caplen, Becky Hothersall, Joanna C. Murrell, and Christine J. Nicol. 2013. "Positive Affective State Induced by Opioid Analgesia in Laying Hens with Bone Fractures." *Applied Animal Behaviour Science* 147 (1/2): 127–31. doi:10.1016/j.applanim.2013.04.015.

O'Connell, Lauren A., and Hans A. Hofmann. 2011. "The Vertebrate Mesolimbic Reward System and Social Behavior Network: A Comparative Synthesis." *The Journal of Comparative Neurology* 519 (18): 3599–639. doi:10.1002/cne.22735.

O'Donohue, William T., and James H. Geer. 1985. "The Habituation of Sexual Arousal." *Archives of Sexual Behavior* 14 (3): 233–46. doi:10.1007/BF01542106.

O'Donohue, William T., and Joseph J. Plaud. 1991. "The Long-Term Habituation of Sexual Arousal in the Human Male." *Journal of Behavior Therapy and Experimental Psychiatry* 22 (2): 87–96. doi:10.1016/0005-7916(91)90003-N.

Paredes, Raúl G. 2009. "Evaluating the Neurobiology of Sexual Reward." *ILAR Journal/ National Research Council, Institute of Laboratory Animal Resources* 50 (1): 15–27. doi:10.1093/ilar.50.1.15.

Paredes, Raúl G. 2014. "Opioids and Sexual Reward." *Pharmacology, Biochemistry, and Behavior* 121 (June): 124–31. doi:10.1016/j.pbb.2013.11.004.

Paredes, Raúl G., and Anders Ågmo. 2004. "Has Dopamine a Physiological Role in the Control of Sexual Behavior? A Critical Review of the Evidence." *Progress in Neurobiology* 73 (3): 179–226. doi:10.1016/j.pneurobio.2004.05.001.

Paredes, Raúl G., and Berenice Vazquez. 1999. "What Do Female Rats Like about Sex? Paced Mating." *Behavioural Brain Research* 105 (1): 117–27. doi:10.1016/ S0166-328(99)00087-X.

Pfaus, James G., and Boris B. Gorzalka. 1987. "Opioids and Sexual Behavior." *Neuroscience and Biobehavioral Reviews* 11 (1): 1–34. doi:10.1016/S0149-7634(87)80002-7.

Pfaus, James G., François Giuliano, and Hélène Gelez. 2007. "Bremelanotide: An Overview of Preclinical CNS Effects on Female Sexual Function." *The Journal of Sexual Medicine* 4 (Suppl 4): 269–79. doi:10.1111/j.1743-6109.2007.00610.x.

Pfaus, James G., Geert Damsma, Danielle G. Wenkstern, and Hand C. Fibiger. 1995. "Sexual Activity Increases Dopamine Transmission in the Nucleus Accumbens and Striatum of Female Rats." *Brain Research* 693 (1/2): 21–30. doi:10.1016/0006-8993(95)00679-K.

Pfaus, James G., Tod E. Kippin, Genaro A. Coria-Avila, Hélène Gelez, Veronica M. Afonso, Nafissa Ismail, and Mayte Parada. 2012. "Who, What, Where, When (and Maybe Even Why)? How the Experience of Sexual Reward Connects Sexual Desire, Preference, and Performance." *Archives of Sexual Behavior* 41 (1): 31–62. doi:10.1007/ s10508-012-9935-5.

Phillips-Farfán, Bryan V., and Alonso Fernández-Guasti. 2009. "Endocrine, Neural and Pharmacological Aspects of Sexual Satiety in Male Rats." *Neuroscience and Biobehavioral Reviews* 33 (3): 442–55. doi:10.1016/j.neubiorev.2008.11.003.

Phoenix, Charles H., and Kathleen C. Chambers. 1988. "Old Age and Sexual Exhaustion in Male Rhesus Macaques." *Physiology & Behavior* 44 (2): 157–63. doi:10.1016/00319384(88)90132-1.

Pruessner, Jens C., Frances Champagne, Michael J. Meaney, and Alain Dagher. 2004. "Dopamine Release in Response to a Psychological Stress in Humans and Its Relationship to Early Life Maternal Care: A Positron Emission Tomography Study Using [11C]raclopride." *The Journal of Neuroscience: The Official Journal of the Society for Neuroscience* 24 (11): 2825–31. doi:10.1523/JNEUROSCI.3422-03.2004.

Puglisi-Allegra, Stefano, Assunta Imperato, Luciano Angelucci, and Simona Cabib. 1991. "Acute Stress Induces Time-Dependent Responses in Dopamine Mesolimbic System." *Brain Research* 554 (1/2): 217–22. doi:10.1016/0006-8993(91)90192-X.

Rivas, Francisco J., and D. Mir. 1990. "Effects of Nucleus Accumbens Lesion on Female Rat Sexual Receptivity and Proceptivity in a Partner Preference Paradigm." *Behavioural Brain Research* 41 (3): 239–49. http://www.ncbi.nlm.nih.gov/pubmed/2288675.

Rivas, Francisco J., and D. Mir. 1991. "Accumbens Lesion in Female Rats Increases Mount Rejection without Modifying Lordosis." *Revista Española De Fisiología* 47 (1): 1–6. http://www.ncbi.nlm.nih.gov/pubmed/1871414.

Rodríguez-Manzo, Gabriela, and Alonso Fernández-Guasti. 1994. "Reversal of Sexual Exhaustion by Serotonergic and Noradrenergic Agents." *Behavioural Brain Research* 62 (2): 127–34. doi:10.1016/0166-4328(94)90019-1.

Rössler, Anne-Sophie, James G. Pfaus, Hossein Kami Kia, Jacques Bernabé, Laurent Alexandre, and François Giuliano. 2006. "The Melanocortin Agonist, Melanotan II, Enhances Proceptive Sexual Behaviors in the Female Rat." *Pharmacology, Biochemistry, and Behavior* 85 (3): 514–21. doi:10.1016/j.pbb.2006.09.023.

Rougé-Pont, Françoise, Pier-Vincenzo Piazza, Martine Kharouby, Michel Le Moal, and Hervé Simon. 1993. "Higher and Longer Stress-Induced Increase in Dopamine Concentrations in the Nucleus Accumbens of Animals Predisposed to Amphetamine Self-Administration. A Microdialysis Study." *Brain Research* 602 (1): 169–74. doi:10.1016/0006-8993(93)90260-T.

Sathe, Rajendra S., Barry R. Komisaruk, Alice K. Ladas, and Shreerang V. 2001. "Naltrexone-Induced Augmentation of Sexual Response in Men." *Archives of Medical Research* 32(3): 221–6. doi:10.1016/S0188-4409(01)00279-X.

Södersten, Per A. T., and Knut Larsson. 1976. "Sexual Behavior in Castrated Male Rats Treated with Monoamine Synthesis Inhibitors and Testosterone." *Pharmacology, Biochemistry, and Behavior* 5 (3): 319–27. doi:10.1016/0091-3057(76)90084-8.

Suschinsky, Kelly D., and Martin L. Lalumière. 2011. "Prepared for Anything?: An Investigation of Female Genital Arousal in Response to Rape Cues." *Psychological Science* 22 (2): 159–65. doi:10.1177/0956797610394660.

Szechtman, Henry, Moshe Hershkowitz, and Rabi Simantov. 1981. "Sexual Behavior Decreases Pain Sensitivity and Stimulated Endogenous Opioids in Male Rats." *European Journal of Pharmacology* 70 (3): 279–85. doi:10.1016/0014-2999(81)90161-8.

Tzschentke, Thomas M. 1998. "Measuring Reward with the Conditioned Place Preference Paradigm: A Comprehensive Review of Drug Effects, Recent Progress and New Issues." *Progress in Neurobiology* 56 (6): 613–72. doi:10.1016/S0301-0082(98)00060-4.

Ventura-Aquino, Elisa, and Alonso Fernández-Guasti. 2013a. "Reduced Proceptivity and Sex-Motivated Behaviors in the Female Rat after Repeated Copulation in Paced and Non-Paced Mating: Effect of Changing the Male." *Physiology & Behavior* 120: 70–76. doi:10.1016/j.physbeh.2013.07.006.

Ventura-Aquino, Elisa, and Alonso Fernández-Guasti. 2013b. "The Antidepressants Fluoxetine and Bupropion Differentially Affect Proceptive Behavior in the Naturally Cycling Female Rat." *The Journal of Sexual Medicine* 10 (11): 2679–87. doi:10.1111/jsm.12280.

Wee, Beth E., David R. Weaver, and Lynwood G. Clemens. 1988. "Hormonal Restoration of Masculine Sexual Behavior in Long-Term Castrated B6D2F1 Mice." *Physiology & Behavior* 42 (1): 77–82. doi:10.1016/0031-9384(88)90263-6.

Wenkstern, Danielle G., James G. Pfaus, and Hand Christian Fibiger. 1993. "Dopamine Transmission Increases in the Nucleus Accumbens of Male Rats during Their First Exposure to Sexually Receptive Female Rats." *Brain Research* 618 (1): 41–46. doi:10.1016/0006-8993(93)90426-N.

Whipple, Beverly, and Barry R. Komisaruk. 1985. "Elevation of Pain Threshold by Vaginal Stimulation in Women." *Pain* 21 (4): 357–67. doi:10.1016/0304-3959(85)90164-2.

Winland, Carissa, Jessica L. Bolton, Brittany Ford, Sumith Jampana, Jennifer R. Tinker, Russell J. Frohardt, Fay A. Guarraci, and Maha Zewail-Foote. 2012. "'Nice Guys Finish Last': Influence of Mate Choice on Reproductive Success in Long-Evans Rats." *Physiology & Behavior* 105 (3): 868–76. doi:10.1016/j.physbeh.2011.10.022.

Wrangham, Richard W. 1993. "The Evolution of Sexuality in Chimpanzees and Bonobos." *Human Nature* 4 (1): 47–79. doi:10.1007/BF02734089.

Zipse, Lauryn R., Erica M. Brandling-Bennett, and Ann S. Clark. 2000. "Paced Mating Behavior in the Naturally Cycling and the Hormone-Treated Female Rat." *Physiology & Behavior* 70 (1/2): 205–9. doi:10.1016/S0031-9384(00)00242-0.

Section II

Neuroendocrinology of Maternal Behavior and Brain Development

9 The Rabbit Doe as a Model of Neuroendocrine Synchronization

Mario Caba, Teresa Morales, and Enrique Meza

CONTENTS

9.1 Introduction 149
9.2 The Master Clock 150
9.3 Suckling Stimulation by Kits during Lactation 151
9.4 Synchronization of Neuroendocrine Brain Areas during Lactation 152
9.5 Oxytocin (OT) and Dopaminergic Cells are Entrained by Suckling 152
9.6 Rhythmic Pattern in Structures Related to Maternal Behavior 156
Acknowledgments 159
References 159

9.1 INTRODUCTION

Rhythms are ubiquitous in nature, from single-celled organisms to complex species, and they are thought to help individuals adapt to a particular ecological environment and physiological situation. The rabbit is an excellent example of how the neuroendocrine system changes to fulfill the specific needs of reproduction, particularly lactation. Hormone levels in blood have a rhythm that reaches peaks and troughs at specific hours during the day. For example, in the rabbit maintained in normal light/dark conditions, the plasma glucocorticoid level is highest in the evening, before the onset of the period of maximal locomotor activity (Szeto et al., 2004), while lowest values occur at 6:00 in the morning. The adaptive significance of the high glucocorticoid concentration is to promote alertness to daily activities of this nocturnal animal during its active period. Similar changes are found across mammalian species, both diurnal and nocturnal. As a result, the levels of corticosteroids and other hormones fluctuate at particular hours to support daily, and even seasonal, activity patterns of individuals and populations. These rhythmic changes are controlled in an orderly way by a clock mechanism located in the brain, specifically in the suprachiasmatic nucleus (SCN).

9.2 THE MASTER CLOCK

The SCN is composed of paired nuclei situated above the optic chiasm and to the ventral sides of the third ventricle in the hypothalamus (Figure 9.1d). Similar to several other brain regions, the SCN presents daily oscillations in electrical activity. However, when neural structures are dissected and explored *in vitro*, some of these structures are arrhythmic or their oscillations steadily decrease, except for the SCN (Abe et al., 2002). It is well established that the oscillations, named circadian oscillations, of the SCN are self-sustained and have a cycle of approximately 24 h. Through humoral and synaptic connections, this nucleus influences rhythmic patterns in other brain regions (Welsh et al., 2010). Consistent with this, ablation of the SCN results in the disruption of rhythms of physiological parameters and behavior, and the animal becomes arrhythmic (Welsh et al., 2010). From this evidence, the SCN is considered to be the master clock in mammals. To maintain its oscillations in phase with the

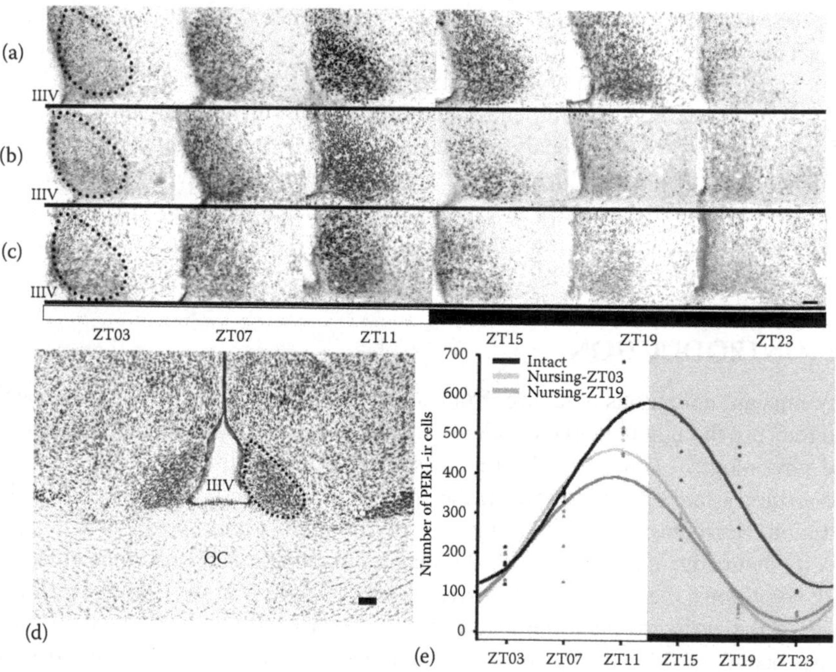

FIGURE 9.1 PER1 expression in the suprachiasmatic nucleus (SCN) in intact and nursing does. (a–c) Photomicrographs of representative sections illustrating the expression of PER1 protein at the level of the middle portion of the SCN in (a) Intact, (b) Nursing-ZT03, and (c) Nursing-ZT19 groups at six different time points throughout a complete 24-h cycle. (d) Photomicrograph (thionine stain) denotes the level at which analyses were performed. Dotted line delimits the SCN. IIIV, third ventricle; OC, optic chiasm. Scale bar = 100 μm. (e) Cosinor analysis indicates a significant rhythmicity of PER1 cells in both intact and nursing groups. (Modified from Meza, E. et al., *Eur J Neurosci*, *28*(7), 1394–1403, 2008.)

environmental and geographical conditions, the SCN needs to be entrained every day by sunlight, which is the main synchronization cue for this nucleus.

The self-sustaining properties of the SCN are supported by a molecular translation/transduction molecular mechanism of several genes known as clock genes and their protein products. In Figure 9.1a we present the rhythmic pattern over a 24-h cycle of the PER1 protein, product of the *Per1* clock gene in the SCN of an adult female rabbit. As shown, the expression of this protein is low early in the day (i.e., at ZT03). (Note: "ZT" is "zeitgeber," literally: time giver; ZT0 = 7 a.m., time of lights on; ZT03 = 10 a.m.), reaches a peak at ZT11, just before the onset of night (ZT12, time of lights off), and then decreases to minimal levels again before the next light period begins. Figure 9.1b and c shows that a similar pattern is observed in lactating does maintained in the same environmental light/dark condition compared to non-lactating rabbits (Figure 9.1e).

Even though the master clock drives rhythmic activity of other brain regions, in specific circumstances some of them, such as neuroendocrine-related brain areas and tissues, adjust their oscillations to a pattern different from that of the SCN to support specific physiological challenges, as during lactation. The rabbit is an excellent model to study synchronization of neuroendocrine structures by cues other than light, as the mother rabbit is exposed only once a day to the suckling of her kits (see below).

9.3 SUCKLING STIMULATION BY KITS DURING LACTATION

Every day at around the same hour, the rabbit doe nurses her kits in the nest for a single period of around 5 min (Zarrow et al., 1965; Jilge, 1995; González-Mariscal et al., 2013). Kits readily attach to the nipple, suck milk, and despite the brevity of the nursing bout, ingest around 30% of their body weight in milk by lactation day 7 (Caba et al., 2003). This daily food pulse is sufficient to synchronize the circadian physiology and behavior of rabbit kits. Corticosterone and other hormones in the blood fluctuate according to the time of nursing (Rovirosa et al., 2005; Morgado et al., 2008, 2010). Additionally, neural structures like the dorsomedial, paraventricular (PVN) and supraoptic nucleus (SON) hypothalamic nuclei show a rhythm in PER1 protein or c-FOS protein, a marker of neuronal activation, that shifts in parallel with the time of suckling of either mother's milk (Caba et al., 2008) or artificial milk (Morgado et al., 2011). Thus, it is clear that rabbit kits are synchronized by the periodic ingestion of food, and they have been considered a natural model of food entrainment (Menaker and Mohawk, 2008; Antle and Silver, 2009; Caba and Gonzalez-Mariscal, 2009). The main consequence of this synchronization is that it supports a state of arousal, just before the arrival of either food or the nursing mother. In agreement, corticosterone blood concentration and locomotor behavior are high during arousal, but after food ingestion kits remain mostly quiescent, and corticosterone and locomotor behavior sharply decrease. During the arousal period the kits vigorously stimulate the mother's nipples during suckling of milk. This stimulus is a powerful signal and we propose that it synchronizes neuroendocrine structures in the lactating rabbit.

9.4 SYNCHRONIZATION OF NEUROENDOCRINE BRAIN AREAS DURING LACTATION

As already mentioned, light is the main cue for the synchronization of circadian rhythms in the SCN, and oscillations of other central and peripheral structures and organs are in phase with those of the SCN. As a result there is a complex hierarchical organization of physiological and behavioral rhythms, with the SCN as the master clock. In the brain, many regions and nuclei including neuroendocrine cells have functional clock genes, which constitute the time-keeping machinery (Kriegsfeld and Silver, 2006; Guilding and Piggins, 2007). Thus, if the SCN is in control of all rhythmic oscillations, why do other cells in different brain regions also have a molecular clock? It had been proposed that these local clocks provide fine-tuning to specific local conditions (Kriegsfeld and Silver, 2006). That is, besides receiving signals from the SCN, the rhythmicity of local clocks in these cells can be adjusted during a particular physiological condition. We have explored this possibility in the lactating rabbit.

9.5 OXYTOCIN (OT) AND DOPAMINERGIC CELLS ARE ENTRAINED BY SUCKLING

During nursing in the rabbit doe, a massive release of OT and prolactin (PRL) into the peripheral blood is elicited by the kits' suckling (Fuchs et al., 1984). Considering that this behavior has a periodicity of around 24 h under both light/dark and constant light conditions (Jilge 1993, 1995), we hypothesized that neural populations that control the release of these hormones could be entrained by the daily suckling stimulus. This idea was explored by the expression of the PER1 protein in two groups of lactating does, scheduled to nurse their kits either at ZT03 or at ZT19 and one control group of females that were not pregnant or lactating.

To determine the effect of suckling on neuroendocrine cells, we analyzed the expression of the PER1 protein as an index of rhythmicity (Feillet et al., 2008) in six populations of neurons: two that produce the hormone OT located in the PVN and SON of the hypothalamus, and four dopaminergic regions: tuberoinfundibular dopaminergic (TIDA), periventricular hypophyseal dopaminergic (PHDA), A15 ventral (A15v), and incertohypothalamic dopaminergic (IHDA) cells in the hypothalamus. The TIDA neurons are located in the arcuate (ARC) nucleus, PHDA in the PVN hypothalamic nucleus, and A15v in the supraoptic neurons. These three populations produce dopamine involved in the control of PRL secretion (Fuxe et al., 1964; Gayrard et al., 1995; Thiery et al., 1995; DeMaria et al., 1998; Ben-Jonathan et al., 2002). IHDA cells are in the dorsal hypothalamus and are not related to the control of PRL (Cheung et al., 1998) and, therefore, they can be used as a control for the effect of suckling. These four regions were analyzed for their coexpression of tyrosine hydroxylase (TH) since this is a good marker for neurons that produce catecholamines (Bjorklund and Lindvall, 1984).

As shown in Figures 9.2a (PVN) and 9.3a (SON), in nonlactating does there is a rhythmic pattern of PER1 protein expression in OT and DA cells in both nuclei

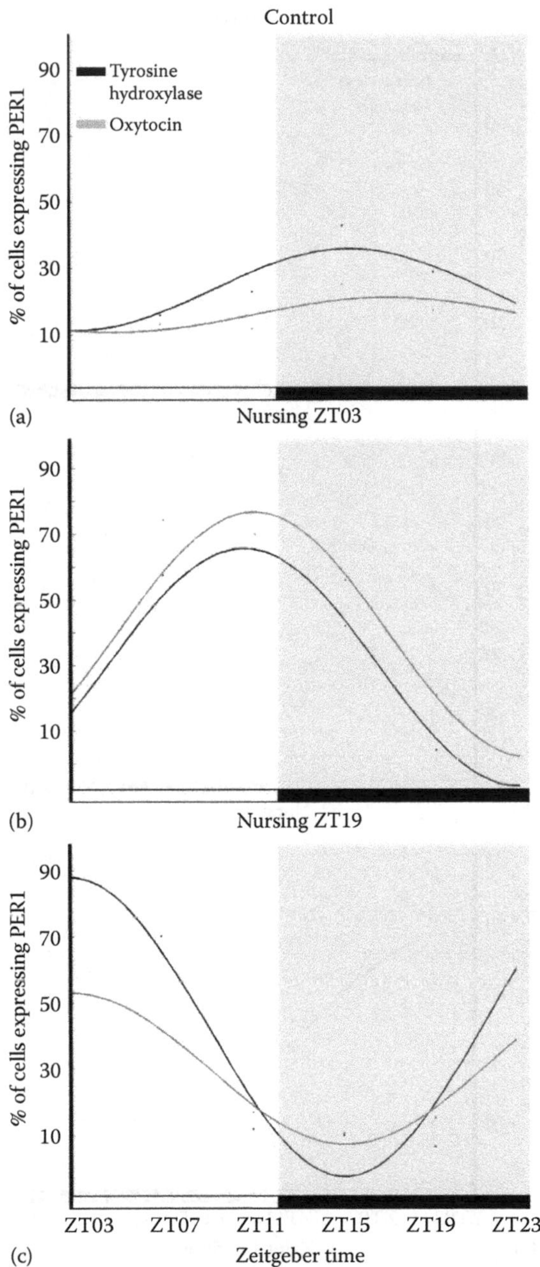

FIGURE 9.2 Rhythmic expression of PER1 in oxytocin and tyrosine hydroxylase cells (A14) in the paraventricular nucleus in intact and nursing does. Cosinor analysis indicates a significant rhythmicity of PER1-ir in both neuroendocrine cells; note that in the control group (a) achrophase is after the onset of night; however, in nursing groups (b, c) achrophase shifts in parallel to the timing of nursing. The black and white bar at the bottom represents the LD condition. ZT, zeitgeber time (ZT0 = 07:00 h/time of lights on).

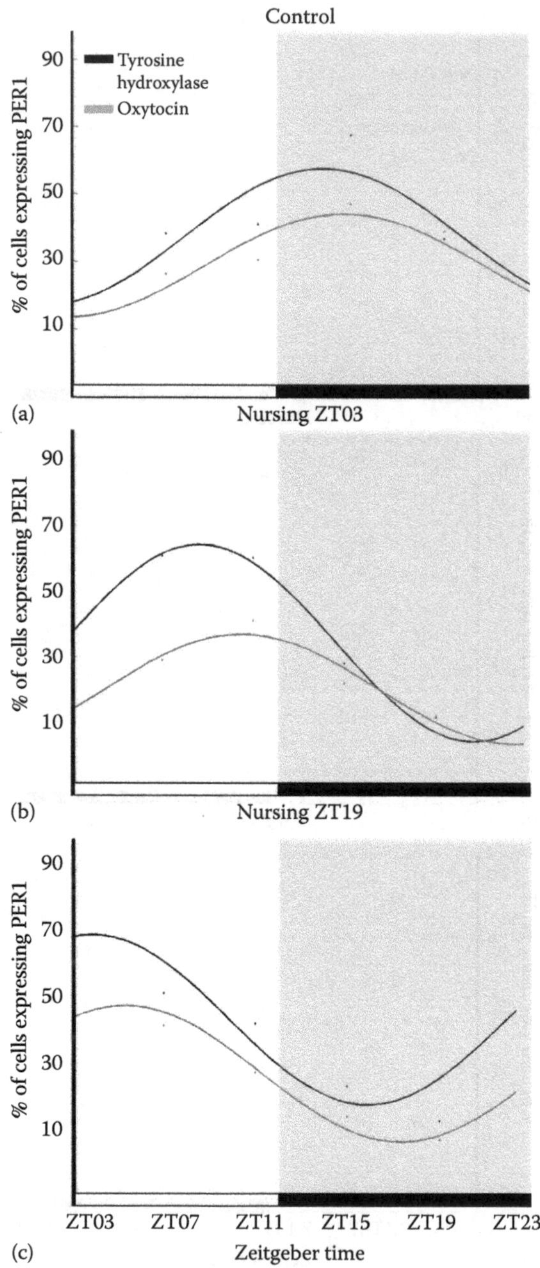

FIGURE 9.3 Rhythmic expression of PER1 in oxytocin and tyrosine hydroxylase cells (A15v) in the supraoptic nucleus in intact and nursing does. Cosinor analysis indicates a significant rhythmicity of PER1-ir in both neuroendocrine cells; note that in the control group (a) achrophase is after the onset of night; however, in nursing groups achrophase shifts in parallel to the timing of nursing (b, c). The black and white bar at the bottom represents the LD condition. ZT, zeitgeber time (ZT0 = 07:00 h/time of lights on).

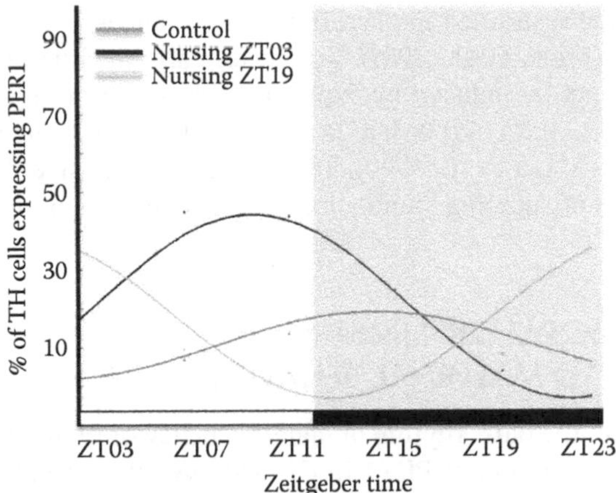

FIGURE 9.4 Rhythmic expression of tyrosine hydroxylase (TH) cells with PER1 protein (PER1/TH) in tuberoinfundibular dopaminergic cells (TIDA) in Nursing-ZT03 and Nursing-ZT19 groups compared to control subjects. Cosinor analysis indicates a significant rhythmicity of PER1-ir cells; note that in the control group achrophase is after the onset of night; however, in nursing groups, achrophase shifts in parallel to the timing of nursing. The black and white bar at the bottom represents the LD condition. ZT, zeitgeber time. (ZT0 = 07:00 h/time of lights on; modified from Meza, E. et al., *J Neuroendocrinol, 23*(6), 472–480, 2011.)

with a maximal value around ZT15. In lactating does, however, the rhythm of PER1 shifts in parallel with the time of nursing, reaching a peak 4 h after nursing in both structures (Figures 9.2b, 9.2c, 9.3b, and 9.3c, respectively; Meza et al., 2008). In TIDA cells a similar pattern was observed (Figure 9.4; Meza et al., 2011) while very few IHDA cells expressed PER1 and there was no detectable rhythm in them (Meza et al., 2011). The persistence of PER1 in OT cells in the PVN and SON and in DA cells in TIDA and PHDA cells at the time of their maximal expression was explored further in nursing females deprived of kits for 24 or 48 h. Skipping one or two nursing bouts significantly decreased the expression of PER1 in these cells, which strongly suggests a role of suckling stimulation for maintaining the rhythm of PER1 expression in them (Meza et al., 2008, 2011). Thus, we conclude that both OT and DA cells are able to show a rhythmic pattern in response to an external stimulus, different from light, specifically, the daily suckling from the kits. This is not a general effect in the brain, as it is not observed in the IHDA, the one dopaminergic population not involved in the neuroendocrine circuit related to the production and ejection of milk.

The importance of the periodic suckling stimulation in the lactating rabbit doe was investigated by Flavio Mena, a student, former collaborator and long-term friend of Dr. Carlos Beyer. Mena and colleagues described that altering the normal once-a-day pattern of suckling by an additional enforced nursing episode led to increased milk yields during early lactation (postpartum day 11), had no effect during mid-lactation (day 21), and decreased milk yields in late lactation (day 31). A reduced milk synthesis and ejection were observed when two extra sucklings were applied to the restrained mother, indicating changes in both the TIDA/PRL and OT systems (Mena et al., 1981). Moreover, early decline or an extension of milk production can be induced by exposing early lactation mothers to suckling by "old" kits (20–23 days old) or late-lactation mothers to "young" kits (4–6 days old), respectively (Mena et al., 1990). These results emphasize the importance of the characteristics of suckling stimulation for the regulation of milk output (Mena et al., 1991).

9.6 RHYTHMIC PATTERN IN STRUCTURES RELATED TO MATERNAL BEHAVIOR

The neural basis of the daily rhythm of maternal behavior in the rabbit has scarcely been explored. The rhythm of PER1 in OT and DA cells provides an insight into the neuroendocrine effects related to the periodic production and release of milk. Therefore, we explored other areas in the brain that had been linked to the control of maternal behavior in different species (Numan and Insel, 2003; Olazábal et al., 2013; Meza et al., 2015), namely, the preoptic area (POA), the bed nuclei of the stria terminalis (BNST), and the lateral septum (LS, dorsal, and ventral subdivisions). Our hypothesis was that these areas would also show a rhythmic pattern, indicated by PER1 protein, entrained by the time of nursing. As shown in Figure 9.5, the control doe shows no rhythm at all, but a robust rhythm is observed in lactating females in the POA, and it shifts in parallel with the time of nursing (Meza et al., 2015).

A similar result was found in the dorsal (Figure 9.6) and ventral (Figure 9.7) subdivisions of the LS (Meza et al., 2015). In contrast, no rhythmic pattern was found under any condition in the BNST (Meza et al., 2015). Moreover, suckling also seems to be an important cue for the effect on the POA and LS (Meza et al., 2015). Our results in the POA and LS agree with those in a variety of species, including the rabbit, indicating the critical role of these structures for the onset and display of maternal behavior (Cruz and Beyer, 1972; Numan and Insel, 2003; Olazabal et al., 2013; González-Mariscal et al., 2009, 2015). The BNST has also been implicated in the display of maternal behavior in rodents, but we did not find any effect. This is interesting as this area is mainly related to the retrieval of kits (Numan and Numan, 1996), a behavior not observed in the maternal rabbit.

The presence of clock genes and oscillation of their protein levels has been demonstrated in the PVN, SON, and ARC nucleus, among many others (Sellix and Freeman, 2003; Guilding and Piggins, 2007), but very little is known about their functional significance. Our results illustrate how the oscillations of these areas can be affected by an external stimulus, offering a new perspective from which to explore the importance of these extra SCN rhythmic patterns as well as the neural basis of maternal rabbit behavior, namely, a neural circuit of clock gene oscillation, that seems to rely on suckling. Indeed, the regular once-a-day nursing behavior seems to rely on a threshold of suckling stimulation as does provided with litters smaller than four kits show multiple entrances into the nest box throughout the

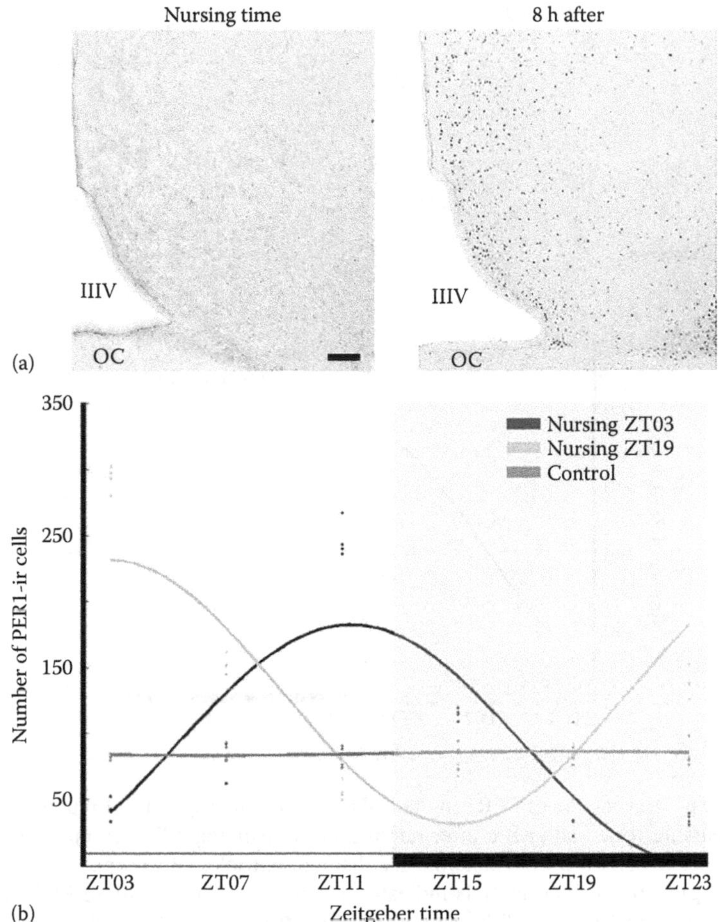

FIGURE 9.5 Expression of PER1 in the POA. (a) Photomicrographs of representative sections at nursing time and at 8 h after nursing in the Nursing-ZT03 group. Note the density of PER1-ir cells 8 h after nursing. IIIV, third ventricle, OC, optic chiasm. Scale bar = 100 μm. (b) Cosinor analysis indicates a significant rhythmicity of PER1-ir cells in Nursing-ZT03 and Nursing-ZT19, but not in the control group. The black and white bar at the bottom represents the LD condition, ZT, Zeitgeber time. (ZT0 = 07:00 h/time of lights on; modified from Meza, E. et al., *Eur J Neurosci*, 41, 196–204, 2015.)

day, that is, without a regular pattern (González-Mariscal et al., 2013). In agreement, prevention of suckling in lactating does decreases the expression of PER1 protein, an index of rhythmicity, in neuroendocrine cells (Meza et al., 2008, 2011) and neural structures important for the onset and display of maternal behavior (Meza et al., 2015).

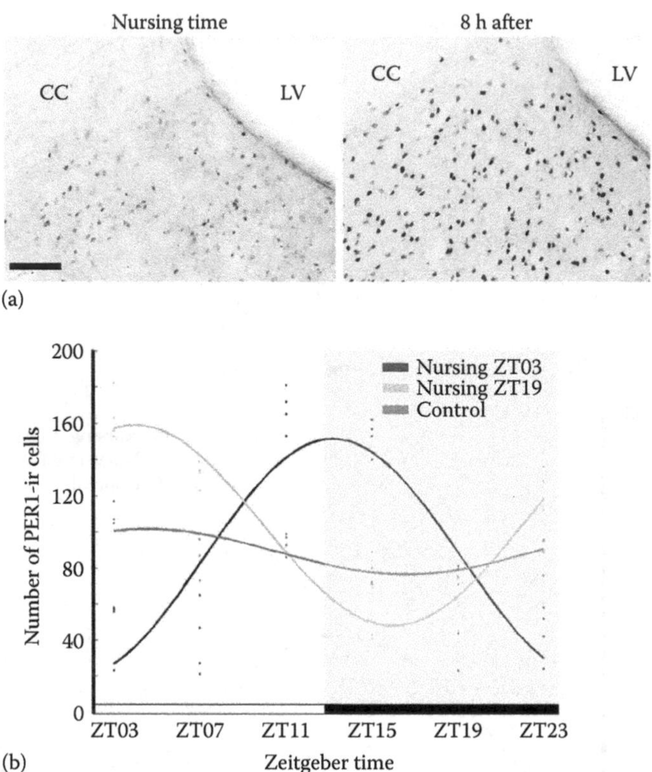

FIGURE 9.6 Expression of PER1 in the LSD. (a) Photomicrographs of representative sections at nursing time and at 8 h after nursing in the Nursing-ZT03 group. Note the density of PER1-ir cells 8 h after nursing. LV, lateral ventricle, CC, corpus callosum. Scale bar = 100 μm. (b) Cosinor analysis indicates a significant rhythmicity of PER1-ir cells in Nursing-ZT03 and Nursing-ZT19, but not in the control group. The black and white bar at the bottom represents the LD condition. (Modified from Meza, E. et al., *Eur J Neurosci*, 41, 196–204, 2015.)

Finally, even though the SCN controls the circadian rhythms in physiology and behavior, this nucleus does not shift its oscillations to the rhythm of lactation, perhaps because this behavior is indeed daily, not circadian, maintained by daily suckling from the kits (Meza et al., 2015).

This line of research on rabbit reproduction started under the influence of Dr. Carlos Beyer, a preeminent scientist in the field of neuroendocrinology, and Dr. Rae Silver, an expert in circadian rhythms. They provided the basic guidance and their contributions established a foundation from which to explore the physiological and neural basis of the rhythmic behavior of lactation in the dyad mother/kits in the rabbit, a model of lactation that is characteristic of lagomorphs.

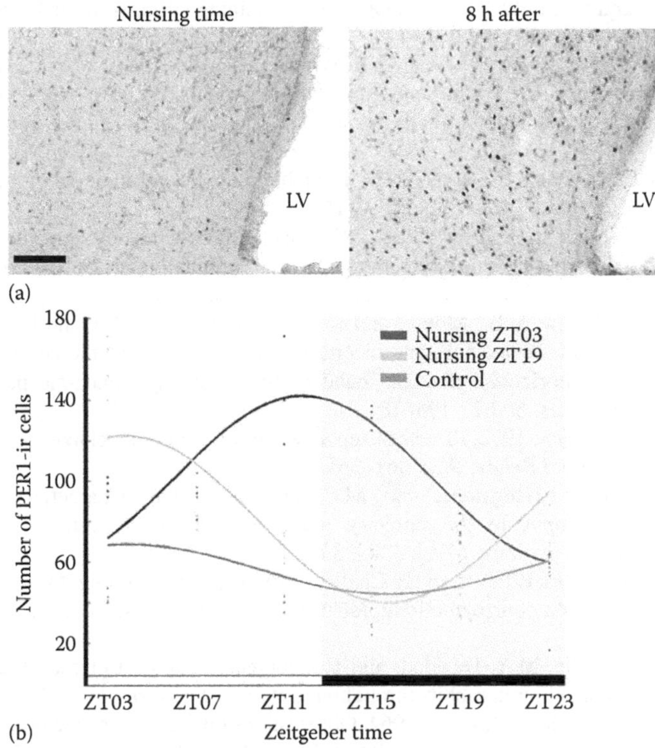

FIGURE 9.7 Expression of PER1 in the LSV. (a) Photomicrographs of representative sections at nursing time and at 8 h after nursing in the Nursing-ZT03 group. Note the density of PER1-ir cells 8 h after nursing. LV, lateral ventricle. Scale bar = 100 μm. (b) Cosinor analysis indicates a significant rhythmicity of PER1-ir cells in Nursing-ZT03 and Nursing-ZT19, but not in the control group. The black and white bar at the bottom represents the LD condition. (Modified from Meza, E. et al., *Eur J Neurosci*, 41, 196–204, 2015.)

ACKNOWLEDGMENTS

The authors thank Dr. Dorothy Pless for the grammar review and Juan Aguirre for technical assistance.

REFERENCES

Abe, M., E.D. Herzog, S. Yamazaki, et al. 2002. Circadian rhythms in isolated brain regions. *J Neurosci*, *22*(1), 350–356.

Antle, M.C., and R. Silver. 2009. Neural basis of timing and anticipatory behaviors. *Eur J Neurosci*, *30*(9), 1643–1649.

Ben-Jonathan, N., K. Liby, M. McFarland, and M. Zinger. 2002. Prolactin as an autocrine/paracrine growth factor in human cancer. *Trends Endocrin Met*, *13*(6), 245–250.

Björklund, A., and O. Lindvall. 1984. Dopamine-containing systems in the CNS. *Handbook Chem Neuroanat*, *2*(Part 1), 55–122.
Caba, M., A. Tovar, R. Silver, E. Morgado, E. Meza, Y. Zavaleta, and C. Juárez. 2008. Nature's food anticipatory experiment: entrainment of locomotor behavior, suprachiasmatic and dorsomedial hypothalamic nuclei by suckling in rabbit pups. *Eur J Neurosci*, *27*(2), 432–443.
Caba, M., and G. González-Mariscal. 2009. The rabbit pup, a natural model of nursing anticipatory activity. *Eur J Neurosci*, *30*(9), 1697–1706.
Caba, M., M.J. Rovirosa, and R. Silver. 2003. Suckling and genital stroking induces Fos expression in hypothalamic oxytocinergic neurons of rabbit pups. *Dev Brain Res*, *143*(2), 119–128.
Cheung, S., J.R. Ballew, K.E., Moore, and K.J. Lookingland. 1998. Contribution of dopamine neurons in the medial zona incerta to the innervation of the central nucleus of the amygdala, horizontal diagonal band of Broca and hypothalamic paraventricular nucleus. *Brain Res*, *808*(2), 174–181.
Cruz, M.L., and C. Beyer. 1972. Effects of septal lesions on maternal behavior and lactation in the rabbit. *Physiol Behav*, *9*(3), 361–365.
DeMaria, J.E., J.D. Livingstone, and M.E. Freeman. 1998. Characterization of the dopaminergic input to the pituitary gland throughout the estrous cycle of the rat. *Neuroendocrinology*, *67*(6), 377–383.
Feillet, C.A., J. Mendoza, P. Pévet, and E. Challet. 2008. Restricted feeding restores rhythmicity in the pineal gland of arrhythmic suprachiasmatic lesioned rats. *Eur J Neurosci*, *28*(12), 2451–2458.
Fuchs, A.R., L. Cubile, M.Y. Dawood, and F.S. Jorgensen. 1984. Release of oxytocin and prolactin by suckling in rabbits throughout lactation. *Endocrinology*, *114*(2), 462–469.
Fuxe, K., T. Hökfelt, and O. Nilsson. 1964. Observations on the cellular localization of dopamine in the caudate nucleus of the rat. *Zeitschriftfür Zellforschung und Mikroskopische Anatomie*, *63*(5), 701–706.
Gayrard, V., J.C. Thiéry, J. Thibault, and Y. Tillet. 1995. Efferent projections from the retrochiasmatic area to the median eminence and to the pars nervosa of the hypophysis with special reference to the A15 dopaminergic cell group in the sheep. *Cell Tissue Res*, *281*(3), 561–567.
González-Mariscal, G., A. Jiménez, R. Chirino, and C. Beyer. 2009. Motherhood and nursing stimulate c-FOS expression the rabbit forebrain. *Behav Neurosci*, *123*(4), 131–139.
González-Mariscal, G., A.C. Lemus, A. Vega-González, and R. Aguilar-Roblero. 2013. Litter size determines periodicity of nursing in rabbits. *Chronobiol Int*, *30*, 711–718.
González-Mariscal, G., M. Caba, M. Martínez-Gómez, A. Bautista, and R. Hudson. 2015. Mothers and offspring: the rabbit as a model system in the study of mammalian maternal behavior and sibling interactions. *Horm Behav*. doi:10.1016/j.yhbeh.2015.05.011.
Guilding, C., and H.D. Piggins. 2007. Challenging the omnipotence of the suprachiasmatic timekeeper: are circadian oscillators present throughout the mammalian brain? *Eur J Neurosci*, *25*(11), 3195–3216.
Jilge, B. 1993. The ontogeny of circadian rhythms in the rabbit. *J Biol Rhythms*, *8*(3), 247–260.
Jilge, B. 1995. Ontogeny of the rabbit's circadian rhythms without an external zeitgeber. *Physiol Behav*, *58*(1), 131–140.
Kriegsfeld, L.J., and R. Silver. 2006. The regulation of neuroendocrine function: timing is everything. *Horm Behav*, *49*(5), 557–574.
Mena, F., C. Clapp, D. Aguayo, M.T. Morales, and G. Martinez de la Escalera. 1991. Stimulatory and inhibitory effects of suckling on lactation. *Endocr Regulat*, *25*, 25–35.
Mena, F., C. Clapp, and G. Martínez-Escalera. 1990. Age-related stimulatory and inhibitory effects of suckling regulate lactation in rabbits. *Physiol Behav*, *48*(2), 307–310.

Mena, F., G. Martínez-Escalera, C. Clapp, D. Aguayo, G. Anguiano, and C.E. Grosvenor. 1981. Effect of acute increases in suckling frequency upon food intake and milk secretion in the rabbit. *Proc Soc Exp Biol Med, 168*: 373–377.

Menaker, M., and J. Mohawk. 2008. Nature's food anticipatory experiment: entrainment of locomotor behavior, suprachiasmatic and dorsomedial hypothalamic nuclei by suckling in rabbit pups. http://f1000.com/prime/1124420 (accessed September 21, 2015); doi:10.3410/f.1124420.581603.

Meza, E., C. Juárez, E. Morgado, Y. Zavaleta, and M. Caba. 2008. Brief daily suckling shifts locomotor behavior and induces PER1 protein in paraventricular and supraoptic nuclei, but not in the suprachiasmatic nucleus, of rabbit does. *Eur J Neurosci, 28*(7), 1394–1403.

Meza, E., J. Aguirre, S. Waliszewski, and M. Caba, M. 2015. Suckling induces a daily rhythm in the preoptic area and lateral septum but not in the bed nucleus of the stria terminalis in lactating rabbit does. *Eur J Neurosci, 41*, 196–204.

Meza, E., S.M. Waliszewski, and M. Caba, M. 2011. Circadian nursing induces PER1 protein in neuroendocrine tyrosine hydroxylase neurones in the rabbit doe. *J Neuroendocrinol, 23*(6), 472–480.

Morgado, E., C. Juárez, A.I. Melo, et al. 2011. Artificial feeding synchronizes behavioral, hormonal, metabolic and neural parameters in mother-deprived neonatal rabbit pups. *Horm Behav, 34*(11), 1807–1816.

Morgado, E., E. Meza, M.K. Gordon, F.K. Pau, C. Juárez, and M. Caba. 2010. Persistence of hormonal and metabolic rhythms during fasting in 7- to 9-day-old rabbits entrained by nursing during the night. *Horm Behav, 58*(3), 465–472.

Morgado, E., M.K. Gordon, M. del Carmen Minana-Solis, et al. 2008. Hormonal and metabolic rhythms associated with the daily scheduled nursing in rabbit pups. *Am J Physiol-Reg I, 295*(2), 690–695.

Numan, M., and M. Numan. 1996. A lesion and neuroanatomical tract-tracing analysis of the role of the bed nucleus of the stria terminalis in retrieval behavior and other aspects of maternal responsiveness in rats. *Dev Psychobiol, 29*(1), 23–51.

Numan, M., and T.R. Insel. 2003. *The Neurobiology of Parental Behavior* (Vol. 1). Springer Science and Business Media. New York: Springer-Verlag.

Olazábal, D.E., M. Pereira, D. Agrati, et al. 2013. Flexibility and adaptation of the neural substrate that supports maternal behavior in mammals. *Neurosci Biobehav R, 37*(8), 1875–1892.

Rovirosa, M. J., S. Levine, M.K. Gordon, and M. Caba 2005. Circadian rhythm of corticosterone secretion in the neonatal rabbit. *Dev Brain Res, 158*(1), 92–96.

Sellix, M.T., and M.E. Freeman. 2003. Circadian rhythms of neuroendocrine dopaminergic neuronal activity in ovariectomized rats. *Neuroendocrinology, 77*(1), 59–70.

Szeto, A., J.A. Gonzales, S. B. Spitzer, et al. 2004. Circulating levels of glucocorticoid hormones in WHHL and NZW rabbits: circadian cycle and response to repeated social encounter. *Psychoneuroendocrinology, 29*(7), 861–866.

Thiery, J. C., V. Gayrard, S. Le Corre, C. Viguie, et al. 1995. Dopaminergic control of LH secretion by the A15 nucleus in anoestrous ewes. *J Reprod Fertil Suppl, 49*, 285–296.

Welsh, D.K., J.S. Takahashi, and S.A. Kay. 2010. Suprachiasmatic nucleus: cell autonomy and network properties. *Annu Rev Physiol, 72*, 551.

Zarrow, M.X., V.H. Denenberg, and C.O. Anderson. 1965. Rabbit: frequency of suckling in the pup. *Science, 150*(3705), 1835–1836.

10 The Domestic Rabbit (*Oryctolagus cuniculus*) as a Model Animal

From Reproductive Neurobiology to Developmental Psychobiology

Kurt Leroy Hoffman, Robyn Hudson, Margarita Martínez-Gómez, and Amando Bautista

CONTENTS

10.1 Introduction 164
10.2 The Rabbit as a Model Animal in Reproductive Neuroendocrinology 164
10.2.1 Ovulation 164
10.2.2 Oxytocin and Gonadal Steroid Hormones 165
10.3 Behavioral Neuroendocrinology of the Rabbit 166
10.3.1 Lactation 166
10.3.2 Hormonal Modulation of Female Sexual Behavior 167
10.3.3 Hormonal Modulation of Male Sexual Behavior 168
10.3.4 Hormonal Modulation of Other Sexually Dimorphic Behavior Patterns 169
10.3.5 Maternal Nest Building Behavior 170
10.4 Beyond Behavioral Neuroendocrinology: From Physiology to Personality 172
10.4.1 The Female Rabbit as a Model for Investigating the Reflex Activity of the Pelvic Region that Underlies Reproductive Processes and Micturition 172
10.4.2 Using Rabbit Behavior to Understand Processes Fundamental to Certain Neuropsychiatric Disorders: Motivation, Memory, and Repetitive Stereotyped Behavior 174

10.4.3 The Rabbit as a Model Animal for Studying the Ontogeny of Individual Differences in Phenotype, Including "Personality"......... 176
10.5 Final Thoughts.. 178
Acknowledgments.. 178
References.. 178

10.1 INTRODUCTION

In 1962, Carlos Beyer, John Tindal, and Charles Sawyer published the study "Electrophysiological study of projections from mesencephalic central gray matter to forebrain in the rabbit" in the journal *Experimental Neurology*. For Carlos, it would be the first of many (nearly 50) published studies involving the European rabbit (*Oryctolagus cuniculus*) as a model animal. Carlos and Charles Sawyer collaborated in three more electrophysiological studies in the rabbit: on the milk ejection reflex and maintenance of lactation, on factors that modulate responsiveness to acoustic stimuli, and on the effects of hormones on the electrical activity of the brain, both in the rat and in the rabbit (Tindal et al. 1963; Beyer and Sawyer 1964; Beyer et al. 1967). For Carlos, this latter paper was one of his first addressing the effects of gonadal hormones on brain activity, a field to which he would continue making important contributions throughout his career.

10.2 THE RABBIT AS A MODEL ANIMAL IN REPRODUCTIVE NEUROENDOCRINOLOGY

10.2.1 Ovulation

Throughout the years, the rabbit has been a consistently important model for studying reproductive neuroendocrinology, most notably with respect to mechanisms underlying ovulation and the regulation of physiological events of pregnancy by estrogen and progesterone. Unlike most of the other mammalian species (e.g., the rat, mouse, guinea pig, and monkey) that were used in laboratory studies at the beginning of the 20th century, the rabbit is an induced ovulator. That is, ovulation is triggered by copulation, thereby presenting an opportunity to experimentally control its timing. Walton and Hammond (1928) were among the first physiologists to take advantage of this aspect of rabbit reproduction, which at the time was considered a "peculiarity" among mammals. In female rabbits that had been anesthetized just before ovulation was expected to occur (about 9–13 hours after mating), they were able to observe this process as it occurred in real time, by exposing the ovaries by means of a ventral incision and viewing them under a dissecting microscope. They described in detail the formation of the pimple-like *macula pellucida* at the apex of the ripe follicle (approximately 9 hours after coitus), its subsequent rupture, and the slow "oozing out" of the follicular fluid down the side of the follicle, carrying the ovum along with it. Soon after, the endocrine induction of ovulation was investigated in a series of seminal studies by Fee, Parkes, and Hill. These studies revealed that removal of the pituitary (hypophysectomy) within the first hour after mating prevented ovulation, but had no effect when performed 1.5 hours or more after mating (Fee and Parkes

1929). Subsequently, they demonstrated that ovulation could be artificially induced in hypophysectomized female rabbits by intravenous administration of acid–alcohol extracts of the anterior pituitary from the ox (Hill and Parkes 1931). The active hormone would later be isolated and termed, "luteinizing hormone" (LH). Marshall and Verney (1936) investigated the neural component of the ovulatory mechanism. Based on the then recent studies in the male guinea pig showing that electrical stimulation of the brain triggered seminal ejaculation, they hypothesized that electrical stimulation of the female brain might similarly provoke an orgasm-like response that, in induced ovulators such as the rabbit, would trigger ovulation. As predicted, ovulation was induced in female rabbits after electrical stimulation of either the brain or the lumbar region of the spinal cord, and, based on the studies of Fee, Parks, Hill, and others, this effect was proposed to be mediated by neural activation of the anterior pituitary and the release of an ovulation-inducing substance. A later series of pharmacological experiments, which employed a variety of drugs with stimulatory or convulsive actions on the nervous system, found that only picrotoxin (now known to be a GABA-A receptor antagonist) consistently induced ovulation (Marshall et al. 1939). These and other studies formed the foundation for later work by Markee et al. (1946a,b) and Harris (1948a,b,c; 1951) showing that brief electrical stimulation of the posterior part of the *tuber cinereum* of the hypothalamus provoked ovulation in estrous rabbits, while stimulation of the pituitary gland itself (pars distalis, pars intermedia, or infundibulum) had no effect. These results provided key support for the model that a substance, derived from neurons in the hypothalamus, is transmitted to the anterior pituitary by the hypophysial portal vessels, where it provokes the release of LH from the anterior pituitary. This substance would later be identified and named luteinizing hormone releasing hormone (LHRH), or gonadotropin releasing hormone (GnRH). Soon after, Harris and colleagues showed that electrical stimulation of the *tuber cinereum* also provoked the release of adrenocorticotropin hormone (ACTH) from the anterior pituitary in the rabbit (Harris 1951), eventually leading to the discovery of corticotropin releasing hormone (CRH). Thus, studies of the "peculiar" means by which ovulation is induced in the rabbit played a key role in elucidating the general mechanism that controls the release of LH, follicle stimulating hormone (FSH), ACTH, thyroid-stimulating hormone (TSH), and growth hormone (GH) from the anterior pituitary. These early studies in the rabbit also began to reveal mechanisms by which the activity of the endocrine system is modulated by environmental stimuli, as well as by emotional and psychological states.

10.2.2 Oxytocin and Gonadal Steroid Hormones

In the 1930s, J. M. Robson carried out a series of studies on the responsiveness of the uterus of estrous, pregnant, and lactating rabbits to estrogens, extracts of the corpus luteum, and extracts of the posterior pituitary. These studies were inspired by the studies of Knaus in 1930, cited in Robson (1933), showing that uterine contractions could be induced by "pituitrin" (an extract of the posterior lobe of the pituitary) only at the very end of pregnancy. Robson tested the capacity of the oxytocic (then called "pitocin," now known as oxytocin) and the pressor (then called "pitressin," now vasopressin) fractions of pituitrin to induce uterine contractions in vitro. During the first

2 weeks of pregnancy, the uterus was unresponsive to oxytocin and vasopressin, but the final days of pregnancy were associated with increased sensitivity to oxytocin, peaking around the time of parturition. Uterine sensitivity to oxytocin then decreased across the first few days postpartum (Robson 1933). Robson attributed the oxytocin insensitivity of the uterus during the first half of pregnancy, to a hormone produced by the corpus luteum. By contrast, treatment with ketohydroxyoestrin (a synthetic estrogen) was found to induce uterine sensitivity to oxytocin in ovariectomized or hypophysectomized rabbits (Robson 1933, 1935a,b). Subsequently, a series of studies carried out in ovariectomized rabbits demonstrated that the then recently isolated hormone progesterone and estrone exerted opposing effects on both progestational proliferation of the endometrium (induced by progesterone and inhibited by estrogens) and uterine sensitivity to oxytocin (inhibited by progesterone, augmented by estrogens) (Robson 1936). With respect to endometrial proliferation, Robson concluded that the *ratio* of the blood concentrations of these two hormones, rather than the absolute levels of either of the individual hormones, was the important factor in determining the type of response. This general mechanism of hormonal control would later be found to operate during maternal nest building by the female rabbit (Zarrow et al. 1963).

10.3 BEHAVIORAL NEUROENDOCRINOLOGY OF THE RABBIT

10.3.1 Lactation

Similar to the process of ovulation, lactation (the long-term production and secretion of milk by the mammary glands, mediated by prolactin) and milk ejection (reflexive ejection of milk from the nipple, mediated by oxytocin) in the rabbit are exquisitely sensitive to sensory stimuli, in this case, suckling by the pups. In 1963, Tindal, Beyer, and Sawyer published a study that tested the hypothesis that activation of the milk ejection reflex by suckling-associated stimuli was necessary in order to maintain lactation in the rabbit. The rabbit is a particularly useful species for the impact of mother–pup contact on lactation and maternal behavior, because of its "peculiar" nursing behavior: the lactating rabbit nurses her pups just once every 24 hours, for only 3–5 minutes (Zarrow et al. 1965a; González-Mariscal et al. 2013). In the study by Tindal et al. (1963), the lactating rabbit was deeply anesthetized at the time of each daily nursing bout while the pups suckled, thereby eliminating suckling-induced neural impulses from the mammary glands to the central nervous system. In order to induce milk ejection, oxytocin was administered systemically immediately before the nursing bout. Tindal et al. found that lactation was maintained under these circumstances, suggesting that sensory stimuli associated with suckling, although necessary to acutely stimulate neurohypophysial oxytocin release and, consequently, milk ejection, were not necessary to maintain milk production. Based on these data, they suggested that, in the rabbit, some mechanism associated with milk ejection itself (which in this study was elicited by exogenous oxytocin administration) was necessary for maintaining prolactin secretion, and, consequently, milk production. This conclusion was supported by two additional studies (Mena and Beyer 1963; Beyer and Mena 1970), in which the spinal cord of pregnant or lactating rabbits was transected at the T2 level, thereby severing the neural connections between the mammary gland and the brain. The milk ejection reflex was eliminated in these rabbits, but lactation was maintained if milk ejection was evoked during nursing by exogenous oxytocin administration. Exogenously applied prolactin or ACTH increased milk production in oxytocin-treated, T2-transected rabbits, but neither treatment was *necessary* to maintain lactation. The spinal pathway responsible for the milk ejection reflex itself was later determined to be ipsilateral and uncrossed, passing through the ventrolateral funiculi (Mena and Beyer 1968a). A series of lesion experiments indicated that input arising from the telencephalon, and more specifically the amygdala and entorhinal cortex regions, exerted a tonic inhibition over milk secretion, most probably via the tonic inhibition of prolactin release (Mena and Beyer 1963, 1968b). Septal lesions, on the other hand, had no effect on mating behavior, ovulation, pregnancy, parturition, or lactation, but severely disrupted motivational components of maternal behavior such as nest building, care of the young, and willingness to nurse (Cruz and Beyer 1972). These rabbit studies, along with those of Zarrow, Denenberg, Sawin, Farooq, Ross, and colleagues (Sawin et al. 1960; Zarrow et al. 1961, 1962, 1963, 1965b), were among the first to examine in detail the neural and neuroendocrine control of lactation, milk ejection, and maternal behavior in a mammalian species.

10.3.2 Hormonal Modulation of Female Sexual Behavior

After publishing their study that characterized projections from the periaqueductal gray to the forebrain (Beyer et al. 1962), Beyer, Sawyer, and colleagues published two more electrophysiological studies in rabbits. These examined arousal-related changes in the evoked EEG potentials, and the effects of various hormones on EEG activity (Beyer and Sawyer 1964; Beyer et al. 1967). This latter study foreshadowed the path that Carlos would follow across the next decades, during which his research efforts were dedicated to characterizing the modulatory effects of gonadal steroid hormones on sexual behavior, and elucidating their underlying cellular mechanisms in the rabbit and rat. The next several studies that Carlos carried out in the rabbit focused on the hormonal control of female sexual receptivity and pseudomale behavior (the display by females of the mounting *pattern* of males). This body of work established that both sexual receptivity and pseudomale behavior in adult ovariectomized rabbits were stimulated by the administration of estradiol and inhibited by progesterone (Yaschine et al. 1967; Beyer et al. 1969). In intact females, sexual receptivity and pseudomale behavior were inhibited during pregnancy as well as during lactation, the former effect being attributed to high circulating levels of progesterone (Beyer and Rivaud 1968). Surprisingly, these investigators found that sexual receptivity and pseudomale behavior in ovariectomized rabbits were also stimulated by testosterone, but not by its active androgenic metabolite, dihydrotestosterone (DHT) (McDonald et al. 1970). Pursuing their hypothesis that testosterone stimulates female sexual behavior via its aromatization to estradiol, Beyer et al. (1970) tested the ability of several androgens—some aromatizable, others nonaromatizable—to stimulate sexual behavior in ovariectomized rabbits. Only aromatizable androgens induced female sexual behavior, supporting their hypothesis that behavioral effects of testosterone can be mediated by its conversion to estradiol within the brain. This seminal concept was ingeniously supported by a subsequent study showing that MER-25, an estrogen receptor antagonist that lacks antiandrogenic activity, blocked testosterone-induced sexual behavior in ovariectomized rabbits (Beyer and Vidal 1971). Thus, these studies carried out in the female rabbit revealed another general mechanism of the hormonal control of behavior: many effects of "male" steroid hormones—the androgens—are actually mediated by their aromatization to estradiol, a "female" hormone. Since then, the aromatization of testosterone has been shown by many other investigators to underlie many of the physiological effects of this hormone, including masculinization and defeminization of brain and behavior during early development.

10.3.3 Hormonal Modulation of Male Sexual Behavior

Beyer and colleagues then began to investigate the role of aromatization (i.e., conversion of testosterone to estradiol) or nonaromatization (using DHT), on sexual behavior in male rabbits. They found that, in prepubertally castrated rabbits, chronic treatment with testosterone propionate (TP) was highly effective in initiating sexual behavior (mounting and intromission), while chronic treatment with DHT propionate (DHTp) was much less effective (Beyer and Rivaud 1973). A second study confirmed that DHT supported sexual behavior in only a small proportion of prepubertally castrated rabbits, while those castrated postpubertally were slightly more responsive to this androgen. However, in both groups, administration of DHT + estradiol benzoate significantly increased mounting and intromissions compared to DHT alone. Estradiol benzoate by itself had no stimulatory effect on male sexual behavior. Taken together, these results suggested that testosterone induced sexual behavior in castrated rabbits through the combined action of its metabolites, DHT and estradiol (Beyer et al. 1975).

Testosterone was also shown to have clear effects on the characteristics of the motor mounting pattern. The male rabbit motor mounting pattern consists of a series of pelvic thrusts performed at a frequency of 11–14 per second, for approximately 1–5 seconds. The male's pelvic thrusting stimulates lordosis in the female, allowing the male's intromission and immediate ejaculation (Contreras and Beyer 1979). Castration had the effect of decreasing the number of mounts that the males displayed (indicating reduced sexual motivation) as well as disorganizing the normal mounting pattern. Thus, the pelvic thrusts of castrated rabbits tended to be weak and were performed at a significantly lower frequency (about 8 per second), and with a more variable interthrust interval. Exogenous testosterone treatment partially restored the motivation to mount, as well as the normal thrusting frequency in castrated rabbits (Beyer et al. 1980). In the case of female rabbits that spontaneously displayed mounting behavior, the motor mounting pattern was characterized by weak thrusting in a very irregular frequency pattern. When these females were ovariectomized and treated with Tp, the motivation to mount was significantly increased compared to preovariectomy levels, and the motor mounting pattern (with respect to frequency of pelvic thrusting) became very similar to that of intact males (Soto et al. 1984). Taken together, these results indicated that a proportion of female rabbits are masculinized sometime during early development with respect to the neural substrate that underlies the motivation to mount and the motor mounting pattern, but normally, they do not fully display these behaviors, due to low circulating androgens.

10.3.4 Hormonal Modulation of Other Sexually Dimorphic Behavior Patterns

The experiments of Beyer and colleagues in maternal and sexual behavior in rabbits laid the foundation for several new lines of investigation in behavioral neuroendocrinology. In the late 1980s, a collaboration involving Robyn Hudson, Gabriela González-Mariscal, Jay Rosenblatt, and Carlos Beyer began to investigate the hormonal modulation of production of the nipple-search pheromone in lactating females, scent marking behavior in males and females, and maternal nest building in females. Hudson and colleagues had already identified and characterized the behavioral response to the nipple-search pheromone—a volatile substance present on the ventrum of lactating rabbits that provokes stereotyped searching behavior in newborn pups—and had reported first evidence for the control of its emission by hormones associated with the female reproductive cycle (Hudson and Distel 1983, 1984; Distel and Hudson 1985; Hudson 1985). Likewise, Mykytowycz had previously described stereotyped scent marking behavior called "chinning," in which the rabbit rubs the undersurface of its chin on objects in its environment and thereby deposits scent gland secretions (Mykytowycz 1965). In natural conditions, male rabbits were observed to chin-mark at a much higher frequency than females, indicating that this behavior is modulated in a sexually dimorphic manner, possibly by gonadal steroid hormones. In the laboratory, a reduction in chinning frequency during pregnancy had been reported by Soares and Diamond (1982).

In another arena, Zarrow, Denenberg, Sawin, Farooq, Ross, and colleagues (Zarrow et al. 1961, 1963, 1965b) had previously shown that maternal nest building could be induced in ovariectomized or intact female rabbits by administering a hormonal treatment regimen of estrogen and progesterone that mimicked hormonal changes that normally occur during pregnancy, but this line of investigation was not pursued further after Zarrow's untimely death.

A series of experiments that used ovariectomized rabbits were carried out by Beyer and colleagues in order to determine the effect of exogenously administered estradiol benzoate, progesterone, or combinations of the two hormones on production of the nipple-search pheromone, chinning, and sexual receptivity. In these studies, the production of the pheromone was induced in females by estradiol treatment, and further increased by adding progesterone or prolactin (Hudson et al. 1990; González-Mariscal et al. 1994a). Estradiol also increased chinning and sexual receptivity, but subsequent treatment with estradiol plus progesterone resulted in the inhibition of both of these behavior patterns, in addition to increasing male-directed aggressive behavior (Hudson et al. 1990). Chinning in both male and female rabbits was later shown to be decreased by castration or ovariectomy, and reinstated by the respective administration of testosterone or estradiol (González-Mariscal et al. 1990, 1992, 1993; Hudson and Vodermeyer 1992; Chirino et al. 1993; Martínez-Gómez et al. 1997). Estradiol benzoate alone did not promote chinning in castrated males but its combination with DHTp was as effective as Tp for restoring chinning (González-Mariscal et al. 1993).

In females, the frequency of chinning changes dramatically with time of year and across the reproductive cycle. It is elevated during the long photoperiod (low melatonin levels) of spring and summer, and during estrus (when circulating levels of estradiol are present in the absence of progesterone) and decreases immediately following mating, throughout pregnancy, and during lactation (González-Mariscal et al. 1990, 1997; Hudson and Distel 1990; Hudson et al. 1994; García-Dalmán and González-Mariscal 2012). The immediate inhibition of chinning (0–4 hours after mating), early inhibition (1–4 days after mating), and long-term inhibition (during the 4 weeks of pregnancy) are mediated by distinct mechanisms, as is the inhibition of chinning during lactation. While the immediate inhibition is probably correlated with mating-induced changes in EEG activity, early inhibition appears to be due to the combined impact of progesterone receptor activation and a decrease in circulating estradiol, and the long-term inhibition of chinning during pregnancy can be attributed to elevated levels of circulating progesterone, acting on the progesterone receptor (Hoffman and González-Mariscal 2006, 2007; Hoffman et al. 2010). During lactation, chinning is temporarily inhibited after each nursing bout. The duration of this inhibition is related to the intensity of the suckling stimulation (García-Dalmán and González-Mariscal 2012). Thus, this "simple" stereotyped behavior of the female rabbit is modulated in a complex and fascinating manner, by a variety of neural, endocrine, and sensory cues.

10.3.5 Maternal Nest Building Behavior

The rabbit is one of the very few (perhaps the only?) mammalian species that is widely used as a subject for biological research, a farm animal raised for meat, and a pet, in addition to being a particularly troublesome invasive species and agricultural pest. Luckily for science, these latter characteristics of *Oryctolagus cuniculus* motivated a number of studies on the natural history of this species, including detailed descriptions of its behavior and social structure in the wild. Many of these investigations were carried out in Australia, the continent so famously plagued by this species. Mykytowycz, Myers, and Poole were some of the pioneers in this area, and their work should be required reading for anyone involved in the study of rabbit behavior (Mykytowycz 1958, 1959; Myers and Poole 1959, 1961, 1962; Mykytowycz and Gambale 1965). Myers and Poole (1961) published a detailed description of the maternal behavior of the rabbit in seminatural conditions. They observed that sometime during the last 14 days of pregnancy, the female rabbit dug a short nesting burrow with a hollowed-out nest chamber at the end, or cleaned out and extended a previously used burrow and nest chamber. Upon leaving the burrow, the rabbit concealed its entrance by covering it with soil. One or two nights before parturition, the female returned to the burrow, opened it, and built a grass nest within the nest chamber. She did this by collecting grass in her mouth and carrying it back to the burrow, repeating this behavior approximately 20 to 40 times. When the nest was finished, the rabbit left the burrow, again concealing its entrance with soil. On the night of parturition, she returned to the burrow, opened it, lined the grass nest with hair that she pulled from her ventrum and thighs, and gave birth inside the nest chamber. After cleaning and nursing the young, she left the nest burrow, concealed its entrance with soil, and thereafter returned only once per day to nurse, each time opening its entrance and covering it again upon leaving (Myers and Poole 1961).

This stereotyped sequence of behaviors (digging, collecting, carrying dry grass, and hair pulling) is faithfully replicated by pregnant rabbits in the laboratory. Zarrow, Denenberg, Sawin, Farooq, Ross, and colleagues were among the first to study rabbit nest building behavior in the lab, initially describing the temporal characteristics of hair loosening during pregnancy, then investigating the endocrine control of this behavior (Sawin et al. 1960; Zarrow et al. 1961, 1962, 1963, 1965b). Thus, ovariectomy after day 14 or 15 of pregnancy resulted in the onset of nest building behavior in most females, but this behavior was not induced in pregnant females ovariectomized before day 14 (Zarrow et al. 1962), indicating that ovarian hormonal signals across the first 2 weeks of pregnancy were necessary to "prime" nest building behavior. Around the same time, Mikhail and colleagues (1961) published a study describing blood progesterone concentrations across pregnancy in the rabbit, showing that levels of this hormone rose steadily until approximately day 16, and then declined until parturition. Given this result, Mikhail et al. (1961) surmised that declining levels of circulating progesterone normally trigger nest building behavior in the preparturient doe. This idea was tested in ovariectomized rabbits by Zarrow et al. (1963), who showed that maternal nest building was induced by the withdrawal of progesterone, while maintaining estradiol treatment, after two or more weeks of having been treated with both hormones. Taken together, these results strongly suggested that the decrease in the ratio of the concentration of circulating progesterone to estradiol is an endocrine signal that triggers maternal nest building. Since then, several studies in the rabbit have described the temporal characteristics of estradiol and progesterone secretion by the ovaries or their levels in the plasma across pregnancy (Mikhail et al. 1961; Hilliard et al. 1967, 1973; Challis et al. 1973; González-Mariscal et al. 1994b).

During the 1990s–2000s, the collaborative efforts of Beyer, González-Mariscal, Rosenblatt, and colleagues resulted in a series of elegant studies of the behavioral neuroendocrinology of rabbit maternal behavior. These investigators developed simple, yet clever, means of quantifying the three behavioral components of maternal nest building in the laboratory: digging, straw collecting and carrying, and hair pulling. Their findings showed that digging increased and decreased during pregnancy concomitant with the rise and fall of circulating progesterone, while straw carrying and hair pulling were consecutively expressed when progesterone levels declined (González-Mariscal et al. 1994b). In ovariectomized rabbits, exogenous treatment with estradiol benzoate plus progesterone promoted digging, while the subsequent removal of progesterone (while maintaining estradiol treatment) triggered straw carrying and hair pulling (González-Mariscal et al. 1996, 1998; Hoffman and González-Mariscal 2006). Intracerebral implants of estradiol benzoate into the nucleus accumbens or medial preoptic area allowed the expression of digging or straw carrying, respectively, when progesterone was administered daily for several days and then withdrawn (González-Mariscal et al. 2005). Further studies implicated prolactin as a hormonal signal that promotes the expression of straw carrying, hair pulling, and maternal care (González-Mariscal et al. 2000, 2004). These studies reveal the dramatic impact that a few key hormones have over the behavior of the female rabbit: the rise and fall of blood progesterone (against constant background levels of circulating estradiol) virtually transform the animal into a digging and straw carrying "robot."

10.4 BEYOND BEHAVIORAL NEUROENDOCRINOLOGY: FROM PHYSIOLOGY TO PERSONALITY

The "peculiar" characteristics of the rabbit continue to stimulate the investigation of physiological and neurobehavioral questions that could not readily be addressed using other animal models. In the following sections, we discuss three specific examples of research areas in which studying the rabbit can give important insight into fundamental biological processes.

10.4.1 The Female Rabbit as a Model for Investigating the Reflex Activity of the Pelvic Region that Underlies Reproductive Processes and Micturition

In mammalian females, many reproductive processes, such as parturition, are mediated by reflexive neuronal activity initiated in the pelvic region. The domestic rabbit gives birth to as many as 10 or more pups across an average time span of around 10 minutes. We have described the behavior of the female rabbit during parturition, as well as several characteristics of her pups, for example, birth latencies, vaginal retention, and the presence or absence of placentas and membranes at the time of birth (Hudson et al. 1999). We found that the pups are very active, are nursed rapidly and briefly, and survive the parturition process with only a limited amount of direct maternal care (e.g., licking each kit as it is being born). We also found that lactating females simultaneously pregnant with a litter from the typical postpartum mating, nurse their young at a time of day when the uterine myometrium is least sensitive to the high level release of oxytocin accompanying milk letdown. In this way, premature birth of the subsequent litter is avoided (Hudson et al. 1995; Ninomiya-Alarcón et al. 2004).

Parturition takes place via a urogenital apparatus that has certain peculiar characteristics: the urethra opens directly into a very long vagina (in proportion to the size of the female), which is surrounded by prominent striated muscle (Martínez-Gómez et al. 1997; Cruz et al. 2002). These characteristics make the distal-most portion of the vagina a structure that serves to mediate parturition, copulation, and urination, and indicate that this portion of the vagina has a complex neural regulation.

The terminal part of the spinal cord includes the lumbar, sacral, and coccygeal segments, and provides the sensorimotor innervation to the pelvic, pudendal, and caudal regions in quadruped mammals. The musculature of these regions is complex and intricate, and comprises delicate layers that subserve or are associated with reproductive organs and urinary and fecal excretion (Thor and de Groat 2010; Martínez-Gómez et al. 2011). We had shown in rats that vaginal stimulation triggers viscerosomatic reflexes that are mediated by several nerves, including the hypogastric and pelvic nerves, the somatomotor branch of the pelvic nerve (Martínez-Gómez et al. 1992), the motor branch of the clitoral nerve, and the ischiocavernosus nerve, producing reflexive activity of the muscles that surround the vagina, such as the ischiocavernosus and the bulbospongiosus (Cruz et al. 2002, 2010). The contraction of each muscle increases vaginal pressure, suggesting that they play a significant role in reproductive processes, such as parturition. Thus, the female rabbit provides an exceptional opportunity to extend our knowledge of the neurophysiology of the female reproductive tract.

Reproduction can have many costs for the female. Multiple pregnancies, dystocic parturitions, and simultaneous pregnancy and lactation imply considerable energetic costs that can put the health of the mother at risk. In women, multiparity (reproductive experience that includes more than two parturitions), despite having clear advantages in terms of biological fitness (increased number of viable offspring), has been associated with urinary pathologies such as incontinence. Urinary incontinence, defined as the involuntary loss of urine, is twice as common in women as in men; age and multiple parturitions are principal risk factors. In addition to the suffering that is caused by this disorder, its management is costly (Abrams et al. 2010).

We view the female rabbit as an advantageous model in which to study basic mechanisms of urinary incontinence. In the breeding season, female rabbits can give birth to two or three litters and, like females of many mammalian species, can be pregnant and lactating simultaneously. Moreover, female rabbits have prominent pelvic and perineal musculature (Martínez-Gómez et al. 1997), and show four distinct urinary behavior patterns: squatting, spraying, squirting, and spotting. These urinary forms can be regulated by various viscerosomatic reflexes and involve the participation of pelvic and perineal striated muscles (Martínez-Gómez et al. 2004).

Using young multiparous rabbits, we tested the hypothesis that multiparity affects the process of micturition by altering the morphology of the urinary apparatus and the surrounding striated muscle. It was necessary to first describe the histological characteristics of the vagina and urethra in young virgin females. We found that the histological organization of the distinct components that comprise each of the tissue layers varies along the length of these organs (Rodríguez-Antolín et al. 2009). When the histological organization of these structures was compared in young virgin versus young multiparous females (having had four parturitions), we found that the organization of the vaginal and urethral walls had changed in the latter animals (Xelhuantzi et al. 2014). The overall thickness of both organs had decreased, which coincided with a decrease in the smooth and striated musculature and an increase in connective tissue (Xelhuantzi et al. 2014). Notably, the histological organization of the vaginal autonomic ganglia was also altered in multiparous females, compared to virgins (Castelán et al. 2013).

Similar to the changes that occur in the vagina, the morphology of the pubococcygeus and bulbospongiosus muscles are also modified in multiparous female rabbits, showing a decreased cross-sectional area of their fibers, changes in the proportions of fiber type—specifically, Type II and IIA—and decreased force generated during simple and tetanic contraction (Fajardo et al. 2008). When bladder pressure and electrophysiological activity of the muscles that surround the urogenital apparatus were measured, we found that the muscles activate synchronously and differentially during micturition in young female rabbits (Corona-Quintanilla et al. 2009). The differential blocking of this muscular activity promotes changes in urodynamic function. Multiparity alters bladder pressure and the temporal pattern of activity in the associated muscles (Martínez-Gómez et al. 2011).

Based on our findings on the innervation of the urogenital apparatus and the effects of differential stimulation on this structure, and those of other investigators, we have proposed a model for the regulation of micturition in virgin female rabbits. We have shown that multiparity affects the process of micturition, altering the anatomy of the urogenital apparatus and the surrounding striated muscle, perhaps also in women. Thus, the domestic rabbit has become an advantageous model for studying the effects of multiple births on urinary function.

10.4.2 Using Rabbit Behavior to Understand Processes Fundamental to Certain Neuropsychiatric Disorders: Motivation, Memory, and Repetitive Stereotyped Behavior

Many neuropsychiatric symptoms and disorders involve alterations in motivation and the persistent expression of repetitive, stereotyped behaviors. These include disorders of the obsessive–compulsive spectrum such as trichotillomania (compulsive hair pulling), obsessive–compulsive disorder (OCD) (compulsive hand-washing, checking, and hoarding), drug addiction (compulsive drug seeking and use), as well as anhedonia and the lack of drive associated with schizophrenia and depression. The female rabbit is a superb model animal for studying the neurobiological substrates that underlie these processes, because hormonal changes that normally occur during pregnancy trigger striking and predictable changes in motivation and behavior. Although the hormonal signals appear remarkably simple, the cognitive processes that are necessary for nest building to be successful should not be overlooked. Constructing a nest burrow, for example, would first require the selection of an appropriate site and then digging at the site for many hours, necessarily involving both a fixed attention on the nest site as well as the maintenance of the motivation to dig. A spatial memory of the site must be formed so that, on the night of nest building, the site can be found again. This spatial map would be repeatedly called upon each time female goes out to collect dry grass and carry it back to the nest burrow. Perhaps most amazing are the predictable changes—initially controlled by estrogen and progesterone—that occur in the animal's responsiveness to dry grass (or straw), a stimulus that the wild rabbit would probably encounter every day. Outside of pregnancy, dry grass would be treated either as something to eat, or perhaps simply ignored. However, when the ratio of progesterone to estrogen in the blood reaches a permissive level during the final week of pregnancy, dry grass abruptly becomes a sought-after material to be collected and carried back to the nest site. Thus, in addition to allowing the behavioral pattern to be expressed, the hormonal changes of pregnancy must also directly or indirectly affect systems that underlie attention, motivation, and spatial memory, as well as those processes that determine the interpretation and incentive value of a very specific stimulus. Finally, once each of the nest building phases has been completed, the underlying motivation to perform these behaviors must be "turned off," so that the animal can return to its normal routine. Late pregnant rabbits provided with an external source of hair will collect it and introduce it into the nest box only if they are shaved but not if they have already built a straw-nest and lined it with their own body hair (González-Mariscal et al. 1998). We (Hoffman and Rueda-Morales 2009, 2012) propose that the rabbit's persistent attention to the nest building task and its prolonged motivation to perform the behaviors (digging, straw carrying) might involve some of the same neural processes that underlie compulsive behavior. Moreover, mechanisms that normally "turn off" nest building behaviors might include those that are altered in obsessive–compulsive and related disorders, in which certain goal-directed behaviors are initiated normally but can be stopped only with extreme difficulty (Szechtman and Woody 2004).

Once the pregnant rabbit begins to collect and carry straw, the continued expression of this behavior is sensitive to external cues associated with the status of nest completion. In the laboratory, a series of experiments indicated that sensory cues associated with a finished nest inside the nest box serve as an external signal to quench the motivation to carry straw (Hoffman and Rueda-Morales 2009). Dopamine signaling appears to participate in maintaining the motivation to carry straw, as dopamine D1 and D2 receptor antagonists significantly shortened the duration of straw carrying bouts, without otherwise altering the expression of this behavior (Hoffman and Rueda-Morales 2012). Thus, the motivation to perform straw carrying, which is initiated by the internal hormonal state and maintained by dopamine signaling, is subsequently quenched by the perception of specific external cues.

The neural mechanisms that underlie this "motivational quenching" are likely to be relevant to understanding OCD. Current models of OCD pathophysiology have begun to implicate dysfunctions in the mechanisms that underlie the person's decision to stop a behavior (hand washing, for example). Studies of persons who suffer from compulsive hand washing indicate that they often relied on subjective internal criteria as signals to stop washing, that is, when it "felt right" (Wahl et al. 2008). By contrast, healthy controls reported either that they did not use any criteria or that they used objective criteria: they stopped automatically or when their hands were visibly clean. Identifying the neural mechanisms that are associated with the rabbit's "decision" to stop collecting straw, as well as those that normally quench the motivation to collect straw, might provide important clues to the identity of neural mechanisms that go awry in OCD.

Schizophrenia, like all neuropsychiatric disorders, encompasses a complex constellation of psychiatric symptoms, generally categorized as positive symptoms (psychosis), negative symptoms (anhedonia, lack of motivational drive) and cognitive symptoms (specific deficits in working and episodic memory). While it is impossible to fully replicate schizophrenia (or, indeed, any neuropsychiatric disorder) in an animal model, nevertheless certain aspects of this disorder can be modeled and studied in nonhuman species. Recently, we described a behavioral test to assess object recognition memory in the adult rabbit, based on the rabbit's tendency to preferentially chin mark unfamiliar objects, compared to objects that had been previously encountered (Hoffman and Basurto 2013, 2014). We found that ketamine or MK-801 (*N*-methyl-D-aspartate [NMDA] receptor antagonists) disrupted the rabbit's ability to recognize a novel object, without affecting the overall frequency of chin marking, and that these deficits were prevented by previous treatment with clozapine, an atypical antipsychotic (Hoffman and Basurto 2014). Similar experimental paradigms have been developed in rats and mice, and are proposed to be valid pharmacological models for schizophrenic symptoms (especially cognitive symptoms), since the pathophysiology of this disorder appears to involve hypofunction of the NMDA receptor. In the rabbit model, we found that glycinamide, a glycine prodrug that passes into the central nervous system and is converted to glycine (Beyer et al. 2013), was as effective as clozapine at preventing MK-801-induced deficits in object recognition memory (Hoffman and Basurto 2014). Subsequently, we showed that glycinamide also prevented MK-801-induced deficits in object recognition memory in the rat (Basurto et al. 2015). Glycine is an obligatory coagonist of the NMDA receptor, and is expected to augment the activity of this receptor. Currently, there is interest in identifying pharmacological strategies that increase extrasynaptic glycine concentrations in the brain, as possible novel pharmacotherapies for schizophrenia.

10.4.3 The Rabbit as a Model Animal for Studying the Ontogeny of Individual Differences in Phenotype, Including "Personality"

Interest has grown rapidly in recent years among behavioral and theoretical biologists in individual differences in behavioral phenotypes of a kind generally referred to as "animal personality." Once considered to be the exclusive domain of human psychologists, differences in animal personality are now considered to be the result of adaptive evolutionary processes, and to occur across a wide range of taxa (Wolf et al. 2007; Briffa and Weiss 2010; Trillmich and Hudson 2011; Hudson et al., in press). However, there have been few studies of the ontogeny of such differences—when they arise during development, if and how they relate to differences in morphology and physiology, or the mechanisms driving such processes (Stamps and Groothuis 2010; Trillmich and Hudson 2011), particularly in the case of mammals. Close observation and manipulation of mammalian young is often made difficult by their being hidden from view in nests, burrows, or pouches; experimental manipulation may lead to their being abandoned by the mother, and they are often defended by mothers or other care givers.

The European rabbit, due to its unusual and highly stereotyped system of "absentee" mothering outlined above (see also Rödel et al. 2012), offers a good experimental opportunity. Starting in Tlaxcala but now also together with colleagues in other countries, we have been investigating the ontogeny of individual behavioral phenotypes in both domestic and wild European rabbits. We began this line of work by focusing on the contribution of differences in body mass and of interactions among littermates to early growth and survival (Drummond et al. 2000; Hudson and Trillmich 2008; Trillmich and Hudson 2011; Hudson et al. 2011a).

In the rabbit, as in other altricial mammals, body mass at birth is closely associated with early postnatal survival and body mass at weaning (Rödel et al. 2008a,b; Hudson et al. 2011a). As in other species, this is a good predictor of postweaning growth and survival. Heavier pups at birth are also generally heavier at weaning than their lighter littermates, they occupy more central thermally advantageous positions in the litter huddle, obtain more milk, and are more efficient at converting this into body mass (Bautista et al. 2008; Rödel et al. 2008a,b; Hudson et al. 2011a). Within-litter differences in body mass at birth are, at least in part, due to the site of implantation of fetuses along the uterine horns. Those implanted at the ovarian end are generally heavier at birth and subsequently show greater weight gain and a higher probability of survival until weaning than their lighter littermates (Bautista et al. 2015a).

Being born into a nest of littermates confronts the young with an interrelated array of challenges and developmental possibilities. These can be broadly grouped according to two functional contexts: suckling and interactions within the litter huddle. Suckling is obviously essential for the young to obtain sufficient milk for survival and growth. Litter interactions within the huddle relate to the need to maintain an adequate body temperature and to obtain somatosensory stimulation necessary for normal neural, motor, and social development (Hudson et al. 2011a).

Considering first suckling, competition among littermates for milk is severe. A high percentage of young fail to obtain milk during at least one nursing event and up to 20% die of starvation within the first postnatal week even under the relatively favorable conditions of the laboratory or farm (Couread et al. 2000; Drummond et al. 2000; Schlolaut et al. 2013). Competition is particularly severe given that the pups only obtain significant amounts of milk during the second minute of nursing (Bautista et al. 2005). Viewing the behavior of the young during nursing in a glass-bottomed nest box showed that they compete for nipples in a vigorous scramble but without obvious signs of overt aggression. Body mass is a good predictor of suckling success and, independent of sex, the heaviest young at birth obtain more milk, are more likely to survive and are heavier at weaning than their lighter sibs (Drummond et al. 2000; Bautista et al. 2005; Rödel et al. 2008a; Hudson et al. 2011a).

Between suckling episodes the young also compete for well-insulated, central positions within the litter huddle, and expend considerable energy climbing over and burrowing under each other in a continuous effort to achieve and maintain such positions (Bautista et al. 2008). Body mass is a good predictor of the outcome of such struggles, with heavier young of either sex generally being in body contact with more littermates than their lighter sibs. Consequently, they also have higher body temperatures, lower expression of uncoupling protein-1 important in altricial young for burning brown fat for nonshivering thermogenesis (Bautista et al. 2013), and they are more efficient at converting milk into body mass (Bautista et al. 2008, 2013; Reyes-Meza et al. 2011; Rödel et al. 2008a,b; Hudson et al. 2011a; García-Torres et al. 2015). They also have higher serum levels of testosterone and corticosterone at the end of the first postnatal week than their lighter sibs (Hudson et al. 2011b).

Aside from such competition, for any individual, littermates represent a significant thermoregulatory resource (Rödel et al. 2008a,b,c). Young raised together, even with a single littermate, have higher mean body temperatures, are more efficient at converting milk into body mass, and have a higher probability of survival than littermates of similar body mass at birth and kept under the same conditions but alone (Bautista et al. 2003). Early sibling presence also appears to provide the young with somatosensory stimulation enabling them to anticipate and prepare for the daily nursing visit (cf. Hudson and Distel 1982; Jilge and Hudson 2001) and enhancing their motor development. Littermates raised together obtained more milk and had better postural control than randomly chosen littermates raised alone (Nicolás et al. 2011; cf. Muciño et al. 2009). Early littermate presence might also contribute to later social competence, as littermates raised together competed more successfully for food and water postweaning than isolation-raised littermates of similar body mass (Nicolás et al. 2011).

In conclusion, the rabbit's system of absentee maternal care means that pups spend the early postnatal period almost exclusively in the company of their siblings. The findings outlined above show that relations among them help shape the early formation of individual profiles in morphological, physiological, and behavioral characteristics (Hudson et al. 2011a; c.f. Sulloway 2010; Bautista et al. 2015b). These considerations suggest that such differences during early development have long-term consequences for individuals' survival and reproductive fitness, and for individual differences in behavior of a kind that can be considered an animal's personality (Rödel and von Holst 2009; Reyes-Meza et al. 2011; Rödel and Monclús 2011).

10.5 FINAL THOUGHTS

Through the years, *Oryctolagus cuniculus* has been an important model animal in a number of distinct research areas. Certain characteristics of the rabbit might be "peculiar," but no more so than the unique characteristics of any other species, including the ubiquitously-studied rat and mouse. The physiology and behavior of every animal species were shaped through evolution, enabling that species to thrive in the face of a unique set of environmental and reproductive challenges. Thus, certain aspects of the physiology and behavior of any species, when considered with a comparative perspective and within the context of its phylogeny and natural history, can provide insight into fundamental biological processes. The "peculiarities" of the European rabbit can be viewed as providing exciting research opportunities, alongside the elegantly "simple" (and thoroughly mapped) nervous system of the nematode *Caenohabditis elegans*, the transparent embryos of the zebrafish *Danio rerio*, the monogamous lifestyle of the prairie vole (*Microtus ochrogaster*), the ability of the zebra finch (*Taenio pygiaguttata*) to learn songs, olfactory imprinting and homing behavior in the Pacific salmon (*Oncorhynchus* spp.), and, indeed, the incredible cognitive capacities of *Homo sapiens*.

ACKNOWLEDGMENTS

Preparation of this chapter was partially supported by a grant from the UNAM funding agency DGAPA-PAPIIT (IN212416).

REFERENCES

Abrams, P., Andersson, K.E., Birder, L., et al. 2010. Fourth international consultation on incontinence recommendations of the international scientific committee: Evaluation and treatment of urinary incontinence, pelvic organ prolapse, and fecal incontinence. *Neurourol Urodyn* 29:213–40.

Basurto, E., González-Flores, O., Hoffman K.L. 2015. Glycinamide prevents MK-801-induced hyperactivity and deficits in object recognition memory in an animal model of positive and cognitive symptoms of schizophrenia. *Schizophr Res* 166:349–50.

Bautista, A., Castelán, F., Pérez-Roldán, H., Martínez-Gómez, M., Hudson, R., 2013. Competition in newborn rabbits for thermally advantageous positions in the litter huddle is associated with individual differences in brown fat metabolism. *Physiol Behav* 118:189–94.

Bautista, A., Drummond, H., Martínez-Gómez, M., Hudson, R. 2003. Thermal benefit of sibling presence in the newborn rabbit. *Dev Psychobiol* 43:208–15.

Bautista, A., García-Torres, E., Martínez-Gómez, M., Hudson, R. 2008. Do newborn domestic rabbits *Oryctolaguscuniculus* compete for thermally advantageous positions in the litter huddle? *Behav Ecol Sociobiol* 62:331–9.

Bautista, A., Mendoza-Degante, M., Coureaud, G., Martínez-Gómez, M., Hudson, R. 2005. Scramble competition in newborn domestic rabbits for an unusually restricted milk supply. *Anim Behav* 70:1011–21.

Bautista, A., Rödel, H.G., Monclús, R., et al. 2015a. Intrauterine position as a predictor of postnatal growth and survival in the rabbit. *Physiol Behav* 138:101–6.

Bautista, A., Zepeda, J.A., Reyes-Meza, V., Martínez-Gómez, M., Rödel, H.G., Hudson, R. 2015b. Contribution of within-litter interactions to individual differences in early postnatal growth in the domestic rabbit. *Anim Behav* 108:145–53.

Beyer, C., de la Torre, L., Larsson, K., Pérez-Palacios, G. 1975. Synergistic actions of estrogen and androgen on the sexual behavior of the castrated male rabbit. *Horm Behav* 6:301–6.

Beyer, C., Komisaruk, B.K., González-Flores, O., Gómora-Arrati, P. 2013. Glycinamide, a glycine pro-drug, induces antinocioception by intraperitoneal or oral ingestion in ovariectomized rats. *Life Sci* 92:576–81.

Beyer, C., McDonald, P., Vidal, N. 1970. Failure of 5α-dihydrotestosterone to elicit estrous behavior in the ovariectomized rabbit. *Endocrinology* 86:939–41.

Beyer, C., Mena, F. 1965. Induction of milk secretion in the rabbit by removal of the telencephalon. *Am J Physiol* 208:289–92.

Beyer, C., Mena, F. 1970. Parturition and lactogenesis in rabbits with high spinal cord transection. *Endocrinology* 87:195–7.

Beyer, C., Ramirez, V.D., Whitmoyer, D.I., Sawyer, C.H. 1967. Effects of hormones on the electrical activity of the brain in the rat and rabbit. *Exp Neurol* 18:313–26.

Beyer, C., Rivaud, N. 1968. Sexual behavior in pregnant and lactating domestic rabbits. *Physiol Behav* 4:753–7.

Beyer, C., Rivaud, N. 1973. Differential effect of testosterone and dihydrotestosterone on the sexual behavior of prepubertally castrated male rabbits. *Horm Behav* 4:175–80.

Beyer, C., Sawyer, C.H. 1964. Effects of vigilance and other factors on nonspecific acoustic responses in the rabbit. *Exp Neurol* 10:156–69.

Beyer, C., Tindal, J.S., Sawyer, C.H. 1962. Electrophysiological study of projections from mesencephalic central gray matter to forebrain in the rabbit. *Exp Neurol* 6:435–50.

Beyer, C., Velazquez, J., Larsson, K., Contreras, J.L. 1980. Androgen regulation of the motor copulatory pattern in the male New Zealand white rabbit. *Horm Behav* 14:179–90.

Beyer, C., Vidal, N. 1971. Inhibitory action of MER-25 on androgen-induced oestrous behaviour in the ovariectomized rabbit. *J Endocrinol* 51:401–2.

Beyer, C., Vidal, N., McDonald, P.G. 1969. Interaction of gonadal steroids and their effect on sexual behaviour in the rabbit. *J Endocrinol* 45:407–13.

Beyer, C., Vidal, N., Mijeres, A. 1970. Probable role of aromatization in the induction of estrous behavior by androgens in the ovariectomized rabbit. *Endocrinology* 87:1386–9.

Briffa, M., Weiss, A.A. 2010. Animal personality. *Curr Biol* 20:R912.

Castelán, F., Xelhuantzi, N., Hernández-Aragón, L.G., Rodríguez-Antolín, J., Cuevas, E., Martínez-Gómez, M. 2013. Morphometry of paravaginal ganglia from the pelvic plexus: impact of multiparity, primarity, and pregnancy. *Eur J Obstet Gynecol Reprod Biol* 170:286–92.

Challis. J.R.G., Davies, I.J., Ryan, K.J. 1973. The concentrations of progesterone, estrone and estradiol-17β in the plasma of pregnant rabbits. *Endocrinology* 93:971–6.

Chirino, R., González-Mariscal, G., Carrillo, P., Pacheco, P., Hudson R. 1993. Effect of removing the chin gland on chin-marking behaviour in male rabbits of the New Zealand race. *Mammal Biol* 58:116–21.

Contreras, J.L., Beyer, C. 1979. A polygraphic analysis of mounting and ejaculation in the New Zealand white rabbit. *Physiol Behav* 23:939–43.

Corona Quintanilla, D.L., Castelán, F., Fajardo, V., Manzo, J., Martínez-Gómez, M. 2009. Temporal coordination of pelvic and perineal striated-muscle activity during micturition in female rabbits. *J Urol* 181:1452–8.

Cruz, M.L., Beyer, C. 1972. Effects of septal lesions on maternal behavior and lactation in the rabbit. *Physiol Behav* 9:361–5.

Cruz, Y., Hudson R., Pacheco, P., Lucio, R.A., Martínez-Gómez, M. 2002. Anatomical and physiological characteristics of perineal muscles in the female rabbit. *Physiol Behav* 74:1–8.

Cruz, Y., Rodríguez-Antolín, J., Nicolás, L., Martínez-Gómez, M., Lucio, R.A. 2010. Components of the neural circuitry of the vaginocavernosus reflex in rabbits. *J Comp Neurol* 518:199–210.

Distel, H. Hudson, R. 1985. The contribution of the olfactory and tactile modalities to the nipple-search behaviour of newborn rabbits. *J Comp Physiol A* 157:599–605.

Drummond, H., Vázquez, E., Sánchez-Colón, S., Martínez-Gómez, M., Hudson, R. 2000. Competition for milk in the domestic rabbit: survivors benefit from littermate deaths. *Ethology* 106:511–26.

Fajardo, V., Pacheco, P., Hudson, R., Jiménez, I., Martínez-Gómez, M. 2008. Differences in morphology and contractility of the bulbospongiosus and the pubococcygeus muscles in nulliparous and multiparous rabbits. *Int Urogynecol J Pelvic Floor Dysfunct* 19:843–9.

Fee, A.R., Parkes, A.S. 1929. Studies on ovulation. I. The relation of the anterior pituitary body to ovulation in the rabbit. *J Physiol* 67:415–32.

García-Dalmán, C., González-Mariscal, G. 2012. Major role of suckling stimulation for inhibition of estrous behaviors in lactating rabbits: acute and chronic effects. *Horm Behav* 61:108–13.

García-Torres, E., Hudson, R., Castelán, F., Martínez-Gómez, M., Bautista, A. 2015. Differential metabolism of brown adipose tissue in newborn rabbits in relation to position in the litter huddle. *J Thermal Biol* 51:33–41.

González-Mariscal, G., Albonette, M.E., Cuamatzi, E., Beyer, C. 1997. Transitory inhibition of scent marking by copulation in male and female rabbits. *Anim Behav* 53:323–3.

González-Mariscal, G., Chirino, R., Flores-Alonso, J.C., Rosenblatt, J.S., Beyer, C. 2004. Intracerebroventricular injections of prolactin counteract the antagonistic effect of bromocriptine on rabbit maternal behaviour. *J Neuroendocrinol* 16:949–55.

González-Mariscal, G., Chirino, R., Hudson, R. 1994a. Prolactin stimulates emission of nipple pheromone in ovariectomized New Zealand white rabbits. *Biol Reprod* 50:373–6.

González-Mariscal, G., Chirino, R., Rosenblatt, J.S., Beyer, C. 2005. Forebrain implants of estradiol stimulate maternal nest-building in ovariectomized rabbits. *Horm Behav* 47:272–9.

González-Mariscal, G., Cuamatzi, E., Rosenblatt, J.S. 1998. Hormones and external factors: are they "on/off" signals for maternal nest-building in rabbits? *Horm Behav* 33:1–8.

González-Mariscal, G., Díaz-Sánchez, V., Melo, A.I., Beyer, C., Rosenblatt, J.S. 1994b. Maternal behavior in New Zealand white rabbits: quantification of somatic events, motor patterns, and steroid plasma levels. *Physiol Behav* 55:1081–9.

González-Mariscal, G., Lemus, A.C., Vega-González, A., Aguilar-Roblero, R. 2013. Litter size determines circadian periodicity of nursing in rabbits. *Chronobiol Int* 30:711–18.

González-Mariscal, G., Melo, A.I., Zavala, A., Beyer, C. 1990. Variations in chin-marking behavior of New Zealand female rabbits throughout the whole reproductive cycle. *Physiol Behav* 48:361–5.

González-Mariscal, G., Melo, A.I., Zavala, A., Beyer, C. 1992. Chin-marking behavior in male and female New Zealand rabbits: onset, development, and activation by steroids. *Physiol Behav* 52:889–93.

González-Mariscal, G., Melo, A.I., Jiménez, P., Beyer, C., Rosenblatt, J.S. 1996. Estradiol, progesterone, and prolactin regulate maternal nest-building in rabbits. *J Neuroendocrinol* 8:901–7.

González-Mariscal, G., Melo, A.I., Parlow, A.F., Beyer, C., Rosenblatt, J.S. 2000. Pharmacological evidence that prolactin acts from late gestation to promote maternal behavior in rabbits. *J Neuroendocrinol* 12:983–92.

González-Mariscal, G., Melo, A.I., Zavala, A., Chirino, R., Beyer, C. 1993. Sex steroid regulation of chin-marking behavior in male New Zealand rabbits. *Physiol Behav* 54:1035–40.

Harris, G. W. 1948a. Electrical stimulation of the hypothalamus and the mechanism of neural control of the adenohypophysis. *J Physiol* 107:418–29.

Harris, G.W. 1948b. The excretion of an antidiuretic substance by the kidney, after electrical stimulation of the neurohypophysis in the unanaesthetized rabbit. *J Physiol* 107:430–5.

Harris, G.W. 1948c. Stimulation of the supraopticohypophysial tract in the conscious rabbit with currents of different wave form. *J Physiol* 107:412–17.

Harris, G.W. 1951. Neural control of the pituitary gland: II. The adenohypophysis. *Br Med J* 15:627–34.

Hill, M., Parkes, A.S. 1931. Studies on ovulation. IV. Induction of ovulation in the hypophysectomized rabbit by administration of anterior lobe extracts. *J Physiol* 71:36–9.

Hilliard, J., Scaramuzzi, R.J., Penardi, R., Sawyer, C.H. 1973. Progesterone, estradiol and testosterone levels in ovarian venous blood of pregnant rabbits. *Endocrinology* 93:1235–8.

Hilliard, J., Spies, H.G., Sawyer, C.H. 1967. Cholesterol storage and progestin secretion during pregnancy and pseudopregnancy in the rabbit. *Endocrinology* 82:157–65.

Hoffman, K.L., Basurto, E. 2013. One-trial object recognition memory in the domestic rabbit (*Oryctolaguscuniculus*) is disrupted by NMDA receptor antagonists. *Behav Brain Res* 250:62–73.

Hoffman, K.L., Basurto, E. 2014. Clozapine and glycinamide prevent MK-801-induced deficits in the novel object recognition (NOR) test in the domestic rabbit (*Oryctolaguscuniculus*). *Behav Brain Res* 271:203–11.

Hoffman, K.L., González-Mariscal, G. 2006. Progesterone receptor activation signals behavioral transitions across the reproductive cycle of the female rabbit. *Horm Behav* 50:154–68.

Hoffman, K.L., González-Mariscal, G. 2007. Relevance of ovarian signaling for the early behavioral transition from estrus to pregnancy in the female rabbit. *Horm Behav* 52:531–9.

Hoffman, K.L., Rueda-Morales, R.I. 2009. Toward an understanding of the neurobiology of "just right" perceptions: nest building in the female rabbit as a possible model for compulsive behavior and the perception of task completion. *Behav Brain Res* 204:182–91.

Hoffman, K.L., Rueda-Morales, R.I. 2012. D1 and D2 dopamine receptor antagonists decrease behavioral bout duration, without altering the bout's repeated behavioral components, in a naturalistic model of repetitive and compulsive behavior. *Behav Brain Res* 230:1–10.

Hoffman, K.L., Rueda-Morales, R.I., González-Mariscal, G. 2010. Relevance of mating-associated stimuli, ovulation and the progesterone receptor for the post-coital inhibition of estrous behavior in the female rabbit. *Horm Behav* 58:747–53.

Hudson, R. 1985. Do newborn rabbits learn the odor stimuli releasing nipple-search behavior? *Dev Psychobiol* 18:575–85.

Hudson, R., Bautista, A., Reyes-Meza, V., Morales Montor, J., Rödel, H.G. 2011a. The effect of siblings on early development: a potential contributor to personality differences in mammals. *Dev Psychobiol* 53:564–74.

Hudson, R., Cruz, Y., Lucio, R.A., Ninomiya, J., Martínez-Gómez, M. 1999. Temporal and behavioral patterning of parturition in rabbits and rats. *Physiol Behav* 66:599–604.

Hudson, R., Distel, H. 1982. The pattern of behaviour of rabbit pups in the nest. *Behaviour* 79:255–71.

Hudson, R., Distel, H. 1983. Nipple location by newborn rabbits: behavioural evidence for pheromonal guidance. *Behaviour* 85:260–75.

Hudson, R., Distel, H. 1984. Nipple-search pheromone in rabbits: dependence on season and reproductive state. *J Comp Physiol A* 155:13–17. Hudson, R., Distel, H. 1990. Sensitivity of female rabbits to change in photoperiod as measured by pheromone emission. *J Comp Physiol A* 167:225–30.

Hudson, R., González-Mariscal, G., Beyer, C. 1990. Chin marking behaviour, sexual receptivity, and pheromone emission in steroid-treated, ovariectomized rabbits. *Horm Behav* 24:1–13.

Hudson, R., Maqueda, B., Velázquez Moctezuma, J., Morales Miranda, A., Rödel, H.G. 2011b. Individual differences in testosterone and corticosterone levels in relation to early postnatal development in the rabbit *Oryctolaguscuniculus*. *Physiol Behav* 103:238–41.

Hudson, R., Melo, A., González-Mariscal, G. 1994. Effect of photoperiod and exogenous melatonin on correlates of estrus in the domestic rabbit. *J Comp Physiol A* 175:573–9.

Hudson, R., Müller, A., Kennedy, G. 1995. Parturition in the rabbit is compromised by daytime nursing: the role of oxytocin. *Biol Reprod* 53:519–24.

Hudson, R., Rangassamy, M., Saldaña, A., Bánszrgi, O., Rödel, H.G. 2015. Stable individual differences in separation calls during early development in cats and mice. *Front Zool* doi: 10.1186/1742-9994-12-S1-S12.

Hudson, R. Trillmich, F. 2008. Sibling competition and cooperation in mammals: challenges, developments and prospects. *Behav Ecol Sociobiol* 62:299–307.

Hudson, R., Vodermeyer, T. 1992. Spontaneous and odour-induced chin-marling in domestic female rabbits. *Anim Behav* 43:329–36.

Jilge, B., Hudson, R. 2001. Diversity and development of circadian rhythms in the European rabbit. *Chronobiol Int* 18:1–26.

Knaus, H. 1930. Zur Physiologie des corpus luteum. *Arch Gynäkol* 141:374–403.

Markee, J.E., Sawyer, C.H., Hollinshead, W.H. 1946a. Activation of the anterior hypophysis by electrical stimulation in the rabbit. *Endocrinology* 38:345–57.

Markee, J.E., Sawyer, C.H., Hollinshead, W.H. 1946b. Electrical stimulation of the ovulatory mechanism in the rabbit. *Anat Rec* 94:521.

Marshall, F.H.A., Verney, E.B. 1936. The occurrence of ovulation and pseudopregnancy in the rabbit as a result of central nervous stimulation. *J Physiol* 86:327–36.

Marshall, F.H.A., Verney, E.B., Vogt, M. 1939. The occurrence of ovulation in the rabbit as a result of stimulation of the central nervous system by drugs. *J Physiol* 97:128–32.

Martínez-Gómez, M., Chirino, R., Beyer, C., Komisaruk, B.R., Pacheco, P. 1992. Visceral and postural reflexes evoked by genital stimulation in urethane-anesthetized female rats. *Brain Res* 575:279–84.

Martínez-Gómez, M., Corona Quintanilla, D.L., Fajardo, V., García, L., Hudson, R. 2004. Patterns of urination in female rabbits of different ages and reproductive state. Paper presented at the 2nd World Lagomorph Conference, Vairão, Portugal.

Martínez-Gómez, M., Guarneros, M., Zempoalteca, R., Hudson, R. 1997. A comparison of spontaneous and odor-induced chin marking in male and female domestic rabbits (*Oryctolaguscuniculusdomestica*). *Ethology* 103:893–901.

Martínez-Gómez, M., Lucio, R.A., Carro, M., Pacheco, P., Hudson, R. 1997. Striated muscles and scent glands associated with the vaginal tract of the rabbit. *Anat Rec* 247:486–95.

Martínez-Gómez, M., Mendoza, G., Corona-Quintanilla, D.L., Fajardo, V., Rodríguez-Antolín, J., Castelán, F. 2011. Multiparity causes uncoordinated activity of pelvic and perineal striated muscle and urodynamic changes in rabbits. *Reprod Sci* 18:1246–52.

McDonald, P.G., Vidal, N., Beyer, C. 1970. Sexual behavior in the ovariectomized rabbit after treatment with different amounts of gonadal hormones. *Horm Behav* 1:161–72.
Mena, F., Beyer, C. 1963. Effect of high spinal section on established lactation in the rabbit. *Am J Physiol* 205:313–16.
Mena, F., Beyer, C. 1968a. Effect of spinal cord lesions on milk ejection in the rabbit. *Endocrinology* 83:615–17.
Mena, F., Beyer, C. 1968b. Induction of milk secretion in the rabbit by lesions in the temporal lobe. *Endocrinology* 83:618–20.
Mikhail, G., Nall, M.W., Allen, W.M. 1961. Progesterone levels in the rabbit ovarian vein blood throughout pregnancy. *Endocrinology* 69:504–9.
Muciño, E., Bautista, A., Jiménez, I., Martínez-Gómez, M., Hudson, R. 2009. Differential development of body equilibrium among littermates in the newborn rabbit. *Dev Psychobiol* 51:24–33.
Myers, K., Poole, W.E. 1959. A study of the biology of the wild rabbit, *Oryctolaguscuniculus (L.)*, in confined populations. I. The effects of density on home range and the formation of breeding groups. *C.S.I.R.O. Wildl Res* 4:14–26.
Myers, K., Poole, W.E. 1961. A study of the biology of the wild rabbit, *Oryctolaguscuniculus* (L.), in confined populations. II. The effects of season and population increase on behaviour. *C.S.I.R.O. Wildl Res* 6:1–41.
Myers, K., Poole, W.E. 1962. A study of the biology of the wild rabbit, *Oryctolaguscuniculus (L.)*, in confined populations. III. Reproduction. *Aust J Zool* 10:225–67.
Mykytowycz, R. 1958. Social behaviour of an experimental colony of wild rabbits, *Oryctolaguscuniculus (L.)*. I. Establishment of the colony. *C.S.I.R.O. Wildl Res* 3:7–25.
Mykytowycz, R. 1959. Social behaviour of an experimental colony of wild rabbits, *Oryctolaguscuniculus (L.)*. II. First breeding season. *C.S.I.R.O. Wildl Res* 4:1–13.
Mykytowycz, R. 1965. Further observations on the territorial function and histology of the submandibular cutaneous (chin) glands in the rabbit, *Oryctolaguscuniculus (L)*. *Anim Behav* 13:400–12.
Mykytowycz, R., Gambale, S. 1965. A study of the inter-warren activities and dispersal of wild rabbits, *Oryctolaguscuniculus (L.)*, living in a 45-Ac paddock. *C.S.I.R.O. Wildl Res* 10:111–23.
Nicolás, L., Martínez-Gómez, M., Hudson, R., Bautista, A. 2011. Littermates presence enhances motor development, weight gain and competitive ability in newborn and juvenile domestic rabbits. *Dev Psychobiol* 53:37–46.
Ninomiya-Alarcón, J.C., Hudson, R., Reyes-Guerrero, G., Barrera-Mera, B., Guevara-Guzmán, R. 2004. Effect of photoperiod on the mechanical response of the pregnant rabbit uterus to oxytocin. *Am J Physiol: Reg Intergr Comp Physiol* 287:R174–80.
Reyes-Meza, V., Hudson, R., Martínez-Gómez, M., Nicolás, L., Rödel, H.G., Bautista, A. 2011. Possible contribution of position in the litter huddle to long-term differences in behavioral style in the domestic rabbit. *Physiol Behav* 104:778–85.
Rideaud, P., Orgeur, P. 2000. Mimicking natural nursing conditions promotes early pup survival in domestic rabbits. *Ethology* 106:207–25.
Robson, J.M. 1933. The reactivity and activity of the rabbit's uterus during pregnancy, parturition and the puerperium. *J Physiol* 79:83–93.
Robson, J. M. 1935a. The action of oestrin on the uterus of the hypophysectomized and of the pregnant rabbit. *J Physiol* 84:148–61.
Robson, J.M. 1935b. The effect of oestrin on the uterine reactivity and its relation to experimental abortion and parturition. *J Physiol* 84:121–32.
Robson, J.M. 1936. The action of progesterone on the uterus of the rabbit and its antagonism by oestrone. *J Physiol* 88:100–11.
Rödel, H.G., Bautista, A., García-Torres, E., Martínez-Gómez, M., Hudson, R. 2008a. Why do heavy littermates grow better than lighter ones? A study in wild and domestic European rabbits. *Physiol Behav* 95:441–8.

Rödel, H.G., Dausmann, K.H., Starkloff, A., Schubert, M., von Holst, D., Hudson, R. 2012. Diurnal nursing pattern of wild-type European rabbits under natural breeding conditions. *Mammal Biol* 77:441–6.

Rödel, H.G., Hudson, R., von Holst, D. 2008c. Optimal litter size for individual growth of European rabbit pups depends on their thermal environment. *Oecologia* 155:677–89.

Rödel, H.G., Monclús, R. 2011. Long-term consequences of early development of personality traits: A study in European rabbits. *Behav Ecol* 22:1123–30.

Rödel, H.G., Prager, G., Stefanski, V., von Holst, D., Hudson, R. 2008b. Separating maternal and litter-size effects on early postnatal growth in two species of altricial small mammals. *Physiol Behav* 93:826–34.

Rödel, H.G., von Holst, D. 2009. Features of the early juvenile development predict competitive performance in male European rabbits. *Physiol Behav* 97:495–502.

Rodríguez-Antolín, J., Xelhuantzi, N., García-Lorenzana, M., Cuevas, E., Hudson, R., Martínez-Gómez, M. 2009. General tissue characteristics of the lower urethral and the vaginal walls in the domestic rabbit. *Int Urogynecol J Pelvic Floor Dysfunct* 20:53–60.

Sawin, P.B., Denenberg, V.H., Ross, S., Hafter, E., Zarrow, M.X. 1960. Maternal behavior in the rabbit: hair loosening during gestation. *Am J Physiol* 198:1099–102.

Schlolaut, W., Hudson, R., Rödel, H.G. 2013. Impact of rearing management on health in domestic rabbits: a review. *World Rabbit Sci* 21:145–59.

Soares, M.J., Diamond M. 1982. Pregnancy and chin marking in the rabbit, *Oryctolaguscuniculus*. *Anim Behav* 30:941–3.

Soto, M. A., Reynoso, M. Beyer, C. 1984. Sexual dimorphism in the motor mounting pattern of the New Zealand white rabbit: steroid regulation of vigor and rhythmicity of pelvic thrusting. *Horm Behav* 18:225–34.

Stamps, J., Groothuis, T.G.G. 2010. The development of animal personality: relevance, concepts and perspectives. *Biol Rev* 85:301–25.

Sulloway, J.F. 2010. Why siblings are like Darwin's finches. Birth order, parental investment, and adaptive divergence within the family. In *The evolution of personality and individual differences,* ed. D.M. Brush, P.H. Hawley, 86–119. Oxford: Oxford University Press.

Szechtman, H., Woody, E. 2004. Obsessive-compulsive disorder as a disturbance of security motivation. *Psychol Rev* 111:111–27.

Thor, K. B., de Groat, W.C. 2010. Neural control of the female urethral and anal rhabdosphincters and pelvic floor muscles. *Am J Physiol Regul Integr Comp Physiol* 299:R416–R38.

Tindal, J.S., Beyer, C., Sawyer, C.H. 1963. Milk ejection reflex and maintenance of lactation in the rabbit. *Endocrinology* 72:720–4.

Trillmich, F., Hudson, R. 2011. The emergence of personality in animals: the need for a developmental approach. *Dev Psychobiol* 53:505–9.

Wahl, K., Salkovskis, P.M. Cotter, I. 2008. "I wash until it feels right" the phenomenology of stopping criteria in obsessive-compulsive washing. *J Anxiety Disord* 22:143–61.

Walton, A., Hammond, J. 1928. Observations on ovulation in the rabbit. *J Exp Biol* 6:190–204.

Wolf, N., Van Doorn, G.S., Leimar, O., Weissing, F.J. 2007. Life-history trade-offs favour the evolution of animal personalities. *Nature* 447:581–4.

Xelhuantzi, N., Rodríguez-Antolín J., Nicolás, L., Castelán, F., Cuevas, E., Martínez-Gómez, M. 2014. Tissue alterations in urethral and vaginal walls related to multiparity in rabbits. *Anat Rec* 297:1963–70.

Yaschine, T., Mena, F., Beyer, C. 1967. Gonadal hormones and mounting behavior in the female rabbit. *Am J Physiol* 213:867–72.

Zarrow, M. X., Denenberg V.H., Anderson C.O. 1965a. Rabbit: frequency of suckling in the pub. *Science* 150:1835–6.

Zarrow, M.X., Denenberg, V.H., Kalberer, W.D. 1965b. Strain differences in the endocrine basis of maternal nest-building in the rabbit. *J Reprod Fertil* 10:397–401.

Zarrow, M. X., Farooq, A., Denenberg, V.H. 1962. Maternal behavior in the rabbit: Critical period for nest building following castration during pregnancy. *Proc Soc Exp Biol Med* 111:537–8.

Zarrow, M.X., Farooq, A., Denenberg, V.H., Sawin, P.B., Ross, S. 1963. Maternal behviour in the rabbit: endocrine control of maternal-nest building. *J Reprod Fertil* 6:375–83.

Zarrow, M.X., Sawin, P.B., Ross, S., Denenberg, V.H., Crary, D., Wilson, E.D., Farooq, A. 1961. Maternal behaviour in the rabbit: evidence for an endocrine basis of maternal nest building and additional data on maternal-nest building in the Dutch-belted race. *J Reprod Fertil* 2:152–62.

11 A View of Rabbit Maternal Behavior from the Perspectives of Complex Systems and Chronostasis

Gabriela González-Mariscal and Raúl Aguilar-Roblero

CONTENTS

11.1 Introduction 187
11.2 What is a Complex System? 189
11.3 How Does a Female (Rabbit) Become a Mother? 189
11.3.1 Nest Building 189
11.3.2 Nursing 191
11.3.2.1 Chronostasis and a Suckling-Entrained Oscillator 192
11.4 Permanent vs Transitory Changes in the Maternal Brain 193
11.5 Conclusions 194
Acknowledgments 195
References 195

11.1 INTRODUCTION

We began to study rabbit maternal behavior because Prof. Carlos Beyer and Prof. Jay Rosenblatt (then director of the Institute of Animal Behavior at Rutgers University) were interested in exploring the underlying neuroendocrine control of the "peculiar" form of parental care displayed by this mammal (see below). Beyer had already performed several studies on the neuroendocrine control of lactation in rabbit does, together with Flavio Mena (see Chapter 10 by Hoffman et al., this volume), but they had hardly touched on the behavioral aspect. Rosenblatt, by contrast, was a pioneer of the studies on maternal behavior and had numerous publications, mostly in rats. Although GGM had never worked on maternal behavior, she was persuaded by both Beyer and Rosenblatt to begin this line of research in the very well-kept colony of rabbits established in the laboratory in Tlaxcala in central Mexico. After several publications devoted to investigating how particular hormones promote specific maternal activities in does (see below), GGM became interested in exploring the time dimension of rabbit nursing. As chronobiology was then an alien field to her, she contacted RAR, a recognized expert on such topic, and tried to interest him in exploring a "new" behavior. Together, we have started a line of research aimed at investigating the underpinnings of the ways by which particular brain regions and somatosensory stimuli orchestrate the timing of rabbit nursing.

Maternal behavior in mammals has been the subject of study in hundreds of papers published during the last 60 years. The approaches used to investigate this topic have ranged from the ethological to the evolutionary, and have employed a multiplicity of experimental methodologies (for reviews, see González-Mariscal and Melo, 2013; González-Mariscal and Poindron, 2002; Numan and Young, 2016; Rosenblatt and Siegel, 1981). Such studies have revealed the intricate web of neural connections that participate in regulating the expression of specific aspects of maternal behavior, the hormones that promote or inhibit particular maternal activities, and the many forms of care that female mammals have adopted to successfully meet the needs of their young. From the analysis of such rich information, specific issues have been found to be common to most mammals studied, for example, a key role of the medial preoptic area (MPOA) for the unfailing occurrence of maternal behavior, a major role of estrogen to promote maternal activities, and a critical role of stimuli coming from the young for maintaining the expression of maternal behavior postpartum. By contrast, other aspects of this behavior vary greatly among mammals, for example, the selectivity—or not—of nursing a female's own vs alien offspring, the frequency and duration of mother-young contact throughout lactation, and the paradoxical action of progesterone (which prevents or stimulates specific maternal activities and triggers others after its removal). Despite their abundance, both types of findings (i.e., the ones revealing common factors and those showing variation among mammals) indicate *immediate factors* of control. They tell us little about the mechanisms by which the elements of the "maternal brain" interact with each other on a higher level to allow the *emergence* of the most complex behavioral task shown by mammalian females during specific periods of their lifespan. It is the purpose of this chapter to present the behavior of mother rabbits in a way that will illustrate the complexities of this activity (for a recent review, see González-Mariscal et al., 2016) and encourage reflection on the general issues, derived from this animal model, that may be valid across mammals. To this end we will use two theoretical frameworks: *complex (or dynamic) systems* and *chronostasis*. The former approach has been used successfully to gain insight on issues as diverse as the mind-brain relationship, the evolution of specific traits, the operation of insect societies, embryology, the production of crops worldwide… (García, 2007; Greenberg et al., 2004; Johnson, 2001). In psychology, this view has been adopted to investigate the development of cognition (Sporns et al., 2004), motor abilities (Smith and Thelen, 2003; Thelen, 1995), and the coordination of perceptual motor-cognitive skills (Schöner and Dineva, 2007). In all these cases, dynamic instabilities (i.e., "the generation of stable patterns of neuronal activation") have been proposed as the common key elements that allow new properties to emerge in the developing organism (Schöner and Dineva, 2007). Moreover, recent progress in the neurosciences is making it possible to ground the concept of emergence in specific neural mechanisms (Sporns et al., 2004) and to test the validity of long-held views on particular issues (e.g., motor development) through the design of incisive experiments (Smith and Thelen, 2003).

Is the process of "becoming a mother" akin to a developmental progression? Does it involve *both* the "unfolding" of a preexisting genetic program and the emergence of new traits as a consequence of interactions among the elements of the system in real time? Similar questions were already posed by Rosenblatt and Lehrman (1963), coinciding with the ideas of Schneirla (1957) on the organization of behavior across stages (levels) of development. This view ("... the events occurring at each stage are, in various ways, relevant to the emergence of the next stage ..."; Rosenblatt and Lehrman, 1963) is in agreement with dynamic systems theory ("... the metric is not whether a given organism has a particular ability. Rather, ... the important dimension is the relative stability of behavior in its particular context over time"; Smith and Thelen, 2003). In the following sections, we will briefly describe the main attributes of complex (or dynamic) systems and will then present evidence, derived from research in rabbits, which supports the notion that the "maternal brain" emerges and operates as a complex system (CS).

11.2 WHAT IS A COMPLEX SYSTEM?

CS is a representation of a slice of reality, an organized entity in which the elements are inseparable from each other and, therefore, cannot be studied in isolation. Moreover, the properties of a CS are determined not by its elements per se but by the relationships among them. Because the structural properties of the system determine its stability (or lack thereof) in relation to the environment, one studies the dynamics of the system (i.e., how it changes across time) rather than its steady state. Thus, the investigation of CS requires interdisciplinary research: CSs are "open"; they lack well-defined limits and are in continuous exchange with the environment. Additional characteristics of CSs are the following: (1) their complexity increases over time; (2) small quantitative increases provoke qualitative discontinuities; (3) the relationships among different levels of organization are nonlinear and probabilistically discontinuous; (4) self-organization: when parts of the system are organized in novel ways, *new properties emerge* as a result of such reorganization. From these properties follows the conclusion that the phenomena at a particular level of organization cannot be explained or predicted by the phenomena or principles that apply at lower levels.

11.3 HOW DOES A FEMALE (RABBIT) BECOME A MOTHER?

11.3.1 Nest Building

Rabbits are gregarious animals, living in colonies inside a series of underground connected tunnels. They emerge from their burrows at dusk to forage communally and they return to their subterranean quarters before sunrise (Mykytowycz, 1958, 1959). These nocturnal habits have been retained in animals housed under laboratory (Jilge, 2001) or farm (González-Mariscal et al., 2007) conditions. Despite their social behavior, females are very exclusive when it comes to maternal care: from mid-pregnancy until parturition, they gradually build a maternal nest totally separated from the colony dwelling. This process begins with digging an underground burrow, continues with the collection of grass (carefully selected to be edible; Hudson and Altbäcker, 1992) and lining of the burrow, and concludes as the doe plucks her body fur (mainly from her ventrum and inner thighs) and covers the "straw nest" with it (Ross et al., 1963; Zarrow et al., 1961). This three-stage nest building process is displayed *only* by pregnant rabbits and it is tightly controlled by specific hormonal combinations. Thus, digging increases under the combined action of estradiol (E) and progesterone (P), a condition characteristic of early to mid-pregnancy (González-Mariscal et al., 1994; Figure 11.1) and mimicked by the exogenous injection of such hormones to ovariectomized (ovx) rabbits (González-Mariscal et al., 1996). Although estrous does can also dig, they do it irregularly and less intensely than do the pregnant ones, and ovx females rarely express this activity.

A decline in the concentration of P triggers a switch between behaviors, that is, a cessation of digging *and* an onset of straw carrying. The predominance of this new behavior over the previous one is *only* observed following P withdrawal (Figure 11.1). Rabbits not exposed to P *never* collect straw and introduce it into the nest box: when given this material, they eat it. Thus, a *behavioral discontinuity* is observed: from being

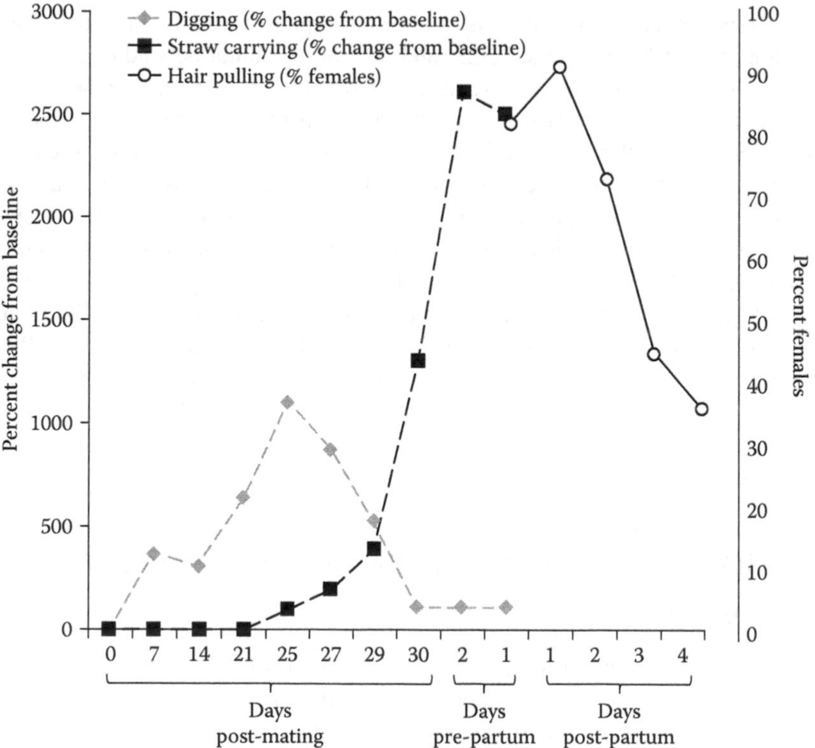

FIGURE 11.1 The expression of the three behaviors necessary to create a maternal nest (digging, straw carrying, and hair pulling) by pregnant rabbits occurs sequentially, that is, the decline in one activity leads to the increase in the next one.

treated as food under the combined influence of E and P, straw becomes "construction material" as the concentration of P declines. These changes are not dependent on the embryos, as the same behavioral switches are observed in ovx rabbits given E + P followed by discontinuing P (González-Mariscal et al., 1996). In the final stages of nest building, when a marked rise in E and prolactin (PRL) occurs, another behavioral discontinuity is observed: straw carrying ceases and mothers pluck their body fur to line the nest. This is the only time when a female rabbit will perform this behavior, and we have obtained evidence that cognition may play a role in it. Thus, shaved females will readily collect fur from a container and introduce it into the straw nest, but the latency to do this depends on the type of fur offered: while they collect their own or synthetic fur almost immediately, it takes them around 20 min to start collecting male fur. Moreover, if shaved pregnant rabbits are given both straw *and* fur, they will collect only the former in late pregnancy and will start collecting fur later on, as parturition approaches (González-Mariscal et al., 1998). The way in which the various stages of nest building are linked to each other, and are stimulated/terminated by specific hormonal combinations, indicates that the relationships between the hormonal and behavioral levels of organization are nonlinear, that is, small quantitative changes provoke qualitative discontinuities.

Although the above evidence supports a major role of E and P in determining the nest building sequence, our knowledge about the substrate(s) on which these hormones act is scarce. The MPOA is a complex telencephalic region with a highly specialized topography (Balthazart and Ball, 2008). In rabbits it participates in controlling functions as diverse as female sexual receptivity, male mounting, scent-marking (Melo et al., 2008), nest building, (González-Mariscal et al., 2005), and ovulation (Ramírez and Beyer, 1988). The MPOA of the rabbit has receptors for E and P (Caba et al., 2003a,b), but their participation in determining the expression of a specific function and the suppression of others is unknown. We have found that the selective implantation of E into the MPOA of ovx rabbits is both necessary and sufficient to stimulate digging (when combined with subcutaneous injections of P) and promote straw carrying (following P withdrawal; González-Mariscal et al., 2005). By contrast, implants of E into the MPOA of ovx rabbits not exposed to P do not promote nest building but, rather, sexual receptivity and scent marking (Melo et al., 2008). Indeed, the latter are behavior patterns characteristic of estrous does because, as rabbits are reflex ovulators, they are not exposed to P unless they mate and a corpus luteum is formed (Ramírez and Beyer, 1988). From the above, we may speculate that a population of MPOA neurons is showing "aggregate" behavior, that is, they are acting as a "unitary whole" toward a specific goal depending on whether or not they are under the influence of P. This speculation is, of course, testable and has the advantage that one can use the presence or absence of P receptors as a means of identifying specific neuronal populations.

11.3.2 Nursing

This activity involves two components: (1) a behavior, that is, the entrance into the nest and the adoption of a posture that will allow suckling by the young, and (2) a neuroendocrine reflex, that is, the massive release of oxytocin, which provokes milk ejection, and the secretion of PRL, which induces milk synthesis. Yet, unlike most mammals, nursing in rabbits occurs only once a day with a periodicity of ca. 24 h (Jilge, 1993, 1995; González-Mariscal et al., 2013) and each bout lasts only around 3 min (Drewett et al., 1982; González-Mariscal et al., 1994; Lincoln, 1974). These characteristics of nursing remain unchanged throughout lactation. This apparent invariability in duration and frequency of nursing is critically dependent on the characteristics of suckling stimulation. Mother rabbits given only one kit to nurse stay a much longer time inside the nest box than do those given four young; however, providing a larger litter does not reduce the time inside the nest box below ca. 3 min (González-Mariscal, 2013b). Similarly, does suckling litters of four or fewer kits show a disruption in nursing periodicity, entering the nest box several times a day (González-Mariscal et al., 2013a). These results indicate that an excitability threshold is required to "turn on" as yet unidentified processes that determine the duration of mother/young contact at ca. 3 min once every 24 h. How is this achieved?

11.3.2.1 Chronostasis and a Suckling-Entrained Oscillator

Recently, Aguilar-Roblero and Muñoz-Díaz (2010) proposed that "… the mammalian timing system is formed by *constitutive clocks* (such as the suprachiasmatic nucleus; SCN) and *emerging clocks* (built up from the coordinated activity of peripheral oscillators)." This timing system underlies a general regulatory process that modulates the "set point" of physiological variables in a periodic manner known as "chronostasis." Thus, the *emergence* of a food-entrained oscillator (FEO) has been proposed to explain the changes in locomotor activity and metabolic rhythms that occur in rodents when food is provided for a restricted period of time, outside the normal schedule of the animal (i.e., diurnal or nocturnal; Stephan, 1983; Escobar et al., 1988; Diaz-Muñoz et al., 2000; Martinez-Merlos et al., 2004; Baez-Ruíz et al., 2005). Moreover, there is evidence that an FEO may also operate in suckling rabbit kits (reviewed in Caba and González-Mariscal, 2009). Specifically, the circadian rhythms of locomotion, corticosterone, and several metabolic indicators shift in relation to the time of the single daily nursing bout (Escobar et al., 2000). Such coordination of behavioral, physiological, and metabolic processes in the organism from SCN control to the assembly of an FEO involves the uncoupling and operation of peripheral oscillators independent of the SCN.

Can we propose the emergence of a suckling-entrained oscillator (SEO) in lactating does? This (putative) entity would, indeed, allow the incorporation of the metabolic changes of lactating does into their "normal" activities and rhythms. For instance, there is a major increase in food intake across lactation (González-Mariscal et al., 2009), which has to be integrated into the neuroendocrine and metabolic changes that support lactation. Such increased metabolic demands may reflect the operation of rheostasis, that is, modifications in the "set point" of processes like food and water intake, energy partitioning, and so on. In addition, the peculiar nursing pattern of rabbit does requires that "the system" be ready at the appropriate time, regarding both milk output and behavior. Chronostatic regulation seems ideally suited to this end, as it allows different configurations of the timing system to deal with changes in the environment and in the organism itself. Accordingly, the circadian system (*constitutive clock*) would allow an optimal timing of metabolic and behavioral events within a 24-h cycle (Aguilar-Roblero, 2015) while an SEO (*emerging clock*) would permit "plasticity" in relation to the number of suckled kits.

Do the hormones of pregnancy-lactation favor the emergence of an SEO? Do they promote plastic changes that favor the engagement of particular brain structures (and peripheral organs) to constitute transitory circuits that "work toward a common end"? Similar questions have been posed and explored regarding the FEO in rodents. Several forebrain areas, for example, the ventromedial, lateral, and dorsomedial hypothalamus, the nucleus accumbens, and the paraventricular thalamic nucleus have been implicated in the display of food anticipatory activity. Yet, none of them is essential (Mistlberger and Rechtschaffern, 1984; Mistlberger and Mumby, 1992; Mistlberger et al., 2003; Landry et al., 2006, 2007a, 2007b; Mistlberger 2011). These findings indicate the operation of a distributed system arranged in a nonhierarchical manner to control food anticipatory activity, consistent with the proposal that the FEO is an emergent system built up in response to daily periodic feeding (Aguilar-Roblero and Díaz-Muñoz, 2010).

Recently, we proposed a model of how nipple stimulation can interact with the circadian system to promote the mother's nursing behavior (González-Mariscal et al., 2013a). It involves three elements: (1) a circadian signal, presumably generated by the SCN, (2) a decaying inhibitory tone that prevents the mother from re-entering the nest after nursing, and (3) a threshold of nipple stimulation ("non-photic zeitgeber"). Although rabbits show circadian rhythmicity in the expression of locomotion, defecation (Jilge, 1991a,b), and corticosteroid secretion (Szeto et al., 2004), the impact of lesions to the SCN—element (1) in the model above—on such activities has not been explored. By contrast, there is experimental evidence supporting the existence of element (2). It is based on the finding that if kits are removed from the maternal cage after suckling and are returned several hours later, does can show the *behavioral component* of nursing much earlier than usual: 60% of mothers that suckled 4 kits entered the nest box and crouched over the litter at 6 h after the previous suckling episode. By contrast, all rabbits that nursed just one kit redisplayed nursing behavior at 6 and even 3 h later (González-Mariscal, 2007). These findings support the existence of a refractory period whose duration depends on the number of kits suckled. Regarding element (3), from the chronostasis framework discussed above, we propose that it corresponds to an SEO.

11.4 PERMANENT VS TRANSITORY CHANGES IN THE MATERNAL BRAIN

Numerous examples in several mammalian species have shown that multiparous females are more resistant to specific challenges, have a lower activation threshold for particular maternal activities, and display better cognitive skills (e.g., memory, learning) than their virgin or primiparous counterparts (for reviews, see González-Mariscal and Kinsley, 2008; González-Mariscal and Melo, 2016). In rabbits, maternal responsiveness is lost in *primiparous* mothers separated from their litter at parturition but is retained in those that are *multiparous* (González-Mariscal et al., 1998).

The mechanisms underlying the permanent modifications in brain function seen in multiparous mammals are varied, and they involve changes in neuronal connections, concentration of neurotransmitters, and number of peptide and steroid receptors, among others (Afonso et al., 2008; Anderson et al., 2006; Barha and Galea, 2011; Gatewood et al., 2005; Kinsley et al., 2007; Leuner and Gould, 2010). Thus, maternal experience illustrates another characteristic of CS: their capacity to self-organize. New properties *emerge* in the maternal brain that lead to adaptive behavior patterns which, in turn, enhance the likelihood of survival of the young and promote a more "finely tuned" maternal behavior.

A different type of reorganization of the maternal brain is illustrated by the responsiveness to specific odors. Nonpregnant rodents, sheep, and rabbits are generally repelled by olfactory signals emanating from newborns, amniotic fluid, and placentas. As parturition approaches, the valence of these stimuli changes from being treated as repulsive to becoming attractive. Thus, females will readily consume the placentas, lick the amniotic fluids, and approach the newborn (Gregg and Wynne-Edwards, 2005; Kristal, 1980; Lévy et al., 1983; Melo and González-Mariscal, 2003; Perea-Rodriguez and Saltzman, 2014; Rosenblatt and Mayer, 1995). Recently we reported that, as early as pregnancy day 7, mother rabbits prefer the odors of kits to "neutral" ones, an effect not seen in virgin animals (Chirino and González-Mariscal, 2015). Both models show that mothers "resolve" the approach/withdrawal conflict under conditions of specific hormonal combinations and somatosensory stimuli. Although these changes also suggest a self-organization of the neural networks underlying placentophagia and the processing of olfactory signals, they differ from the "maternal experience" effects in that they are transitory. That is, only during a limited period will females (even multiparous ones) consume placenta and amniotic fluid or prefer odors emanating from the young.

11.5 CONCLUSIONS

We present evidence herein that supports the notion that rabbit maternal behavior operates as a "complex system." We provide examples, derived from experimental data, illustrating that, (1) small quantitative changes result in qualitative discontinuities, (2) the complexity of the system increases over time and shows self-organization, and (3) specific brain regions are operating in a pattern suggestive of "aggregate" behavior. These properties characterize the anagenetic nature of the process, that is, a trend from the simple (the virgin female) to the complex (the maternal female). From this perspective, "becoming a mother" is a developmental process. Yet, unlike other forms of behavioral development, maternal traits are lost by the end of lactation (weaning), and this loss is an essential condition to allow the reinitiation of a new reproductive cycle. Nonetheless, "the maternal brain" apparently retains characteristics that facilitate the future expression of complex cognitive, motor, and spatial tasks (reviewed in González-Mariscal and Kinsley, 2009; González-Mariscal and Melo, 2016). Thus, the expression of maternal behavior seems to permanently modify the animal, as occurs in other developmental processes.

ACKNOWLEDGMENTS

Our gratitude, respect, and affection for Prof. Carlos Beyer cannot be conveyed adequately in this manuscript. We hope that his many teachings, generosity, and "life wisdom" will encourage us to pursue noble academic ideals and produce scientific findings worthy of that which we received from him.

This work was supported by CINVESTAV (GGM) and CONACYT grant 128528 (to RAR).

REFERENCES

Afonso, V.M., Grella, S.L., Chatterjee, D., and Fleming, A.S. 2008. Previous maternal experience affects accumbal dopaminergic responses to pup-stimuli. *Brain Research* 1198:115–23.

Aguilar-Roblero, R. 2015. Chronostasis: the timing of physiological systems. In *Mechanisms of Circadian Systems in Animals and Their Clinical Relevance*, eds. R. Aguilar-Roblero, M. Díaz-Muñoz, and M. L. Fanjul-Moles. Springer, New York, pp. 221–236.

Aguilar-Roblero, R., and Muñoz-Díaz, M. 2010. Chronostatic adaptations in the liver to restricted feeding: the FEO as an emergent oscillator. *Sleep and Biological Rhythms* 8:9–17.

Anderson, G.M., Grattan, D.R., van den Ancker, W., and Bridges, R.S. 2006. Reproductive experience increases prolactin responsiveness in the medial preoptic area and arcuate nucleus of female rats. *Endocrinology* 147:4688–94.

Baez-Ruiz, A., Escobar, C., Aguilar-Roblero, R., Vazquez-Martinez, O., and Diaz-Munoz, M. 2005. Metabolic adaptations of liver mitochondria during restricted feeding schedules. *Am J Physiol Gastrointest Liver Physiol* 289:G1015–23.

Balthazart, J., and Ball, G.2007. Topography in the preoptic region: differential regulation of appetitive and consummatory male sexual behavior. *Frontiers in Neuroendocrinology* 28:161–78.

Barha, C.K., and Galea, L.A.M. 2011. Motherhood alters the cellular response to estrogens in the hippocampus later in life. *Neurobiology of Aging* 32:2091–5.

Caba, M., Beyer, C., González-Mariscal, G., and Morrell, J.I. 2003a. Immunocytochemical detection of estrogen receptor alpha in the female rabbit forebrain: topography and regulation by estradiol. *Neuroendocrinology* 77:208–22.

Caba, M., and González-Mariscal, G. 2009. The rabbit pup, a natural model of nursing anticipatory activity. *European Journal of Neuroscience* 30:1697–706.

Caba, M., Rovirosa, M.J., Beyer, C., and González-Mariscal, G. 2003b. Immunocytochemical detection of progesterone receptor in the female rabbit forebrain: distribution and regulation by oestradiol and progesterone. *Journal of Neuroendocrinology* 15:855–64.

Chirino, R., and González-Mariscal, G. 2015. Changes in responsiveness to kit odors across pregnancy: relevance for the onset of maternal behavior. *World Rabbit Science* 23:103–109.

Díaz-Muñoz, M., Vázquez-Martínez, O., Aguilar-Roblero, R., and Escobar, C. 2000. Anticipatory changes in liver metabolism and entrainment of insulin, glucagon, and corticosterone in food-restricted rats. *American Journal of Physiology—Regulatory Integrative and Comparative Physiology* 279:R2048–56.

Drewett, R.F., Kendrick, K.M., Sanders, D.J., and Trew, A.M. 1982. A quantitative analysis of the feeding behavior of suckling rabbits. *Developmental Psychobiology* 15:25–32.

Escobar, C., Díaz-Muñoz, M., Encinas, F., and Aguilar-Roblero, R. 1998. Persistence of metabolic rhythmicity during fasting and its entrainment by restricted feeding schedules in rats. *American Journal of Physiology—Regulatory Integrative and Comparative Physiology* 274:R1309–16.

Escobar, C., Hudson, R., Martínez-Gómez, M., Aguilar-Roblero, R., and Escobar, C. 2000. Metabolic correlates of the circadian pattern of suckling-associated arousal in young rabbits. *Journal of Comparative Physiology—A Sensory, Neural, and Behavioral Physiology* 186:33–8.

García, R. 2007. *Sistemas Complejos*. México, DF: Gedisa.

Gatewood, J.D., Morgan, M.D., Eaton, M., McNamara, I.M., Stevens, L.F., Macbeth, A.H., Meyer, E.A.A., Lomes, L.M., Kozub, F.J., Lambert, K.G., and Kinsley, C.H. 2005. Motherhood mitigates aging-related decrements in learning and memory and positively affects brain aging in the rat. *Brain Research Bulletin* 66:91–8.

González-Mariscal, G. 2007. Mother rabbits and their offspring: timing is everything. *Developmental Psychobiology* 49:71–6.

González-Mariscal, G., Caba, M., Martínez-Gómez, M., Bautista, A., and Hudson, R. 2016. Mothers and offspring: the rabbit as a model system in the study of mammalian maternal behavior and sibling interactions. *Hormones and Behavior.* DOI:10.1016/j.yhbeh.2015.05.011.

González-Mariscal, G., Chirino, R., Rosenblatt, J.S., and Beyer, C. 2005. Forebrain implants of estradiol stimulate maternal nest-building in ovariectomized rabbits. *Hormones and Behavior* 47:272–9.

González-Mariscal, G., Cuamatzi, E., and Rosenblatt, J.S. 1998. Hormones and external factors: are they "on/off" signals for maternal nest-building in rabbits? *Hormones and Behavior* 33:1–8.

González-Mariscal, G., Díaz-Sánchez, V., Melo, A.I., Beyer, C., and Rosenblatt, J.S. 1994. Maternal behavior in New Zealand white rabbits: quantification of somatic events, motor patterns, and steroid plasma levels. *Physiology and Behavior* 55:1081–9.

González-Mariscal, G., Gallegos, J.A., Sierra-Ramírez, A., and Garza Flores, J. 2009. Impact of concurrent pregnancy and lactation on maternal nest-building, estradiol and progesterone concentrations in rabbits. *World Rabbit Science* 17:145–52.

González-Mariscal, G., and Kinsley, C.H. 2009. From indifference to ardor: the onset, maintenance and meaning of the maternal brain. In *Hormones, Brain, and Behavior*, eds. D. Pfaff, A. Arnold, A. Etgen, S. Fahrbach, and R. Rubin, 109–36. San Diego, CA: Academic Press.

González-Mariscal, G., Lemus, A.C., Vega-González, A., and Aguilar-Roblero, R. 2013a. Litter size determines circadian periodicity of nursing in rabbits. *Chronobiology International* 30:711–18.

González-Mariscal, G., McNitt, J.I., and Lukefahr, S.D. 2007. Maternal care of rabbits in the lab and on the farm: endocrine regulation of behavior and productivity. *Hormones and Behavior* 52:86–91.

González-Mariscal, G. and Melo, A.I. 2013. Parental behavior. In *Neuroscience in the 21st Century*, ed. D. Pfaff, 2069–100. New York: Springer-Verlag.

González-Mariscal, G., and Melo, A.I. 2016. Bidirectional effects of mother-young contact on the maternal and neonatal brains. In *The Plastic Brain,* eds. R. Von Bernhardi and J. Eugenin. New York: Springer.

González-Mariscal, G., Melo, A.I., Chirino, R., Jiménez, P., Beyer, C., and Rosenblatt, J.S. 1998. Importance of mother/young contact at parturition and across lactation for the expression of maternal behavior in rabbits. *Developmental Psychobiology* 32:101–11.

González-Mariscal, G., Melo, A.I., Jiménez, P., Beyer, C., and Rosenblatt, J.S. 1996. Estradiol, progesterone, and prolactin regulate maternal nest-building in rabbits. *Journal of Neuroendocrinology* 8:901–7.

González-Mariscal, G., and Poindron, P. 2002. Parental care in mammals: immediate internal and sensory factors of control. In *Hormones, Brain, and Behavior*, eds. D. Pfaff, A. Arnold, A. Etgen, S. Fahrbach, and R. Rubin, 215–98. San Diego, CA: Academic Press.

González-Mariscal, G., and Rosenblatt, J.S. 1996. Maternal behavior in rabbits: a historical and multidisciplinary perspective. In *Parental Care: Evolution, Mechanism and Adaptive Significance*, eds. J.S. Rosenblatt and C.T. Snowdon, 333–60. San Diego, CA: Academic Press.

González-Mariscal, G., Toribio, A., Gallegos-Huicochea, J.A., and Serrano-Meneses, M.A. 2013b. The characteristics of suckling stimulation determine the daily duration of mother-young contact and milk output in rabbits. *Developmental Psychobiology* 55:809–17.

Greenberg, G., Partridge, T., Weiss, E., and Pisula, W. 2004. Comparative psychology, a new perspective for the 21st century: up the spiral staircase. *Developmental Psychobiology* 44:1–15.

Gregg, J.K., and Wynne-Edwards, K.E. 2005. Placentophagia in naive adults, new fathers, and new mothers in the biparental dwarf hamster. *Phodopuscampbelli. Developmental Psychobiology* 47:179–88.

Hudson, R., and Altbäcker, V. 1992. Development of feeding and food preference in the European rabbit: environmental and maturational determinants. In *Ontogeny and Social Transmission of Food Preferences in Mammals: Basic and Applied Research*, eds. B.G. Galef, M. Mainardi, and P. Valsecchi, 125–45. London: Harwood Academic Publishing.

Jilge, B. 1991a. The rabbit: a diurnal or a nocturnal animal. *Journal of Experimental Animal Science* 34:170–83.

Jilge, B. 1991b. Restricted feeding: a nonphotic zeitgeber in the rabit. *Physiology and Behavior* 51:157–66.

Jilge, B. 1993. The ontogeny of circadian rhythms in the rabbit. *Journal of Biological Rhythms* 8:247–60.

Jilge, B. 1995. Ontogeny of the rabbit's circadian rhythms without an external zeitgeber. *Physiology and Behavior* 58:131–40.

Jilge, B., and Hudson, R. 2001. Diversity and development of circadian rhythms in the European rabbit. *Chronobiology International* 18:1–26.

Johnson, S. 2001. *Emergence: The Connected Lives of Ants, Brains, Cities and Software*. New York: Simon and Schuster.

Kinsley, C.H., Bardi, M., Karelina, K., Rima, B., Christon, L., Friedenberg, J., and Griffin, G. 2007. Motherhood induces and maintains behavioral and neural plasticity across the lifespan in the rat. *Archives of Sexual Behavior* 37:43–56.

Kristal, M.B. 1980. Placentophagia: a biobehavioral enigma (or De gustibus non disputandumest). *Neuroscience and Biobehavioral Reviews* 4:141–50.

Landry, G.J., Simon, M.M., and Webb, I.C., Mistlberger, R.E., 2006. Persistence of a behavioral food-anticipatory circadian rhythm following dorsomedial hypothalamic ablation in rats. *American Journal of Physiology—Regulatory, Integrative and Comparative Physiology* 290:R1527–34.

Landry, G.J., Yamakawa, G.R., and Mistlberger, R.E. 2007a. Robust food anticipatory circadian rhythms in rats with complete ablation of the thalamic paraventricular nucleus. *Brain Research* 1141:108–18.

Landry, G.J., Yamakawa, G.R., Webb, I.C., Mear, R.J., and Mistlberger, R.E. 2007b. The dorsomedial hypothalamic nucleus is not necessary for the expression of circadian food-anticipatory activity in rats. *Journal of Biological Rhythms* 22:467–78.

Leuner, B., and Gould, E. 2010. Dendritic growth in the medial prefrontal cortex and cognitive flexibility are enhanced during the postpartum period. *Journal of Neuroscience* 30:13499–503.

Lévy, F., Poindron, P., and Le Neindre, P. 1983. Attraction and repulsion by amniotic fluids and their olfactory control in the ewe around parturition. *Physiology and Behavior* 31:687–92.

Lincoln, L.W. 1974. Suckling: a time constant in the nursing behavior of the rabbit. *Physiology and Behavior* 13:711–14.

Martínez-Merlos, M.T., Ángeles-Castellanos, M., Díaz-Muñoz, M., Aguilar-Roblero, R., Mendoza, J., and Escobar, C., 2004. Dissociation between adipose tissue signals, behavior and the food-entrained oscillator. *Journal of Endocrinology* 181:53–63.

Melo, A.I., Chirino, R., Jiménez, A., Cuamatzi, E., Beyer, C., and González-Mariscal, G. 2008. Effect of forebrain implants of testosterone or estradiol on scent-marking and sexual behavior in male and female rabbits. *Hormones and Behavior* 54:676–83.

Melo, A.I., and González-Mariscal, G. 2003. Placentophagia in rabbits: incidence across the reproductive cycle. *Developmental Psychobiology* 43:37–43.

Mistlberger, R.E. 2011. Neurobiology of food anticipatory circadian rhythms. *Physiology and Behavior* 104:535–45.

Mistlberger, R.E., Antle, M.C., Kilduff, T.S., and Jones, M. 2003. Food- and light-entrained circadian rhythms in rats with hypocretin-2-saporin ablations of the lateral hypothalamus. *Brain Research* 980:161–8.

Mistlberger, R.E., and Mumby, D.G. 1992. The limbic system and food-anticipatory circadian rhythms in the rat: ablation and dopamine blocking studies. *Behavioural Brain Research* 47:159–68.

Mistlberger, R.E., and Rechtschaffen, A. 1984. Recovery of anticipatory activity to restricted feeding in rats with ventromedial hypothalamic lesions. *Physiology and Behavior* 33:227–35.

Mykytowycz, R. 1958. Social behaviour of an experimental colony of wild rabbits, *Oryctolagus cuniculus* L. I: Establishment of the colony. *CSIRO Wildlife Research* 3:7–25.

Mykytowycz, R. 1959. Social behaviour of an experimental colony of wild rabbits, *Oryctolagus cuniculus* L. II: First breeding season. *CSIRO Wildlife Research* 4:1–13.

Numan, M., and Young, L.J. 2016. Neural mechanisms of mother-infant bonding and pair bonding: similarities, differences, and broader implications. *Hormones and Behavior*. DOI:10.1016/j.yhbeh.2015.05.015.

Perea-Rodriguez, J. P., and Saltzman, W. 2014. Differences in placentophagia in relation to reproductive status in the California mouse (*Peromyscus californicus*). *Developmental Psychobiology* 56:812–20.

Ramírez, V.D., and Beyer, C. 1988. The ovarian cycle of the rabbit: its neuroendocrine control. In *The Physiology of Reproduction*, eds. E. Knobil and J. Neill, 1873–92. New York: Raven Press.

Rosenblatt, J.S., and Lehrman, D.S. 1963. Maternal behavior of the laboratory rat. In *Maternal Behavior in Mammals*, ed. H.L. Rheingold, 8–57. New York: Wiley.

Rosenblatt, J.S., and Mayer, A.D. 1995. An analysis of approach/withdrawal processes in the initiation of maternal behavior in the laboratory rat. In *Behavioral Development, Concepts of Approach/Withdrawal and Integrative Levels*, eds. K.E. Hood, G. Greenberg, and E. Tobach, 177–230. New York: Garland Publishing.

Rosenblatt, J.S., and Siegel, H.I. 1981. Factors governing the onset and maintenance of maternal behavior among non-primate mammals. In *Parental Care in Mammals*, eds. D.J. Gubernick and P.H. Klopfer, 113–76. New York: Plenum Press.

Ross, S., Zarrow, M.X., Sawin, P.G., Denenberg, V.H., and Blumenfield, M. 1963. Maternal behavior in the rabbit under semi-natural conditions. *Animal Behaviour* 11:283–5.

Schneirla, T.C. 1957. The concept of development in comparative psychology. In *The Concept of Development*, ed. D.B. Harris, 78–108. Minneapolis, MN: University of Minneapolis Press.

Schöner, G., and Dineva, E. 2007. Dynamic instabilities as mechanisms for emergence. *Developmental Science* 10:69–74.

Smith, L.B., and Thelen, E. 2003. Development as a dynamic system. *Trends in Cognitive Sciences* 7:343–8.

Sporns, O., Chiaivo D.R., Kaiser, M., and Hilgetag, C.C. 2004. Organization, development and function of complex brain networks. *Trends in Cognitive Sciences* 8:418–25.

Stephan, F.K. 1983. Circadian rhythm dissociation induced by periodic feeding in rats with suprachiasmatic lesions. *Behavioural Brain Research* 7:81–98.

Szeto, A., Gonzales, J.A., Spitzer, S.B., Levine, J.E., Zaias, J., Saab, P.G., Schneiderman, N., and McCabe, P.M. 2004. Circulating levels of glucocorticoid hormones in WHHL and NZW rabbits: circadian cycle and response to repeated social encounter. *Psychoneuroendocrinology* 29:861–6.

Thelen, E. 1995. Motor development, a new synthesis. *American Psychologist* 50:79–95.

Zarrow, M.X., Sawin, P.B., Ross, S., Denenberg, V.H., Crary, D., Wilson, E.D., and Farooq, A. 1961. Maternal behavior in the rabbit: evidence for an endocrine basis of maternal nest-building and additional data on maternal nest-building in the Dutch-Belted race. *Journal of Reproduction and Fertility* 2:152–62.

12 Multisignaling Approach to the Study of Sexual Differentiation of Brain and Behavior in Mammals

María Cruz Rodríguez Del Cerro, Carmen Pérez-Laso, Francisco Gómez, Antonio Guillamón, and Santiago Segovia

CONTENTS

12.1 Introduction 202
12.2 The Vomeronasal System; An Integrative Model for the Study of Brain and Behavioral Sex Differences 202
12.2.1 Sexual Dimorphism in the VNS 203
12.2.1.1 Vomeronasal Organ 203
12.2.1.2 The Accessory Olfactory Bulb 203
12.2.1.3 The Bed Nucleus of the Accessory Olfactory Tract 204
12.2.1.4 The Medial and Posteromedial Cortical Amygdaloid Nuclei 204
12.2.1.5 The Posteromedial Division of the Bed Nucleus of the Stria Terminalis 204
12.2.1.6 Hypothalamic Nuclei Receiving Vomeronasal Input 205
12.2.2 Maternal Behavior in Mammals 205
12.2.3 Functional Implications of Sexual Dimorphism of the VNS; Neurofunctional Hypothesis of Sex Differences 207
12.2.4 Vomeronasal Input in Human Mothering 209
12.3 Neurotransmitter Role in the Development of Sex Differences in VNS and Parental Behavior 211
12.3.1 GABA-a Effects on Locus Coeruleus and AOB 212
12.3.2 Neuroactive Steroids 213
12.3.3 Endocrine-Disrupting Chemicals 214

12.4 Environmental Perinatal Stress; An Epigenetic Factor Involved in the Multisignaling Process of Brain and Parental Behavior Sexual Differentiation.. 215
12.4.1 Environmental Prenatal Stress Alters Sexual Dimorphism of AOB, BAOT, Maternal Behavior, and HPA and HPG Axis......... 216
12.4.2 Maternal Care as a Preventive Factor in EPS Behavioral Effects; Cross-Fostering/In-Fostering Studies in Rats 218
12.5 Stress Effects on Human MB 220
12.6 Summary 222
In Memoriam: Carlos, "El Amigo"/Dr. Beyer, "El Maestro" 222
Acknowledgments... 223
References.. 223

12.1 INTRODUCTION

In the present chapter, we review our research experience with "el maestro" Beyer during his stay in our laboratory that was funded by the prestigious award "Programa Iberdrola de Profesores Visitantes en Ciencia y Tecnología" (1996). We integrate our research on sexual dimorphism in the vomeronasal system (VNS) and the neuro-functional expression of sex differences in parental behavior (PB), with Prof. Beyer's contributions to the hormonal control of sexual differentiation of the brain.

In this chapter, we: (1) review sex differences in the VNS. We show that these differences between male and female start in the sensory organ and continue with the complete neural network involved in the modulation of PB. We note two patterns (males > females and females > males) that depend on the VNS structures; (2) review involvement of neurotransmitters in the sexual dimorphism of the brain and in PB. This represents the multisignaling mechanism of sexual differentiation of the brain and behavior, and (3) describe an exemplary multisignaling process affecting sexual difference in brain and PB. We demonstrate how the epigenetic factor environmental prenatal stress (EPS) alters not only the hormonal "milieu" of mother and fetuses, but also the neuromorphological and PB sex differences. We conclude with a review of our cross-fostering and in-fostering studies relevant to the understanding of the classic scientific–philosophical dilemma: nature versus nurture/biology versus environment/inheritance versus education.

12.2 THE VOMERONASAL SYSTEM; AN INTEGRATIVE MODEL FOR THE STUDY OF BRAIN AND BEHAVIORAL SEX DIFFERENCES

The olfactory system consists of olfactory receptor organs and the primary and secondary centers to which they project. Most vertebrates from anurans to humans are provided with two types of receptors for responding to olfactory cues: the olfactory epithelium and the vomeronasal organ (VNO) (Raisman and Field 1971; Scalia and Winans 1975). It is generally accepted that the olfactory epithelium responds to odors, whereas the VNO responds to pheromones.

The main olfactory system (MOS) and the VNS differ morphologically and functionally. They have different neural pathways that project to the hypothalamus, although both probably project to some of the same areas, for example, the posteromedial cortical nucleus of the amygdala (Licht and Meredith 1987; Guillamón and Segovia 1996).

The VNS is a series of neural structures that can be found along the phylogenetic scale in vertebrates (Allison 1953; Johnson et al. 1985). It is a chemosensory system, whose primary sensory cells are located in a peripheral organ, the VNO. The first synapse from the VNO to a central brain structure is the accessory olfactory bulb (AOB), which transmits information to other limbic brain structures implicated in the development and expression of reproductive, aggressive, and maternal behavior (MB) (Scalia and Winans 1975; Kevetter and Winans 1981; Lehman and Winans 1982; Halpern 1987).

12.2.1 Sexual Dimorphism in the VNS

12.2.1.1 Vomeronasal Organ

In the mid-1990s, two populations of receptor cells, based on differential expression of G-proteins, were found in the vomeronasal epithelium. They were subsequently found to project to different parts of the AOB, forming segregated pathways with possible different functions. The two populations of VNO neurons express different putative receptors, apparently activated by specific ligands that trigger distinct behavioral patterns (Halpern 2003). In previous years we demonstrated the existence of sexual dimorphism in the VNO (Segovia and Guillamón 1982). These sex differences were organized by the action of gonadal steroids during critical periods of brain development, which are coincident with the testosterone (T) surges in males on prenatal day 18 and postnatal day 21. The VNO of male rats contains more bipolar neurons and has a larger volume than the VNO of female rats. This pattern was reversed by gonadectomy of the males and androgenization of females on postnatal day 1. These initial findings led us to propose VNS as a possible sexually dimorphic neural network.

12.2.1.2 The Accessory Olfactory Bulb

The AOB is a VNS structure embedded in the dorsocaudal main olfactory bulb (MOB). Its cytoarchitecture consists of five concentric layers: (1) external plexiform, formed by fibers from the VNO, (2) glomerular, in which one of the two axons of the bipolar cells of the VNO synapses with the apical dendrites of the mitral cells of the AOB, (3) mitral, containing the mitral cell bodies, (4) inner plexiform, containing the axons of the mitral cells and the dendritic branches of the granule cells, and (5) granular, containing light and dark granules (Allison 1953). The AOB is involved in the control of sexual and PB in both male and female rats (Wysocki and Meredith 1987; Del Cerro 1998). It exerts an inhibitory control on PB in virgin female and male rats (Fleming and Rosenblatt 1974; Fleming et al. 1979) and on feminine sexual behavior in male rats (Schaeffer et al. 1986; Segovia et al. 2009).

The AOB shows the same pattern of sexual dimorphism as the VNO: its total volume, the volume of each of its layers (Segovia et al. 1984), the number of mitral cells, and the light and dark granules are greater in males than in females (Segovia et al. 1986; Valencia 1986). This sexual dimorphism is reversed by androgenization of females and gonadectomy of males on day 1 postpartum (Segovia et al. 1986; Valencia et al. 1986). Furthermore, we reported that aromatization of T to estradiol produces the masculinization (Pérez-Laso et al. 1997).

12.2.1.3 The Bed Nucleus of the Accessory Olfactory Tract

The bed nucleus of the accessory olfactory tract (BAOT) is ventrally located in the lateral olfactory tract (LOT), while its caudal portion is ventrolaterally located adjacent to the anterior medial amygdala (MeA) (Kevetter and Winans 1982; De Olmos et al. 1985). The BAOT maintains direct, reciprocal, projections with the AOB via the accessory olfactory tract, which is a ventral component of the LOT.

The BAOT is sexually dimorphic, males having a greater volume and number of neurons than females (Collado et al. 1990). The BAOT exerts inhibitory control over PB, as bilateral electrolytic lesions in males result in the display of MB similar to that of intact control females (Del Cerro et al. 1991; Izquierdo et al. 1992).

12.2.1.4 The Medial and Posteromedial Cortical Amygdaloid Nuclei

There is sexual dimorphism in the medial amygdaloid nucleus (Me), the male rats having a larger volume than females (Mizukami et al. 1983). This sexual dimorphism can be observed starting on postnatal day 21, and can be abolished by treating the females with estradiol on postnatal days 1 to 30 (Mizukami et al. 1983). The number of shaft synapses is higher in adult male rats than in females, and these sex differences can be reversed by injecting T propionate to females, by performing an orchidectomy of males on postnatal day 5. Our group found sex differences in the posteromedial cortical nucleus of the amygdala, males showing more neurons and a larger volume than females. These differences were reversed with a single injection of T propionate to females or estradiol benzoate to males (Vinader-Caerols et al. 2000).

12.2.1.5 The Posteromedial Division of the Bed Nucleus of the Stria Terminalis

The bed nucleus of the stria terminalis (BST) is a complex forebrain nucleus, in which four divisions can be distinguished along an anterior–posterior gradient: BST medial (BSTM), BST lateral (BSTL), BST ventral (BSTV), and BST intermediate (BSTI). The BSTM is subdivided into three regions: the BSTM posterior (BSTMP), the BSTM central (BSTMC), and the BSTM anterior (BSTMA) regions. The BSTMP receives VNS input, while the olfactory input from the MOS reaches the BSTMA and the BSTL (for review, see Segovia and Guillamón 1993).

Two different patterns of sexual dimorphism have been found in components of the BST: (1) males have more neurons and greater volume than females in the BSTMP; (2) females have more neurons and greater volume than males in the BSTMA (Del Abril et al. 1987). The effects of male orchidectomy and female androgenization on postnatal day 1 are also different in these two regions of the BST: sex differences are reversed in the BSTMP, while they are only suppressed in the BSTMA. That is, in the BSTMA, males showed an increase in volume, whereas female rats remained unchanged (Del Abril et al. 1987). No sex differences were found in the BSTL region.

12.2.1.6 Hypothalamic Nuclei Receiving Vomeronasal Input

The medial preoptic area (MPOA) is the final station of the above-mentioned neural circuit involved in the control of PB. It receives olfactory projections from the AOB via MeA and BST (Leonard and Scott 1971; De Olmos and Ingram 1972; Krettek and Price 1978; Kevetter and Winans 1981; Simerley and Swanson 1986).

Sex differences in the volume of neurons in the MPA-anterior hypothalamus continuum were first described by Dörner and Staudt (1968). Gorski and his collaborators described the sexually dimorphic nucleus of the MPA (sdn-MPA), which has greater volume in males than in females (Arnold and Gorski 1984) and receives direct projections from the BST (Hutton et al. 1998). The volume and number of neurons of the MPA and the sexual dimorphic nucleus of the MPA (sdn-MPA) are greater in males than in females (Madeira et al. 1999). The hypothalamic ventromedial nucleus (VMH), the ventral premamillary nucleus (PMV), the arcuate nucleus (ARCN), and the supraoptic neurons (SON) also receive vomeronasal input (Scalia and Winans 1975; Smithson et al. 1992; Larriva-Sahd et al. 1993). The number of neurons and volume of VMH are greater in males than in females (Matsumoto and Arai 1983). However, when analyzing the three subdivisions of the VMH, we found that there are significant differences in the number of neurons in its dorsomedial and ventrolateral subdivisions and in the volume of the ventrolateral subdivision, males showing higher values than females; this is reflected in the total number of neurons and total volume of the VMH (Segovia et al. 2009). The VMH ultrastructure is also sexually dimorphic: when examining its ventrolateral part (VMHL), a region in which estrogen receptors are abundant, more dendritic and shaft spines were found in female rats (Matsumoto and Arai 1986).

The VNS is considered the main neuroendocrine pathway involved in reproductive behavior, since Powers and Winans (1975) demonstrated that the VNO was necessary for the onset of sexual behavior in naïve hamsters (see for review, Del Cerro 1998; Keverne 2004). Subsequent studies clarified the function of the VNO as a pheromone detector for triggering sexual behavior (Wysocki 1981; Singer et al. 1987; Krieger et al. 1999; Halper and Martinez-Marcos 2003). However, sexual behavior of experienced males remained unaltered after VNO lesions. These findings suggest that in the absence of VNO, its function can be performed by the MOS (Meredith 1986) (Figure 12.1).

12.2.2 Maternal Behavior in Mammals

PB has been defined by Numan (1994) as any behavior of a member of a species that increases the likelihood that the immature individual will survive to maturity. MB can be viewed as a specific component of "parental behavior."

The onset of MB is concurrent with several physiological demands, for example, hormonal changes at the end of pregnancy; its maintenance is regulated by internal cues in the mother and external stimulation by the pups. Thus, maternal care is a motivated behavior pattern that involves the integration of these multiple factors (Rosenblatt 1967; Del Cerro 1998).

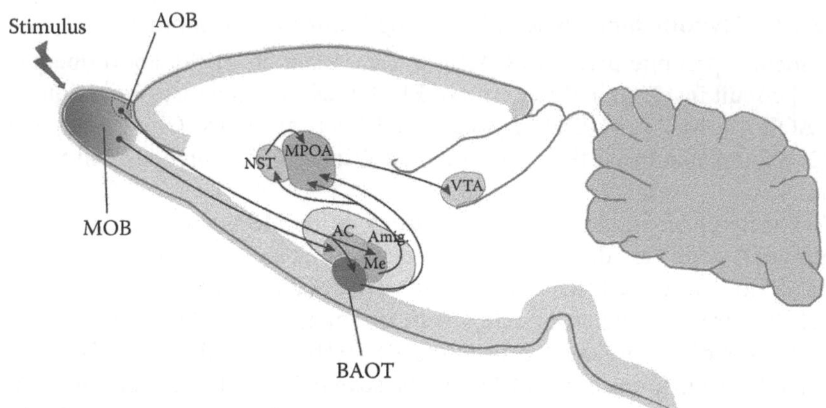

FIGURE 12.1 VNS pathway involved in the control of parental behavior.

Maternal care differs among species, depending on the maturity of the young at birth. In altricial species, exemplified by the rat, the mother builds a nest in which she gives birth to her litter—generally large, up to 12 or more pups—which have limited locomotor and sensory capacities (Del Cerro 1998). Under these conditions, the pups are gathered into the nest, nursed, and kept warm and clean. The study of MB has characterized a complex stereotypic sequence of behavior patterns that include: nest building, grooming of the pups, licking of their anogenital area, retrieving the litter to the nest, and nursing or crouching over them. Frequency, duration, and latency of all these behavioral patterns are recorded and analyzed (Figure 12.2).

Postpartum MB is not initiated by olfactory stimuli, based on: (1) destruction of the olfactory mucosa by zinc sulfate application (Benuck and Rowe 1975; Jirik-Babb et al. 1984; Kolunie and Stern 1995), (2) removal of the vomeronasal nerves (Fleming et al. 1992), or (3) bilateral bulbectomy (Fleming and Rosenblatt 1974). However, once maternal response has been established, olfaction plays a crucial role. Odors from the pups trigger specific behavior that is necessary for their survival (Brouette-Lahlou 1991), that is, licking of the anogenital area, which is essential for the stimulation of reflexive defecation and micturition (Moore 1981; Rosenblatt et al. 1979). On the other hand, the pups use chemosensory cues to orient themselves to the nest as they acquire greater motor control.

Pup odor is aversive to virgin female rats, whose typical initial response is avoidance of the pups. However, after several consecutive days of exposure to pups, they display MB (Rosenblatt 1967). This process of eliciting MB has been called "induction" (Rosenblatt 1967) or "sensitization" (Noirot 1972). Rosenblatt and Mayer (1995) proposed a motivational model in which MB emerges when avoidance of the pups decreases and motivation to approach them increases. Olfactory input plays an essential function in these two motivational systems (approach/avoidance). This procedure has been used with naïve male rats; they display some MB patterns, although they require more days of exposure than females, and they frequently attack the pups (infanticide) (Menella and Moltz 1988), a major sex difference in PB.

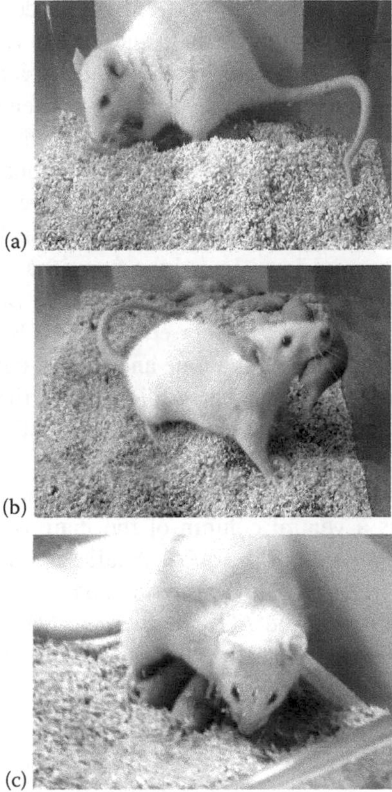

FIGURE 12.2 Maternal behavior patterns: (a) anogenital licking, (b) retrieval, and (c) crouching.

12.2.3 Functional Implications of Sexual Dimorphism of the VNS; Neurofunctional Hypothesis of Sex Differences

The total number of neurons and their relative proportions in different regions of the nervous system appear to be related to their function. The size of several VNS structures also correlates with expression of PB in rodents. In female rats, a significant increment is observed in dendritic bundling of supraoptic neurons (SON) after the induction of MB (Hatton et al. 1992). Gubernick and Alberts (1987, 1989) found in a biparental species of mouse (*Peromyscus californicus*) that the MPA/brain size ratio was significantly larger in naïve males than in virgin females. When females of this species become mothers they do not show changes in MPA size; however, fathers show a decrease in their MPA/brain size ratio compared with naïve males.

As demonstrated in the morphological studies described above by our group, the VNS is a sexually dimorphic neural network, and there are two patterns of sexual dimorphism: male > female, and female > male, in neuron number and volume. Numerous lesion studies have demonstrated that some VNS structures are implicated in mating (Meredith 1986; Keverne 2004) and MB. Thus, deafferentation of the VNO diminished active avoidance of pups by male and female virgin rats (Fleming et al. 1979), and ablation of VNO decreased infanticide behavior by male rats (Menella and Moltz 1988). Facilitation of induced MB in virgin female and naïve male rats was reported after lesion of various VNS structures (Numan et al. 1993; Del Cerro 1998). By contrast, lesion of MPA, which receives vomeronasal input, disrupted MB (Numan et al. 1977; Numan and Callahan 1980) and prevented the facilitation of MB produced by lesions of the amygdala (Fleming and Luebke 1981) (Figure 12.3).

Based on these findings, we investigated the functional significance of quantitative sex differences in VNS structures, specifically in the BAOT and AOB (Collado et al. 1990; Segovia et al. 1986; Valencia et al. 1986). Both structures present a dimorphic pattern: male > female in volume and neuron number. This dimorphism is dependent on the presence of gonadal steroids during the prenatal and postnatal period; for example, aromatization of estradiol induces AOB differentiation (Pérez-Laso et al. 1997).

The BAOT maintains reciprocal connections with the AOB via the accessory olfactory tract, which is a ventral column of the main olfactory tract, consisting of the axons of the mitral cells of the AOB (Scalia and Winans 1975; De Olmos et al. 1978). The BAOT projects to the posteromedial cortical amygdaloid nucleus, and input that it receives from the AOB reaches the MPO via the striaterminalis (De Olmos 1985). After bilateral electrolytic lesions of the BAOT, we found shorter

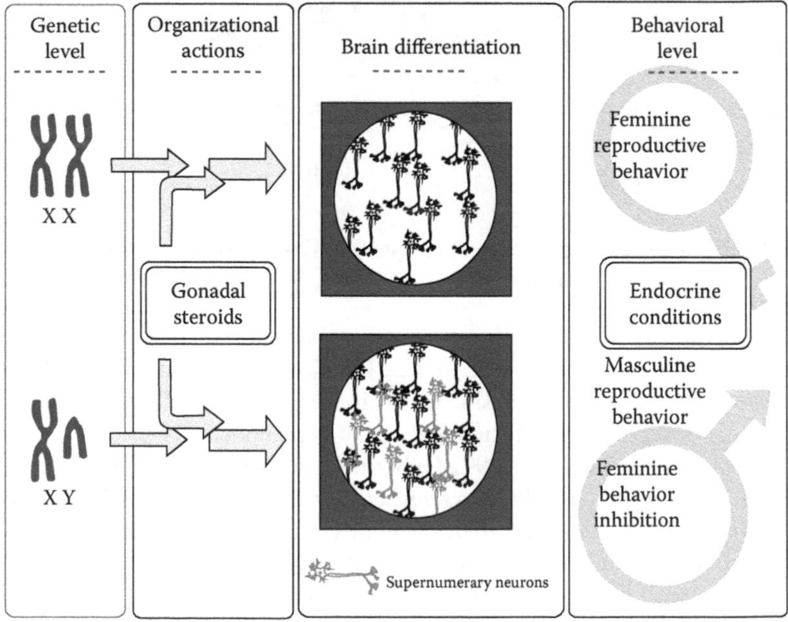

FIGURE 12.3 Neurofunctional hypothesis on the significance of quantitative sex differences in the VNS and their relation to sexual differentiated reproductive behavior. (Modified from Segovia, S. and A. Guillamón, *Behav. Brain Res.*, 18, 51–74, 1993.) Initial hypothesis.

latencies for retrieving pups and becoming maternal (sensitized animals) compared to control and sham surgery groups of virgin females (Del Cerro et al. 1991). In a similar experiment performed with male rats (Izquierdo et al. 1992), we found that significantly more animals in the sham and bilateral-lesion groups became maternal compared with controls; the bilateral-lesion group did not differ from the bilateral-lesion female group of our prior experiment.

Compared to females, naïve males require longer exposure to pups to become parental (Mayer et al. 1979) and perform more attacks and cannibalizing behavior (Menella and Moltz 1988). We suggest that these functional differences can be attributed to the greater number of neurons in the BAOT of males than females (Collado et al. 1990). Thus, our observation of BAOT suggests that bilateral lesions eliminated a greater inhibition in males than in females, which supports the hypothesis that this sexual dimorphism (male > female) has two functional consequences: (1) it allows for a greater tonic inhibition of maternal care and lordosis behavior in males than in virgin females, and (2) it enables males to exhibit masculine sexual behavior. The functional significance of this "extra number of neurons" in the males is depicted in Figure 12.3.

For the first time in the literature, we demonstrated the inhibitory role of the BAOT in PB; prior to our report, there was no known functional role for the BAOT.

12.2.4 Vomeronasal Input in Human Mothering

Anthropological studies and cross-species comparative data suggest that close contact between the mother and her newborn infant is the biological norm that has evolved in our species (Konner and Worthman 1980). There are common behavioral patterns of maternal care across mammalian species (Rosenblatt 1994). In addition to those described above in the rat, other mammalian behavioral elements are: perceptual exploration of the offspring, retrieval and reciprocal calls, grooming, kissing or licking, nursing and lactation and feeding, prolonged physical contact or sleeping together, and aggressive response to perceived threats to their offspring (Swain et al. 2007). Most of these behavior patterns are based on skin-to-skin contact, which promotes exchange of olfactory cues between mother and infant.

There is substantial evidence in humans for the motivating effect of infant crying (Boukydis 1985; LaGasse et al. 2005), and a wide-ranging role of vision in mother–infant interaction (Messer and Vietz 1984; Mäntymaa et al. 2006). Infants perceive and respond to biologically relevant odors within the first minutes after birth. Lack of responsiveness to olfactory stimuli is a valuable diagnostic marker. Mizuno et al. (2004) have shown that a brief period of mother–infant skin-to-skin contact immediately after birth, accompanied by suckling at the breast, enhances neonates' recognition of maternal milk odor and has enduring effects on breastfeeding. The benefits to the mother of early skin-to-skin contact may be less obvious than the benefits to her infant. However, there are findings that illustrate the importance of the combination of multiple sensory systems—tactile, olfactory, and visual. Infants that were placed in a prone position on their mother's abdomen within the first hour after birth displayed intermittent extension and flexion of the lower legs (Varendi et al. 1996). Several mothers reported that they experienced uterine contractions and expulsion of blood when stimulated by these stepping movements. It has been suggested that manipulation of the mother's nipples by the newborn infant may have similar effects (Matthiesen et al. 2001). Such neonatal motor activity therefore appears to serve the same function as common medical interventions (uterine massage or the administration of pharmaceuticals) to expel the placenta and prevent the accumulation of blood in the uterus. However, at present, the biological function of the human sense of smell is not well understood. Although our olfactory sense may be less acute than that of other mammals, it is becoming evident that olfactory cues exert a subtle but meaningful influence on human behavior and physiology. Does vomeronasal sensory input play a role in human mothering? Although the VNS has been studied and described in a variety of nonhuman vertebrates, its existence in adult humans has been established in the last decade of the past century (Morán et al. 1995). It was previously assumed that the VNO was absent in adult humans except in rare cases, in which it was considered vestigial and without function (Meredith 1983). However, there are indications that the human VNO is involved in pheromone detection, on the basis of clinical studies in adults (García-Velasco and Mondragón 1991; Morán et al. 1995) and more consistently, electrophysiological (Monti-Bloch and Grosser 1991), electromicroscopic (Morán et al. 1991), and immunohistochemical studies (Takagi 1989; Morán et al. 1992). The term "vomeropherins" has been substituted for "pheromone" to refer to substances that stimulate the VNO in humans (Stensaas et al. 1991; Monti-Bloch et al. 1994). The reported frequency of occurrence of the VNO in adult humans ranges from 39% of 100 patients (Johnson et al. 1985) to 100% of 200 patients (Morán et al. 1991). This apparent discrepancy may be accounted for by differences in methodology. In one study, visual inspection was used and in the second, microscopic observations of the characteristic vomeronasal pits were reported. There is evidence that in adult humans the VNO: (1) does exist, (2) has a unique electron-microscopic ultrastructure, (3) contains cells with several immunohistochemical markers for neurons, and (4) displays distinct chemoreceptive responses to specific chemical stimulants. However, the presence of AOB has not been demonstrated clearly in adult humans. That is, there is no clear experimental evidence of a neuroanatomical connection between the human VNO and the brain. Nevertheless, specific autonomic changes (galvanic skin response and body temperature) occur during VNO stimulation by vomeropherins, suggesting the possibility that the VNO is, indeed, functionally connected to the brain. The short latencies of these physiological responses indicate that the transmission of information from the VNO to the brain is mediated by a neuronal, rather than a blood-borne mechanism. Some studies suggest that, although the human AOB may disappear as a distinct anatomical structure in the second trimester of gestation, the cells of the AOB may undergo a physical displacement, rather than neuronal degeneration, and persist into adult life (Morán et al. 1995).

Monti-Bloch et al. (1994) have reported sex differences in the specific responses of VNO and VNO-mediated effects of different groups of vomeropherins. However, no sexual dimorphism has been observed for olfactory responses. These reports coincide with the crucial role that the VNS plays in the detection of pheromones in animals (Wysocki 1979). The role of VNS in the onset of MB in rats has been reviewed above. By contrast with rodents, human mothers are highly responsive to visual stimuli from their infants. However, the influence of olfactory cues is less obvious. At present there is no direct evidence for vomeronasal involvement in the early attachment between mother and infant. The largest body of evidence relevant to MB in humans is related to the correlation between psychological and physiological influences in the expression of maternal care in the mothers. Psychobiological processes involved in the control of normal or deviant mothering are complex, as background and situational factors seem to mask the role of biological factors in early mothering (Harkness and Super 1995). Although the role of the VNS in human mothering has not been studied, there is evidence of the sensory, hormonal, and neural factors that regulate PB in human beings (Benoist and Porter 1991; Corter and Fleming 1995; Lonstein et al. 2015). Some studies have provided evidence of possible prenatal olfactory learning related to the attractiveness of amniotic fluid odor, which might involve vomeronasal processing (Varendi et al. 1996; Mizuno and Ueda 2004).

There is also a popular view that during the interval from pregnancy to the first days postpartum, women undergo dramatic changes in emotional lability, and specifically in odor sensitivity. New mothers can recognize their own infants on the basis of olfactory cues; for instance, they can discriminate their own infant's soiled t-shirt from those of same-age infants, requiring very little interaction with their baby to do so (Corter and Fleming 1995). Odor recognition capability may be related to early postpartum experience and circulating hormones. For example, mothers who correctly identified their own infant's odors had experienced earlier and longer contact with their infants after birth, spent more time in close contact with them during interactions, and reported more positive maternal feelings and attitudes, in contrast to mothers who were unable to identify their infant's odor (Fleming et al. 1997).

What is occurring in mothers' brain during auditory (mainly cry), olfactory, and visual stimuli coming from their own or alien infants? Evidence from neuroimaging studies that focused on reward and affect shows considerable overlap with brain regions that are activated in mothers by infant cues. These include the prefrontal cortex, anterior cingulate cortex (ACC), orbitofrontal cortex (OFC), striatum (nucleus accumbens), amygdala, and hypothalamus. Johansen-Berg and colleagues (2008) found a considerable overlap between mothering neural circuits and those involved in affect. They examined the connectivity of the ventral ACC, the anatomical target of deep brain stimulation that has alleviated depression in some patients (Mayberg et al. 2005). The ACC has been implicated in the regulation of mood, including major depression, as well as anxiety (Drevets et al. 2008). Further research is warranted on the relationship between mothering, parenting, and the newborn stimuli that trigger the brain responses. Better understanding of this neurofunctional process might prevent or avoid potentially detrimental alterations that can occur during the very first moments in the critical interaction among mother, father, and infant.

12.3 NEUROTRANSMITTER ROLE IN THE DEVELOPMENT OF SEX DIFFERENCES IN VNS AND PARENTAL BEHAVIOR

Neurotransmitters play an important role in the development of the nervous system, in synaptic events such as changes in excitability of neuronal membranes and in building sex differences in brain structure (Dörner 1979; Lipton 1989; Mattson 1988; Beyer and Feder 1987; Arrati et al. 2006; Tabatadze et al. 2015). De Vries (1990) showed Vasopressin (AVP) immunoreactivity in cells and projections of BST and Me, depending on gonadal steroid circulation in the adult period. The AVP projections from the BST showed fiber staining that was more dense in male than in female rats. The male and female rats were gonadectomized and treated with T before sacrifice to ascertain whether the differences observed were due to the different hormone levels in adulthood. Higher density was seen in the AVP-IR fiber networks expressed in the Me and ventral hippocampus of males than females. However, no differences were found in the anteroventral portion of the periventricular nucleus and the dorsomedial nucleus of the hypothalamus that receive AVP innervation from the suprachiasmatic nucleus. These findings indicated that the sex difference in the steroid-sensitive AVP pathways depends on other factors besides circulating hormone levels in adulthood. Based on this and other studies, in addition to our own research tradition, we decided to investigate some of these "other factors," focusing on the "organizational" period of sexual differentiation of the brain. Thus began a collaborative research adventure between our research group and Professor Beyer.

12.3.1 GABA-a Effects on Locus Coeruleus and AOB

Special attention was focused on the participation of the inhibitory neurotransmitter, GABA, in the developmental process of brain sex differences. We knew that the GABA-a agonist, muscimol, significantly decreased the volume of the sexually dimorphic nucleus of the preoptic area (SDN-POA) in males, but had no comparable effect in females (Bach et al. 1992). We had studied GABA in some of our selected structures related to the neural control of PB, such as AOB (Segovia et al. 1991; Pérez-Laso et al. 1994) or to female > male pattern in locus coeruleus (LC) (Segovia et al. 1991; Rodríguez Zafra et al. 1993), using the GABA-a agonist, diazepam (DZ) and the antagonist, picrotoxin (PTX). In relation to the sexually dimorphic LC, the administration of DZ affected male and female rats differentially, depending on the timing of the treatment (prenatal–postnatal and pre+postnatal). Males were affected just by the prenatal DZ administration, increasing volume and number of neurons in LC, approaching the number in females. Females were affected by pre+postnatal DZ exposure, showing reduced volume and number of neurons in the LC. In relation to AOB, applying the same treatment in dosage and period as in the LC study, we found that all treatments (prenatal, postnatal and pre+postnatal DZ) affected AOB mitral and light and dark granule cells, decreasing their number and volume, approaching the number similar to those of females. By contrast, female AOB cell populations were not affected by perinatal exposure to DZ.

These studies suggested that the GABA-a receptor is implicated in sexual differentiation of the brain. Questions arose after these experiments, related to the peripheral or central site of action of DZ. Since DZ could affect steroidogenesis (Krueger and Papadopoulos 1992), our findings from DZ treatment did not clarify whether our results were indirect, due to changes in the gonadal secretion of sex hormones during critical periods, or whether the GABA-a receptor is involved in sexual differentiation of the brain independent of hormones. In order to elucidate this question, we used the antagonist, PTX, which selectively blocks the CL^- channel of the GABA-a receptor in the CNS, and evaluated the effects on the AOB and MB in male and female rats

(Segovia et al. 1996; Del Cerro 1998). The administration of PTX to the females during the postnatal period induced a masculine pattern in the number of AOB mitral cells, and disrupted the expression of MB. We found an opposite effect on the males: postnatal DZ exposure induced a feminine pattern in the number of AOB mitral cells and facilitated the induction of MB. At the conclusion of the maternal induction test, we found no changes in the plasma levels of estradiol, progesterone, or testosterone. Thus, sex differences in AOB and the induction of MB were reversed by the treatment, while leaving the gonadal function unaltered in both sexes. We concluded that the GABA neurotransmitter, through GABA-a receptors, is involved in the development of sex differences in the brain and reproductive behavior. In both cases, we observed that perinatal DZ treatment induced a feminine pattern in males and a masculine pattern in females in both number of neurons and volume of the brain structures.

12.3.2 Neuroactive Steroids

The sex differences in which males have more neurons than females is a consequence of a preventive mechanism, against natural neuronal death, by T and its aromatization to estradiol (Nordeen et al. 1985). On the other hand, the sexual differentiation of the brain and reproductive behavior is thought to be mainly estrogen-receptor dependent. Estrogen acts on its receptors to transcribe genes capable of inducing brain sex differences (McCarthy et al. 1993). However, based on the above studies and those of the last three decades, genomic estrogen action is not the only process that induces sexual differentiation of the brain (Del Cerro et al. 2010; Pérez-Laso et al. 2008, 2013; Tabatadze et al. 2015) (Figure 12.4).

Progesterone administered neonatally protects against steroid-induced sterility in neonatal male and female rats (Kincl and Maqueo 1965) and the incidence of sterility in females induced by neonatal androgenization was reduced after simultaneous administration of reserpine, chlorpromazine, progesterone, deoxycorticosterone, or pregnanodione, while almost complete protection was seen after barbiturate treatment (Arai and Gorski 1968). In a pioneer study (Erickson et al. 1967), it was demonstrated that progesterone was able to block androgen-induced behavior in male ring doves, and in subsequent studies, Ring-A-reduced anesthetic pregnanes counteracted androgen-induced defeminization in neonatal female rats (González-Mariscal et al. 1982). Metabolically active steroids, such as progesterone, pregnenolone, dehydroepiandrosterone, and ring-A-reduced pregnanes are synthesized in the brain from cholesterol, on which basis they were termed "neurosteroids" (Beaulieu and Robel 1990). Thus, brain cells are targets for endogenous steroids or neurosteroids, ring-A-reduced pregnanes, and other steroids of glandular origin. It is important to point out that ring-A-reduced metabolites of progesterone and deoxycorticosterone are potent modulators of the GABA-a receptor complex, interacting with these receptors like barbiturates (Majewska et al. 1992). Our above-described studies have demonstrated that agonists and antagonists of the GABA-a receptor complex can affect brain and behavioral sex differences. Other neurosteroids are able to modulate the GABA-a receptor activity in fetal forebrain synaptoneurosomes (Kellog et al. 1998). These findings provide evidence that ring-A-reduced metabolites acting on the GABA-a

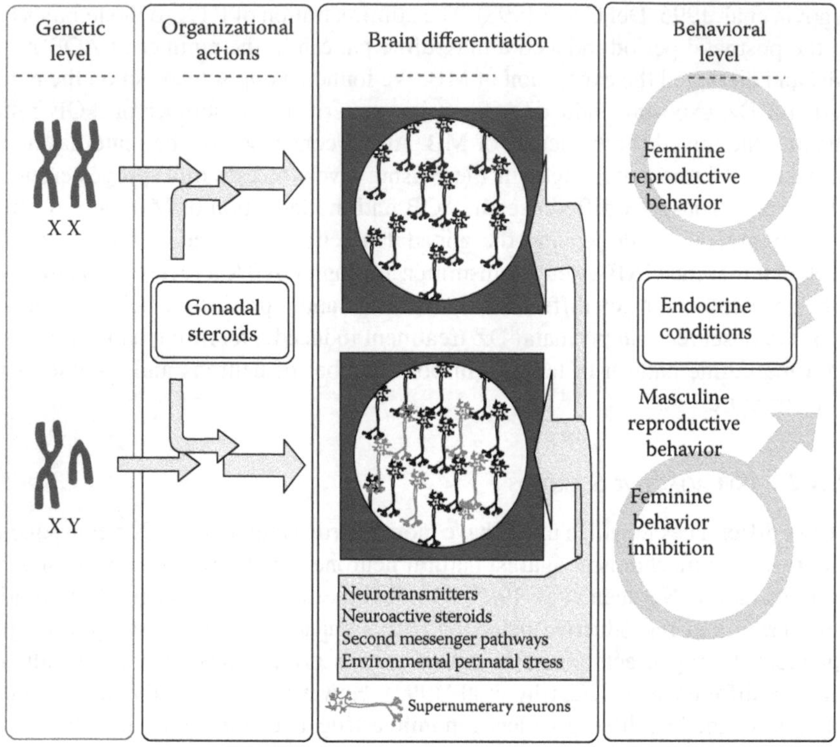

FIGURE 12.4 Multisignaling: neurofunctional hypothesis of the significance of quantitative sex differences in the VNS and their relation to sexually differentiated reproductive behavior patterns, including epigenetic factors modulating the expression of maternal behavior.

receptor complex could be involved in the sexual differentiation of brain and behavior (Beyer et al. 1991; Segovia and Guillamón 1996; Segovia et al. 1996). We conclude that in addition to their genomic action, steroids can influence brain and behavior via their nongenomic activity, altering the membrane permeability for ions and inducing a cascade of intracellular events in a cross-talk mechanism related to transcription.

12.3.3 Endocrine-Disrupting Chemicals

Strong conservation exists in brain development among species from rodents to humans (Rice and Barone 2000; Howdehell 2002). The development of sex differences in the brain has been conceptualized as a multisignaling process (Segovia et al. 1999) in which the action of genomic agents can be altered and/or modulated by neurotransmitters, neurotrophic factors, neuroactive steroids, and environmental agents.

The expression of PB in mature mammals is closely related to the correct development of their VNS. Various neural and/or endocrine factors can dramatically influence MB, for example, exposure to endocrine disrupting chemicals

(EDCs), among which are bisphenol A (BPA) and phthalates. BPA suppresses the steroid-acute-regulatory-protein (StAR) mRNA in male and female gonads of rodents and fishes (D'Cruz et al. 2012; Liu et al. 2012; Savchuk et al. 2013; Horman et al. 2012). StAR is an essential mitochondrial protein for transporting cholesterol into the cell and is thus considered the rate-limiting step for steroid hormone production. BPA and phthalates exposure also disrupts hypothalamic–adrenocortical axis function. Prenatally treated female rats showed increased basal levels of corticosterone (Panagoitidou et al. 2014). In several studies with mice, prenatal BPA had deleterious effects in the expression of MB of their offspring when mature, reducing the time spent nursing the pups and crouching (Panza et al. 2005). Acute exposure of female Sprague-Dawley rats to BPA (40 µ/kg) from mating until weaning of their pups, showed a trend to less anogenital licking and less time crouching over the pups (Della Seta et al. 2005). In Wistar rats, females treated with 5 µ/kg of BPA from the first day of gestation to lactation, also reduced time spent nursing the pups (Boudalia et al. 2014).

12.4 ENVIRONMENTAL PERINATAL STRESS; AN EPIGENETIC FACTOR INVOLVED IN THE MULTISIGNALING PROCESS OF BRAIN AND PARENTAL BEHAVIOR SEXUAL DIFFERENTIATION

Stress related to natural or provoked disasters, or chronic stress during gestation, can induce marked alterations in neural development, neuroendocrine axes, and behavior (Pérez-Laso et al. 2008, 2013; Weinstock 2008, 2011; Del Cerro et al. 2010; Del Giudice 2011; Hepfer et al. 2012). In previous research, we demonstrated that environmental prenatal stress (EPS) increased the number of mitral cells in the AOB of female rats to a level that did not differ from the male control group, estimated in adulthood. EPS also reduced estradiol levels and increased CpdB, ACTH and P in virgin female rats (Pérez-Laso et al. 2008; Del Cerro et al. 2010). Induced MB tested at 90 days of age was also altered, exhibiting a male-like pattern in the prenatally stressed females. These findings support our concept of the inhibition of MB by vomeronasal structures.

There are two morphological patterns of sexual dimorphism in the structures of the VNS in rats: males > females and females > males. These patterns can be reversed by neonatal gonadecomy of males, or administration of T propionate to females. Moreover, a single injection of estradiol benzoate after gonadectomy on postnatal day 1 to male rats counteracts the effects of gonadectomy in VNS structures (AOB, BAOT and VMH), which demonstrates that the male-pattern differentiation in these nuclei is mediated by the aromatization of T to estradiol.

In women, the postpartum period is a time of increased vulnerability to mood disorders, with about 20%–30% of women experiencing spontaneous mood changes within the first six weeks postpartum (O'Hara and Swain 1996; Llewellyn et al. 1997; Mastorakos and Ilas 2001).

Numerous animal studies have demonstrated that stress during pregnancy can cause biological dysfunctions in both dams and offspring and also may result in a high rate of fetal resorptions and abortions (De Catanzaro 1988) in rats and mice. Prenatal stress also reduces birth weight of offspring (Pardon et al. 2000) and reduces fertility of female offspring when adult (Herrenkohl 1979). There is an abundant literature alerting to the persistent deleterious effects of stress experienced during early life on adult neuroendocrine parameters and behavior (Pérez-Laso et al. 2008, 2013; Weinstock 2008, 2011; Del Cerro et al. 2010; Del Giudice 2011; Herpfer et al. 2012). Early stress alters stress coping in both virgin female and male offspring when mature (Weinstock et al. 1992; Levine 2001; Bosch et al. 2007). Moreover, gestational stress can alter maternal care that the dam provides to her offspring (Fanelly et al. 1995; Champagne and Meany 2006).

12.4.1 Environmental Prenatal Stress Alters Sexual Dimorphism of AOB, BAOT, Maternal Behavior, and HPA and HPG Axis

We have demonstrated that: (1) VNS is a sexually dimorphic multisynaptic network, (2) VNS structures are directly involved in the neural control of PB, and (3) MB is affected by environmental stress during the gestational period. Based on this framework, we evaluated a new possible factor in the multisignaling process of sexual differentiation of brain, hormones, and behavior: *peri*natal environmental stress.

It is still controversial as to whether behavioral and neurofunctional anomalies observed in the stressed offspring are exclusively due to: (1) prenatal stress, (2) early postnatal environment (i.e., type of maternal care), or (3) their interaction.

In order to address this experimental question, we have always applied the same EPS procedure: three daily environmental stress sessions of 45 minutes (immobility, heat, and light) (Figure 12.5) to pregnant Wistar rats during the last week of gestation, while their corresponding control group of mothers are left undisturbed throughout their whole pregnancy (for methodological details, Pérez-Laso et al. 2008, 2011, 2013; Del Cerro et al. 2010, 2015). Offspring are reared by their own mothers, weaned at 21 days and when they reach 90 days, we segregate the different groups: control females, control males, and prenatally stressed females and males.

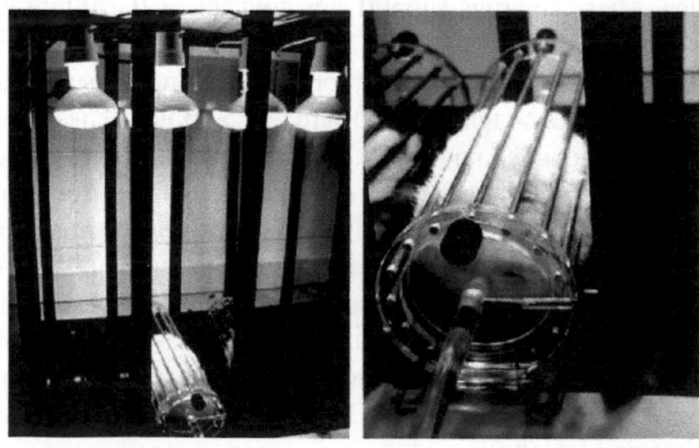

FIGURE 12.5 Stressor device (restraint/heat/light) designed by Del Cerro, M. C. R., used in our EPS studies.

For studying the combination of EPS and type of MB received, we introduced a cross-fostering design, switching entire litters, which generates prenatally stressed females and males reared by nonstressed mothers and nonstressed pups reared by stressed dams. In order to ensure that the exchange of mother "per se" is not affecting the results we introduce the in-fostering condition, we exchange the entire litter of nonstressed control mother with another nonstressed control dam and the same for the stressed mothers. In two of our studies (Pérez-Laso et al. 2008; Del Cerro et al. 2015) we tested the effects of EPS on AOB and BAOT sex differences, hormone levels and induced MB in prenatally stressed female offspring in comparison with control female and male offspring.

Our results demonstrated that EPS altered MB, in that the prenatally stressed females built disorganized nests, spent less time in contact with the pups and did not display full MB. EPS also increased the number of mitral cells in the AOB and the total number of BAOT neurons in female rats to a level that did not differ from the male control group, estimated in adulthood (Figure 12.6). EPS also reduced estradiol and increased CpdB, ACTH, and P levels in virgin female rats (Pérez-Laso et al.

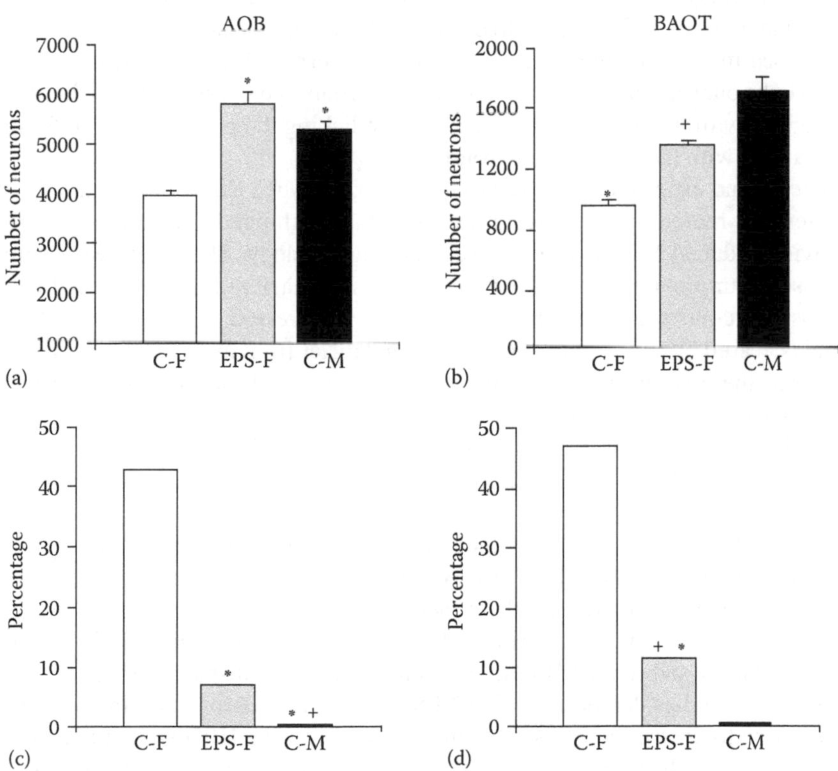

FIGURE 12.6 Effects of EPS on induced maternal behavior and sex differences in the number of neurons in two VNS structures implicated in the control of maternal behavior. (a) number of AOB mitral cells, (b) number of neurons in the BAOT, (c) and (d) percentage of animals expressing sensitization in the AOB and BAOT, respectively.

2008; Del Cerro et al. 2015). In sum, EPS induced a male-like pattern in behavior and the morphology of the AOB and BAOT in the prenatally stressed females, and induced extended duration of dysfunction of the HPA and HPG axes.

12.4.2 Maternal Care as a Preventive Factor in EPS Behavioral Effects; Cross-Fostering/In-Fostering Studies in Rats

In order to show that maternal care can be a protective factor on the possible EPS effects on the litter, we used cross-fostering strategy (Del Cerro et al. 2010; Pérez-Laso et al. 2013). We intended to differentiate the effects due to EPS from those of possible disorganized maternal care received by the pups of gestationally stressed mothers. Using the same EPS procedure, we included in our design the cross-fostering condition. We divided the pregnant rats into four groups: two gestationally stressed and two nonstressed mothers. After parturition, we proceeded to cross the litters as before. That is, we exchanged the litter from a stressed mother to a control mother and vice versa. We also maintained two groups of noncrossed litters: stressed mothers that reared their own pups and nonstressed mothers that reared their own pups. This study consisted of two experiments. First, we tested the groups of lactating mothers for MB (natural MB test) and our results indicated that the group of stressed mothers exhibited a disorganized pattern of MB. That is, they did not retrieve the pups as quickly as the nonstressed group, spent less time showing physical contact with the pups and the time spent in licking the pups was also inferior in comparison with the nonstressed group.

The second experiment referred to the offspring, when adult. The results showed that females reared by their own stressed mothers exhibited a male-like pattern in behavior (induced MB), hormone levels, and morphology. These data were similar to those of our previous study (2008). By contrast, behavior of the two cross-fostered groups: (1) control females that were reared by stressed mothers and, (2) EPS females, reared by nonstressed mothers, did not differ from the control female group. However, there were differences in neuroendocrine and morphological variables: EPS females reared by nonstressed mothers showed higher number of AOB mitral cells, elevated plasma ACTH levels, and reduced estradiol levels in comparison with control females. These measures did not differ from EPS females reared by their own stressed mothers.

In this study, we also included another variable. The c-*fos* expression during MB performance in the AOB and MPOA. We found that in the absence of prenatal stress (the control females reared by their own nonstressed mother or reared by foster stressed mothers), female offspring, when adult, showed Fos-IR cells in the AOB mitral cell layer and in the MPOA, only if and when they displayed MB.

All these studies demonstrate that EPS affects: (1) maternal care in the dams in comparison to control nonstressed dams, (2) the offspring HPG system at maturity, (3) sexual dimorphism of vomeronasal structures, and (4) induced MB exhibited by female offspring when adult. In addition, we found a striking effect using the cross-fostering procedure: "appropriate" maternal care during lactation counteracted the behavioral effects of prenatal stress on induced MB in female offspring when they

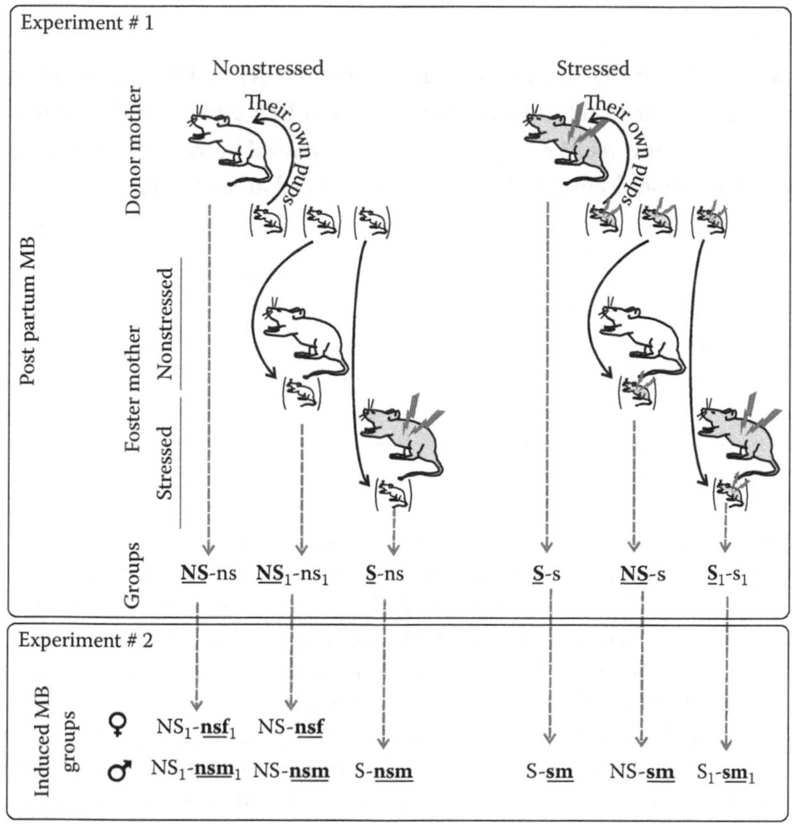

FIGURE 12.7 Experimental design used in cross-fostering and in-fostering procedures for the study of EPS effects on parental behavior.

reached maturity. However, the maternal care did not counteract neuroanatomical or hormonal alterations in those females.

Our next research question addressed was whether EPS and maternal care have a comparable effect on male offspring. Thus, we designed a complex experimental design including offspring male rats and cross-fostering and in-fostering groups (Figure 12.7). We included "in-fostering" groups in order to discriminate the effects due to EPS from those of maternal care.

The gestational stress affected maternal care exhibited by stressed mothers, but we found no effect of the in-fostering procedure on mothers. In other words, the type of MB exhibited was dependent on the mother's condition, stressed or nonstressed, regardless of whether she reared her own or foster litter. When pups reached 90 days of age, they were tested for induction of MB in the second part of the study. Our findings indicated no effect of the in-fostering procedure in females or in the male offspring when adult. Prenatally stressed males reared by their own stressed mother or by a foster stressed mother did not differ in any of the studied parameters

TABLE 12.1
Overall Results of the Cross-Fostering Procedure After EPS. Summary of the Findings of the Several Studies of the Effects of EPS and Cross-Fostering on Maternal Behavior, AOB Morphology, and Gonadal Steroid Plasma Levels, Shown by Male and Female Offspring, Measured at Maturity

	NS-ns		S-s		Ns-s		S-ns	
	♂	♀	♂	♀	♂	♀	♂	♀
MB	---	+++	---	---	+++	+++	+++	+++
AOB	+++	---	++++	+++	---	---	---	---
E_2	---	+++	---	++	+	++	++	++
T	+++	---	++++	---	++	---	+	---

(morphological, hormonal, or behavioral). The same was true for nonstressed pups reared by their own nonstressed mothers or foster nonstressed mothers.

However, cross-fostering had differential effects in males and in females. As in our previous study (Del Cerro et al. 2010), adequate maternal care counteracted the behavioral effects of EPS, but could not prevent neuroendocrine dysfunction, or AOB and BAOT male-like pattern in EPS females.

The EPS males reared by nonstressed mothers presented a greater number of AOB mitral cells and elevated T plasma levels, compared to control males (nonstressed males reared by nonstressed mothers), but their scores in the induction of MB were similar to those obtained by the control female groups. Moreover, the most striking result was that nonstressed males reared by stressed mothers exhibited a feminine-like pattern in the induction of MB, in the number of mitral cells and a significant decrease of plasma T level, when compared with control males (Table 12.1).

We conclude that adequate maternal care, soon after birth, can counteract detrimental behavioral effects of EPS in virgin female and naïve male rats. However, we emphasize that appropriate maternal care has different effects on neuroendocrine and morphological parameters in females versus males.

12.5 STRESS EFFECTS ON HUMAN MB

The prenatal period is characterized by sequential and organized changes during which organs and systems are developing and are susceptible to both organizing and disorganizing influences coming from the mother and the external milieu. These influences on the fetus have been described as "programming," that is, a process by which a stimulus or insult during a vulnerable developmental period has long-lasting effects (Harris and Seckl 2011). As we have seen in work by others, and in our own research with male and female rats, prenatal stress alters sexual dimorphism in the brain, HPA response and PB, when the rats stressed while in utero reach maturity (Weinstock 2008; Pérez-Laso et al. 2013; Del Cerro et al. 2015). The influence on the developing fetal HPA axis, of prenatal exposure to maternal stress, has been proposed as one of the mechanisms that underlies fetal "programming" of health at maturity. Similarly in humans, chronic adverse life situations experienced by the mother during gestation, can alter not only the typical maternal care and mother–infant bond, but also the fetal environment, which can induce deleterious effects on the developmental process, and mental and physical health of the infant (Wadhwa et al. 2001; Koenig et al. 2002; Maccari et al. 2003; Feinberg 2007; Grant et al. 2009). The stress effects on children are present as early as the immediate postnatal period. Some examples are: a higher incidence of preterm birth and lower birth weight (<37th week/<2.5 kg), which in turn is associated with subsequent developmental impairments and motor disabilities (Knoches and Doyle 1993; Ruiz et al. 2002), a slower rate of development of speech, walking, and impairment of other developmental traits (Grant et al. 2009), and HPA axis dysregulation and depressive symptoms during adolescence (O'Connor et al. 2005; Tronick and Reck 2009). Unfortunately, in our society there are too many opportunities for studying the outcomes of chronic stress experience during pregnancy on the mother and the children. Typically, these studies offer correlational data obtained from studies on tragic situations, such as terrorist attacks or maintained domestic violence experienced by the mother, which show deleterious effects on both mother and child. In such cases, maternal care is compromised, associated with postpartum depression and anxiety, and negative behavioral reactivity in the children (Davis et al. 2004; Yehuda et al. 2005). Research using structural MRI (Buss et al. 2009) has shown that high levels of prenatal maternal anxiety at 19 weeks gestation, but not at 25 or 31 weeks, was associated with significant gray matter volume reduction in the prefrontal cortex, the premotor cortex, the medial temporal lobe, the lateral temporal cortex, the postcentral gyrus, and the cerebellum, extending to the middle occipital gyrus and the fusiform gyrus in 6- to 9-year-old children. These results provide evidence of a temporal pattern of pregnancy-related anxiety related to specific changes in morphology of brain regions involved in cognitive performance (in the "medial temporal lobe memory system"). These and related findings may account for the increased prevalence of psychopathology in the offspring (Khashan et al. 2008). Furthermore, the results of reduced gray matter density in the premotor cortex and the cerebellum may sustain the anatomical basis for other previous well documented observations of delayed and impaired motor developmental skills associated with prenatal stress (Huizink et al. 2003).

The above findings by our group and others raises the question as to whether the behavior of the mother toward her offspring can "program" stable changes in gene expression (Meany 2010), which could induce changes in the neuroendocrine and behavioral responses to stress of the offspring later in life. But it is also important to note, based on our findings, the *preventive* effects of appropriate care giving early after birth, independent of chronic stress or anxiety during pregnancy (Del Cerro et al. 2010, 2015; Pérez-Laso et al. 2013). This dramatic effect holds promise for developing a human maternal care program to counteract the possible long-lasting effects of mothers' gestational stress on their children.

12.6 SUMMARY

There are two morphological patterns of sexual dimorphism in the structures of the VNS of rats: males greater than females and females greater than males. These patterns can be reversed by neonatal gonadectomy of the male and T propionate injection to the female. Moreover, a single injection of estradiol benzoate after gonadectomy on postnatal day 1 to male rats counteracts the effects of gonadectomy in VNS structures such as the AOB, BAOT, VMH, evidence that the differentiation to male in these nuclei is mediated by the aromatization of T to estradiol.

EPS has behavioral, neuroendocrine and morphological long-duration effects. EPS alters sexual differentiation of VNS structures such as the AOB and BAOT. EPS has differential long-duration effects on male and female offspring: it induces a male-like pattern in behavior, when tested in the MB induction test, and eliminates the c-*fos* expression concomitant with induced MB in virgin female rats. EPS induces neuroendocrine dysfunction of the HPA and HG axes: increased ACTH and reduced estradiol levels, morphology of the AOB mitral cell layer, and total number of neurons in BAOT. However, it increases T levels and the number of mitral cells, but does not affect behavior in male rats. There is an evident interaction between EPS and early life developmental events. We have demonstrated that adequate maternal care can counteract the effects of EPS in female offspring, but does not prevent the neuroendocrine dysfunction and morphological alterations in the AOB. Human studies on vomeronasal involvement in MB and EPS are presented showing interestingly similar results to those seen in animal models. Gestational stress may be an epigenetic agent to be considered by public health policy as a critical key to social welfare. We emphasize that improvement of our world condition can start in our perinatal life.

IN MEMORIAM: CARLOS, "EL AMIGO"/DR. BEYER, "EL MAESTRO"

Our chapter is a modest contribution to the rich and relevant experimental body of knowledge related to Carlos Beyer's collaborative research activity with different neuroendocrine and behavioral research groups. Additionally, we want to mention some brief, but deep memories of his personal imprint on our way of seeing the world.

Carlos, our friend, was one of the most erudite persons we have ever met. It was such a delightful experience to share with him any conversation, experimental design or, even, a trip. We have fond memories of the weekly seminar that he presented at our Department of Psychobiology during his two long stays with us, of the invitation to him through the prestigious award "IBERDROLA excellence researchers" from UNED, Madrid, of his frequent visits to the Prado Museum (he lived 50 m across the street from it in Madrid), and of our 2010 summer trip to the Montealbán archeological site in the state of Guerrero, México. Carlos always was not only a fine and stimulating conversationalist, but also a clever, insightful observer of the world around us, near and far, with a delightful sense of humor.

We also appreciate very much the extreme generosity of Fina, his beloved wife, who shared with us some of our common trips. She has always been the best host and kindest person during any time we spent together. His daughters, Maria Emilia and Gaby, have always been part of his culturally rich life, which he shared with all of us.

Dr. or Professor Beyer, El Maestro, was one of the most distinguished neuroendocrinological scientists in the past six decades. His contribution to the neurohormonal regulation of reproductive behavior in mammals, including humans, led him to the 2007 National Sciences and Arts Award, the most prestigious Science award in México. Maestro Beyer was one of the world leaders in the area of the mechanism of steroid hormone action on the brain.

Finally, allow us to let you in on one of Carlos' profound metaphoric thoughts on the adaptive ability of our human nature, which gives a clear and concise picture of his extremely intelligent and astute personality: *"...Deberíamos saber estar tanto en la cima como en la sima." ("...We should be able to be at the top as well as at the bottom").*

ACKNOWLEDGMENTS

The research presented in this chapter was supported by grants DGSICBSO2000-19 (M.C.R. del Cerro) and PSI2014-58004 (A. Guillamón and M.C.R. del Cerro). We also thank Dr. Komisaruk (Rutgers University NJ, USA) for his suggestions and editorial help and to L. Carrillo, L. Troca, and G. Moreno for the technical assistance along the experimental process. Our thanks are particularly due to A. Marcos for his artwork support. Finally, we will always be grateful to Dr. Beyer for his valuable scientific insights that underlie most of the work we address in this chapter.

REFERENCES

Allen, L.S. and Gorski, R.A. 1990. Sex differences in the bed nucleus of the stria terminalis of the human brain. *J. Comp. Neurol.* 302:607–706.

Allison, A. 1953. The morphology of the olfactory system in vertebrates. *Bil. Rev.* 28:195–244.

Arnold, A.P. and Gorski, R.A. 1984. Gonadal steroid induction of structural sex differences in the central nervous system. *Annu. Rev. Neurosci.* 7:413–442.

Arratia, P.G., Carmona, C., Dominguez, C., Beyer, C. and Rosenblatt, J.S. 2006. GABA receptor agonist in the medial preoptic area and maternal behavior in lactating rats. *Physiol. Behav.* 87(1):51–65.

Bach, F., Pflugge, G. and Wuttke, W. 1984. GABAergic influences on the development of the sexually dimorphic nucleus of male and female rats. *Brain. Res.* 573:341–344.

Beaulieu, E.E. and Robel, P. 1990. Neurosteroids: a new brain function. *J. Steroid. Biochem. Mol. Biol.* 37:395–403.

Benoist, S. and Porter, R. 1991. *Advances on Study of Behavior*, Vol. 20. Academic Press, New York, pp. 135–199.

Benuck, I. and Rowe, F.A. 1975. Centrally and peripherally induced anosmia: influences on maternal behavior in lactating female rats. *Physiol. Behav.* 14:439–447.

Beyer, C., M. Caba, C., Banas, and B.R. Komisaruk. 1991. Vasoactive intestinal polypeptide (VIP) potentiates the behavioral effect of substance P intrathecal administration. *Pharmacol. Biochem. Behav.* 39:695–698.

Beyer, C. and Feder, H.H. 1987. Sex steroids and afferent input: their roles in brain sexual differentiation. *Annu. Rev. Physiol.* 49:349–365.

Bosch, O. L., Müch, W., Bredewold, R., Slattery, D.A. and Neumann, I.D. 2007. Prenatal stress increases HPA axis activity and impairs maternal care in lactating female offspring: implications for postpartum mood disorder. *Psychoneuroendocrinology*. 32:267–278.

Boudalia, S., Berges, R., Chabanet, C., Folia, M., Decocq, L., Pasquis, B., Abdennebi-Najar, L. and Caniven-Lavier, M.C. 2014. A multigenerational study on low-dose BPA exposure in Wistar rats: effects on maternal behavior, flavor and development. *Neurotoxicol. Teratol.* 41:16–26.

Boukydis, C. 1985. Perception of infant crying as an interpersonal event. In: *Infant Crying*, eds. Lester, B.M. and Boukydis, C.F.Z. New York: Plenum Press, pp. 187–216.

Brouette-Lahlou, I. 1991. Dodecyl propionate, attractant from rat preputial gland: Characterization and identification. *J. Chem. Ecol.* 17(7):1343–1354.

Buss, C., Poggy D.E., Tugan, M.L. and Head, K. 2009. High pregnancy anxiety during gestation is associated with decreased gray matter density in 6–9 year-old children. *Psychoneuroendocrinology.* 35(1):141–153.

Champagne, F.A. and Meany, M.J. 2006. Stress during gestation alters postpartum maternal care and the development of the offspring in a rodent model. *Bil. Psychiatry.* 59:1227–1235.

Collado, P., Guillamón, A. and Segovia, S. 1990. Sexual dimorphism in the bed nucleus of the accessory olfactory tract in the rat. *Dev. Brain. Res.* 56:263–268.

Corter, C.M. and Fleming, A.S. 1995. Psychobiology of maternal behavior in human beings. In: *Handbook of Parenting*, ed. Bornstein, M.H. Mahwah, NJ: Lawrence Erlbaum Associates, pp. 87–116.

D'Cruz, S.C., Jubewndradass, R., Jayakanthan, M., Rani, J.S. and Mathur, P.P. 2012. Bisphemol impairs insulin signaling and glucose homeostasis and decreases steroidogenesis in rat testis: an in vivo and in silico study. *Food Chem. Toxicol.* 50:1124–1133.

Davis, E.P., Snidman, N., Wadhwa, P.D., Dunkel, S.C., Glynn, L. and Sandman, C.A. 2004. Prenatal maternal anxiety and depression predict negative behavioral reactivity in infancy. *Infancy.* 6:319–331.

De Catanzaro, D. 1988. Effect of predator exposure upon early pregnancy in mice. *Physiol. Behav.* 43:691–696.

De Olmos, J.S., Alheid, G. and Beltramino, C.A. 1985. Amygdala. In *The Rat Nervous System*, ed. Paxinos. New York: Academic Press, 223–234.

De Olmos, J.S. Hardy, H. and Heimer, L. 1978. The afferent connections of the main and accessory olfactory bulb formations in the rat: An experimental HRP-study. *J. Comp. Neurol.* 181:213–244.

De Olmos, J.S. and Ingram, W.R. 1972. The projection field of the stria terminalis in the rat brain. An experimental study. *J. Comp. Neurol.* 146:303–334.

De Vries, G.J. 1990. Sex differences in neurotransmitters systems. *J. Neuroendocrinol.* 2:1–13.

Del Abril, A., Segovia, S. and Guillamón, A. 1987. The bed nucleus of the stria terminalis in the rat: regional sex differences controlled by gonadal steroids early after birth. *Dev. Brain. Res.* 32:295–300.

Del Cerro, M.C.R., Izquierdo, M.A.P., Collado, P., Segovia, S. and Guillamón, A. 1991. Bilateral lesions of the bed nucleus of the accessory olfactory tract facilitate maternal behavior in virgin female rats. *Physiol. Behav.* 50:67–71.

Del Cerro, M.C.R., Ortega, E., Gómez, F., Segovia, S. and Pérez-Laso, C. 2015. Environmental prenatal stress eliminates brain and maternal behavioral sex differences and alters hormone levels in female rats. *Horm. and Behav.* 73:142–147.

Del Cerro, M.C.R., Pérez-Laso, C., Ortega, E., Martín, J.L.R., Gómez, F., Pérez-Izquierdo, M.A. and Segovia, S. 2010. Maternal care counteracts behavioral effects of prenatal stress in female rats. *Behav. Brain. Res.* 208:593–602.

Del Giudice, M. 2011. Fetal programming by maternal stress: insights from a conflict perspective. *Psychoneuroendrocrinology.* 37(10):1614–1629.

Della Seta, D., Minder, L., Dessi-Fulgheri, F. and Farabollini, F. 2005. Bisphemol A exposure during pregnancy and lactation affects maternal behavior in rats. *Brain. Res. Bull.* 65:255–260.

Dörner, G. 1979. Hormones and sexual differentiation of the brain. *Sex hormones and Behavior.* Ciba Foundation Symposium 62. Amsterdam. Elsevier. 81–110.

Dörner, G. and Staudt, J. 1968. Structural changes in the preoptic anterior hypothalamus of the mail rat, following neonatal castration and androgen substitution. *Neuroendocrinology.* 3:136–140.

Erickson, C.J., Bruder, R.H., Komisaruk, B.R. and Lehrman, D.S. 1967. Inhibition by progesterone of androgen-induced behavior in male ring doves (Streptopeliarisoria). *Endocrinology.* 81:39–44.

Fanelli, M., Kitraki, E. and Stylianopoulou, F. 1995. Maternal behavior of dams treated with ACTH during pregnancy. *Physiol. Behav.* 57:397–400.

Feinberg, A.P. 2007. Phenotypic plasticity and the epigenetics of human disease. *Nature.* 447:433–440.

Fleming, A.S., Gavanath, K. and Sarker, J. 1992. Effects of transections to the vomeronasal nerves or to the main olfactory bulbs on the initiation and long-term retention of maternal behaviour in primiparous rats. *Behav. Neural. Biol.* 57(3):177–188.

Fleming, A.S and Luebke, C. 1981. Timidity prevents the virgin rat from being a good mother: functionally differences between nulliparous and parturient females. *Physiol. Behav.* 27:863–868.

Fleming, A.S., Miceli, M. and Moretto, D. 1983. Lesions of the medial preoptic area prevent the facilitation of maternal behaviour produced by amygdala lesions. *Physiol Behav.* 32:503–510.

Fleming, A.S. and Rosenblatt, J.S. 1974. Olfactory regulation of maternal behaviour in rats. Effects of olfactory bulb removal in experienced and inexperienced lactating and cycling females. *J. Comp. Physiol. Psychol.* 86:221–232.

Fleming, A.S., Ruble, D., Krieger, H. and Wong, P.Y. 1997. Hormonal and experiential correlates of maternal responsiveness during pregnancy and the puerperium in human mothers. *Horm. and Behav.* 31:145–158.

Fleming, A.S., Vaccarino, F., Tambosso, L. and Chee, P.H. 1979. Vomeronasal and olfactory system modulation of maternal behaviour in the rat. *Science.* 203:372–374.

González-Mariscal, G., Fernández Guasti, A. and Beyer, C. 1982. Anesthetic pregnanes counteract androgen-induced defeminization. *Neuroendocrinology* 34:357–362.

Grant, K., McMahon, C., Austin, M., Reilly, N., Leader, L. and Ali, S. 2009. Maternal prenatal anxiety, postnatal caregiving and infants' cortisol responses to the still-face procedure. *Devel. Psychobiol.* 51:625–637.

Gubernick, D.J. 1989. Prolactin and paternal behaviour in the biparental California, *Peromyscus californicus. Horm. Behav.* 23(2):203–210.

Gubernick, D.J. and Alberts, J.R. 1987. The biparental care system of the California mouse, *Peromyscus californicus. J. Comp. Neurol.* 101(2):69–77.

Gubernick, D.J. and Alberts, J.R. 1989. Postpartum maintenance of paternal behavior in the biparental California mouse, *Peromyscus californicus. Animal Behavior* 37:656–664.

Guillamón, A. and Segovia, S. 1996. Sexual dimorphism in the CNS and the role of steroids. In: *CNS Neurotransmitters and Neuromodulators. Neuroactive Steroids*, ed. Stone, T.W. Boca Raton, FL: CRC Press, pp. 127–152.

Halpern M. 1987. The organization and function of the vomeronasal system. *Annu. Rev. Neurosci.* 10:325–362.

Halpern, M. and Martínez-Marcos. 2003. Structure and function of the vomeronasal system. An update. *Prog. Neurobiol.* 70:245–318.

Harkness, J. and Super, X. 1995. Cultural variation of maternal behavior. In: *Handbook of Parenting*, ed. Bornstein, M.H. Mahwah, NJ: Lawrence Erlbaum Associates.

Harris, A. and Seckl, J. 2011. Glucocorticoids, prenatal stress and the programming of disease. *Horm Behav.* 59:279–289.

Hatton, G.I., Modney, B.K. and Salm, A.K. 1992. Increases in dentritic bundling and dye coupling of supraoptic neurons after the induction of maternal behavior. *Annals of the New York Academy of Science* 652:142–155.

Heepfer, I., Hezel, H., Reichardt, W., Clark, K., Geiger, J., Gross, C.M., Heyer, A., Neagu, V., Bhatia, J., Atas, H.C., Bernd, L., Fubich, J.B., Bischofberger, J., Haas, C.A., Lieb, K. and Norman, C. 2012. Early life stress differentially modulates distinct forms of brain plasticity in young and adult mice. *PloS One*. 7(10):e46004.

Herrenkohl, L.H. 1979. Prenatal stress reduces fertility and fecundity in female offspring. *Science* 206:1097–1099.

Holmes, W.G. 1990. Parent-offspring recognition in mammals: a proximate and ultimate perspective. In: *Mammalian Parenting: Biochemical, Neurobiological and Behavioral Determinations,* ed. Krasnegor, N.A., Bridges, R.S. New York: Oxford University Press, 441–460.

Hortman, K.A., Naciff, J.M., Overmann, G.J., Foertsch, L.M., Richardson, B.D. and Daston, G.P. 2012. Effects of transplacental 17-alpha estradiol or bisphemol A on the developmental profile of steroidogenic acute regulatory protein in the testis. *Birth Defects Res B Dev Reprod Toxicol* 95:318–325.

Howdehell, K.K. 2002. A model of the development of the brain as a construct of the thyroid system. *Environ Health Perspect* 110(suppl.3):337–348.

Huizink, P.G., Robles de Medina, P.G., Mulder, E.J., Visser, G.H. and Buitelaar, J.K. 2003. Stress during pregnancy is associated with developmental outcome in infancy. *J. Child Psychol. Psychiatry.* 44:810–818.

Hutton, L.A. and Gu, G., Simerly, R.B. 1998. Development of a sexually dimorphic projection from the bed nucleus of the stria terminalis to the anteroventral periventricular nucleus in the rat. *J. Neurosci.* 18(8):3003–3013.

Izquierdo, M.A.P., Collado, P., Segovia, S., Guillamón, A. and Del Cerro, M.C.R. 1992. Maternal behavior induced in male rats by bilateral lesions of the bed nucleus of the accessory olfactory tract. *Physiol. Behav.* 52:707–712.

Jirik-Babb, P., Manaker, S., Tucker, A.M. and Hofer, M.A. 1984. The role of the accessory and main olfactory systems in maternal behavior of the primiparous rat. *Behav. Neural. Biol.* 40:170–178.

Johnson, A., Josephson, R. and Hawke, M. 1985. Clinical and histological evidence of the vomeronasal (Jacobson's) organ in adult humans. *J Otolaryngol.* 14:101–122.

Kellog, C.K., Olson, V.G. and Pleger, G.L. 1998. Neurosteroids action at the GABAA receptor in fetal forebrain. *Dev. Brain Res.* 108:131–137.

Keverne, E.B. 2004. Importance of olfactory and vomeronasal system for the male sexual function. *Physiol. Behav.* 83:177–187.

Kevetter, G.A. and Winans, S.S. 1981. Connections of the corticomedial amygdala in the golden Hamsterm I. Efferents of the vomeronasal amygdala. *J Comp Neurol* 197:81–98.

Khashan, A.S., Abel, K.M., McNamee, R., Pedersen, M.G., Webb, R.T., Baker, P.N., Kenny, L.C. and Mortensen, P.B. 2008. Higher risk of offspring schizophrenia following antenatal maternal exposure to severe adverse life events. *Arch. Gen. Psychiatry.* 65:146–152.

Kincl, F.A. and Maqueo, M. 1965. Prevention by progesterone of steroid-induced sterility in neonatal male and female rats. *Endrocrinology.* 77:859–862.

Knoches, A.M. and Doyle, L.W. 1993. Long-term outcome of infants born preterm. *Bailliere's Clin. Obstet. Gynecol.* 7:633–651.

Koenig, J.I., Kirkpatrick, B. and Lee, P. 2002. Glucocorticoid hormones and early brain development in schizophrenia. *Neuropsychopharmacology.* 27:309–318.

Kolunie, J.M. and Stern, J.-J. 1995. Maternal aggression in rats: effects of olfactory bulbectomy, $ZnSO_4$-induced anosmia and vomeronasal organ removal. *Horm. Behav.* 29:492–518.

Konner, M. and Worthman, C. 1980. Nursing frequency, gonadal function and birth spacing among !Kung hunter-gatherers. *Science*, 207:788–791.

Krettek, J.E. and Price, J.L. 1978. Amygdaloid projections to Subcortical structures within the basal forebrain and brainstem in the rat. *J. Comp. Neurol.* 178:255–280.

Krieger, J., Schmidt, A., Lobel, D., Gudermann, T., Schulz, G., Breer, H. and Boekhoff, I. 1999. Selective activation of G-protein subtypes in the vomeronasal organ upon stimulation with urine-derived compounds. *J. Biol. Chem.* 274:465–462.

Krueger, K.E. and Papadopoulos, V. 1992. Mitochondrial benzodiazepine receptors and the regulation of steroid biosynthesis. *Annu. Rev. Pharmacol. Toxicol.* 32:211–237.

LaGasse, L.L., Messinger, D., Lester, B.M., Seifer, R., Tronick, E.Z., Bauer, C.R., Shankaran, S., Bada, H.S., Wright, L.L., Smeriglio, V.L., Finnegan, L.P., Maza, P.L. and Liu, J. 2005. Prenatal drug exposure and maternal and infant feeding behavior. *Arch. of Disease in Childhood: Fetal and Neonatal Edition*, 88:391–399.

Larriva-Sahd, J., Rondan, A., Orozco-Estevez, H. and Sánchez-Robles, M.R. 1993. Evidence of a direct projection from the vomeronasal organ to the medial preoptic nucleus and hypothalamus. *Neurosci. Lett.* 163:45–49.

Lehman, M.M. and Winans, S.S. 1982. Vomeronasal and olfactory pathways to the amygdala controlling male hamster sexual behavior: autoradiographic and behavioral analyses. *Brain. Res.* 240:27–41.

Leonard, C.M. and Scott, J.W. 1971. Origin and distribution of amygdal of ugal pathways in the rat. An experimental neuroanatomical study. *J. Comp. Neurol.* 141:313–330.

Levine, S. 2001. Primary social relationships influence the development of the hypothalamic-pituitary axis in the rat. *Physiol. Behav.* 73:235–260.

Licht, G. and Meredith, M. 1987. Convergence of main and accessory olfactory pathways onto single neuron in the hamster amygdala. *Exp. Brain Res.* 69:7–18.

Liu, S., Qin, F., Wang, H., Wu, T., Zhang, Y. and Zheng, Y et al. 2012. Effects of 17alpha-ethinylestradiol and bisphenol A on steroidogenic messenger ribonucleic acid levels in the rara minnow gonads. *Aquat. Toxicol.* 122/123:19–27.

Llewellyn, A.M., Stowe, Z.N. and Nemeroff, C.B. 1997. Depression during pregnancy and puerperium. *J. Clin. Psychiatry.* 58:26–32.

Lonstein, J.S., Lévy, F. and Fleming, A.S. 2015. Common and divergent psychobiological mechanism underlying maternal behaviors in non-human and human mammals. *Horm Behav.* 73:156–185.

Maccari, S., Darnaudery, M., Morley-Fletcher, S., Zuena, A.R., Cinque, C., Van-Reeth, O. 2003. Prenatal stress and long-term consequences: implications of glucocorticoid hormones. *Neurosci. Biobehav. Rev.* 27:119–127.

Majewska, M.D. 1992. Neurosteroids: endogenous bimodal modulators of the GABAA receptor; mechanisms f action and physiological significance. *Prog. Neurobiol.* 38:379–395.

Mastorakos, G. and Ilias, I. 2000. Maternal hypothalamic-pituitary-adrenal axis in pregnancy and the postpartum period: post-partum-related-disorders. *Ann. NY Acad. Sci.* 105:663–668.

Matsumoto, A. and Arai, Y. 1983. Sex differences in the volume of the Ventromedial nucleus of the hypothalamus in the rat. *Endocrinol. Jpn.* 30(3):277–280.

Matsumoto, A. and Arai, Y. 1986. Male-female difference in synaptic organization of the ventromedial nucleus of the hypothalamus in the rat. *Neuroendocrinology.* 42:232–236.

Mattson, M.P., Lee, R.E. Adams, M.E., Guthrie, P.B. and Karter, S.B. 1988. Interaction between entorhinal axons and target hippocampal neurons: a role for glutamate in the development of hippocampal circuitry. *Neuron.* 1:865–876.

Mayer, A.D., Freeman, N.C.G. and Rosenblatt, J.S. 1979. Ontogeny of maternal behavior in the laboratory rat: Factors underlying changes in responsiveness from 30 to 90 days. *Developmental Psychobiology* 12:425–439.

McCarthy, M.M., Schlenker, E.H. and Pfaff, D.W. 1993. Enduring consequences of neonatal treatment with antisense oligodeoxynucleotides to estrogen-receptor messenger ribonucleic acid on sexual differentiation of rat brain. *Endocrinology.* 133:433–439.

Meaney, M. 2010. Epigenetics and the biological definition of gene x environment interactions. *Child Dev.* 81:41–79.

Menella, J.A. and Moltz, H. 1988. Infanticide in the male rat: Role of the vomeronasal organ. *Physiol Behav.* 42:303–306.

Meredith, M. 1986. Vomeronasal organ removal before sexual experience impairs male hamster mating behavior. *Physiol. Behav.* 36:737–743.

Messer, D. and Vietz, P.1984. Timing and transitions in mother–infant gaze. *Infant Behavior and Development* 7:167–181.

Mizukami, S., Nishizuka, M. and Arai, Y. 1983. Sexual difference in nuclear volume and its ontogeny in the rat amygdala. *Exp. Neurol.* 79:659–675.

Monti-Bloch, L. and Grosser, B.I. 1991. Effect of putative pheromones on the electrical activity of human vomeronasal organ and olfactory epithelium. *J. of Steroid Biochem. and Mol. Biol.* 39: 573.

Monti-Bloch, L., Jennings-White, C., Dolberg, D.S. and Berliner, D.L. 1994. The human vomeronasal system. *Psychoneuroendocrinology.* 19:673–686.

Morán, D.T., Jaffek, B.W. and Rowley, J.C. 1991. The vomeronasal (Jacobson's) organ in man: ultrastructure and frequency of occurrence. *J. of Steroid Biochem. and Mol. Biol.* 39:522–545.

Morán, D.T., Jaffek, B.W. and Rowley, J.C. 1992. Ultrastructure of the human olfactory mucosa. In: The *Human Sense of Smell*, eds. Laing, D.G., Doty, R.L. and Breipohlw, X. Berlin: Springer, pp. 3–28.

Morán, D.T., Monti-Bloch, L., Stensaas, L.J. and Berliner, D.L. 1995. Structure and function of the human vomeronasal organ. In *Handbook of Olfaction and Gustation,* ed. Doty, R.L. New York. Marcel Decker, 793–820.

Noirot, E. 1972. The onset and development of maternal behavior in rats, hamsters and mice. *Adv. Study Behav.* 4:107–145.

Nordeen, E.J., Nordeen, K.W., Sengelaub, D.R. and Arnold, A.P. 1985. Androgens prevent normally occurring cell death in a sexually dimorphic spinal nucleus. *Science.* 229:671–673.

Numan, M. 1994. Maternal behavior. In *The Physiology of Reproduction.* ed. Knobil, E., Neil. J.D., vol. 2, 221–301. New York. Raven Press.

Numan, M. and Callahan, E.C. 1980. The connections of the medial preoptic region and maternal behavior in the rat. *Physiol. Behav.* 25:653–665.

Numan, M., Numan, M.J. and English, J.B. 1993. Excitotoxic amino acid injections into the medial amygdala facilitate maternal behavior in virgin female rats. *Horm. Behav.* 27:56–81.

Numan, M., Rosenblatt, J.S. and Komisaruk, B.R. 1977. Medial preoptic area and the onset of maternal behavior in the rat. *J. Comp. Physiol. Psychol.* 91:146–164.

O'Connor, T.G., Ben-Shlomo, Y, Heron, J., Golding, J., Adams, D. and Glover, V. 2005. Prenatal anxiety predicts individual differences in cortisol in pre-adolescent children. *Biol. Psychiat.* 58:211–217.

O'Hara, M.W. and Swain, A.M. 1996. Rates and risks of postpartum depression—a meta-analysis. *Int. Rev. Psychiatry.* 8:37–54.

Panagoitidou, E., Zerva, S., Mitsiou, D.J., Alexis, M.N. and Kitraki, E. 2014. Perinatal exposure to low-dose bisphenol A affects the neuroendocrine stress response in rats. *J. Endocrinol.* 220:207–218.

Panza, G., Mura, E., Pwssarri, M. and Viglietti-Panzica, C. 2005. Early embryonic administration of xenoestrogens alters vasottocin system and male sexual behavior of the Japanese quail. *Domest. Anim. Endocrinol.* 29:436–445.

Pardon, M., Gerardin, P., Joubert, C., Pérez-ias, F. and Cohen-Salmon, C. 2000. Influence of prepartum chronic ultramild stress on maternal pup care behavior in mice. *Biol. Psychiatry.* 47:858–863.

Pérez-Laso, C., Ortega, E., Martin, J.L.R., Pérez Izquierdo, M.A.P., Gómez, F., Segovia, S. and Del Cerro, M.C.R. 2013. Maternal care interacts with prenatal stress in altering sexual dimorphism in male rats. *Horm. Behav.* 64:624–633.

Pérez-Laso, C., Rubio, S., Martín, J.L.R., Gómez, F., Segovia, S. and Del Cerro, M.C.R. 2011. Differential regional brain responses to induced maternal behavior in rats measured by cytochrome oxidase immunohistochemistry. *Behav. Brain Res.* 223:293–296.

Pérez-Laso, C., Segovia, S., Collado, P., Rodríguez-Zafra, M., Del Abril, A. and Gillamón, A. 1997. Estradiol masculinizes the number of the accessory olfactory tract mitral cells in the rat. *Brain Res Bull* 43(3):227–230.

Pérez-Laso, C., Segovia, S., Martín, J.L.R., Ortega, E., Gómez, F. and Del Cerro, M.C.R. 2008. Environmental prenatal stress alters sexual dimorphism of maternal behavior in rats. *Behav. Brain Res.* 187:284–288.

Pérez-Laso, C., Valencia, A., Rodríguez-Zafra, M., Calés, J.M., Guillamón, A. and Segovia, S. 1994. Perinatal administration of diazepam alters sexual dimorphism in the rat accessory olfactory bulb. *Brain Res.* 634:1–6.

Powers, J.B. and Winnans, S.S. 1975. The vomeronasal organ: critical role in mediating sexual behavior in the male hamster. *Science.* 187:961–963.

Raisman, G. and Field, P.M. 1971. Sexual dimorphism in the preoptic area of the rat. *Science.* 173:731–733.

Rice, D. and Barone, S. Jr. 2000. Critical periods of vulnerability for the developing nervous system evidence from humans and animal models. *Environ. Health. Perspect.* 108(Sppl. 3):511–533.

Rodríguez-Zafra, M., De Blas, R., Pérez-Laso, C., Calés, J.M., Guillamón, A. and Segovia, S. 1993. Effects of perinatal diazepam exposure on the sexually dimorphic rat Locus Coeruleus. *Neurotoxicol. Teratol.* 15:139–144.

Rosenblatt, J.S. 1967. Nonhormonal basis of maternal behaviour in the rat. *Science.* 156:1512–1514.

Rosenblatt, J.S. and Mayer, A.D. 1995. An analysis of approach/withdrawal processes in the initiation of maternal behaviour in the laboratory rat. In *Handbook of Behavioral Neurobiology,* ed. K.E. Hood, G. Greenberg, and E. Tobach. vol. 7. New York: Plenum Press, 229–298.

Rosenblatt, J.S., Siegel, H.I. and Mayer, A.D. 1979. Progress in the study of maternal behaviour in the rat: hormonal, non-hormonal, sensory and developmental aspects. *Adv. Study. Behav.* 10:225–311.

Ruiz, R.J., Fullerton, J., Brown, C.E. and Dudley, D.J. 2002. Predicting risk of preterm birth: the roles of stress, clinical risk factors, and corticotropin-releasing hormone. *Biol. Res. Nur* 4:54.

Runner, L. 1959. Embryological effect of handling pregnant ice and its prevention with progesterone. *Nat. Rec.* 133:330–331.

Savchuk, I., Soder, O. and Svechnikov, K. 2013. Mouse leydig cells with different androgen production potential are resistant to estrogenic effects after maternal exposure during organogenesis via the intraperitoneal route. *Reprod. Toxicol.* 37:6–14.

Scalia, F. and Winans, S.S. 1975. The differential projections of the olfactory bulb and accessory olfactory bulb in mammals. *J. Comp. Neurol.* 181:31–56.

Schaeffer, C., Chabli, A. and Aron, C. 1986. Endogenous progesterone and lordosis behavior in male rats given estrogen alone. *J. Steroid Biochem.* 1:99–102.

Segovia, S., Del Cerro, M.C.R., Ortega, E., Pérez-Laso, C., Rodríguez-Zafra, M., Izquierdo, M.A.P. and Guillamón, A. 1996. The role of $GABA_A$ receptors in the organization of brain and behavioral sex differences. *Neuro. Report.* 7(15):2553–2557.

Segovia, S., García-Falgueras, A., Pérez-Laso, C., Pinos, H., Carrillo, B., Collado, P., Claro, F. and Guillamón, A. 2009. The effects of partial and complete masculinization on the sexual differentiation of nuclei that control lordotic behavior in the male rat. *Behav. Brain Res.* 196:261–267.

Segovia, S. and Guillamón, A. 1982. Effects of sex steroids in the development of the vomeronasal organ in the rat. *Dev. Brain Res.* 5:209–212.

Segovia, S. and Guillamón, A. 1993. Sexual dimorphism in the vomeronasal pathway and sex differences in reproductive behaviors. *Behav. Brain Res.* 18:51–74.

Segovia, S., Guillamón, A., Del Cerro, M.C.R., Ortega, E., Pérez-Laso, C., Rodríguez-Zafra, M. and Beyer, C. 1999. The development of brain sex differences: a multisignaling process. *Behav. Brain. Res* 105:69–80.

Segovia, S., Orensanz, L.M., Valencia, A. and Guillamón, A. 1984. Effects of sex steroids on the development of the accessory olfactory bulb in the rat. A volumetric study. *Dev. Brain Res.* 16:312–314.

Segovia, S., Pérez-Laso, C., Rodríguez-Zafra, M et al. 1991. Early postnatal diazepam exposure alters sex differences in the rat brain. *Brain Res. Bull.* 26:899–907.

Segovia, S., Valencia, A., Calés, J.M. and Guillamón, A. 1986. Effects of sex steroids on the development of two granule cell subpopulations in the accessory olfactory bulb. *Dev. Brain Res.* 30:283–286.

Simerley, R. and Swanson, L. 1986. The organization of neural inputs to the medial preoptic nucleus of the rat. *J. Comp. Neurol.* 246:312–342.

Singer, A. G., Agosta, W.C., Clancy, A.N. and Macrides, F. 1987. The chemistry of vomeronasally detected pheromones characterization of an aphrodisiac protein. *Ann. N.Y. Acad. Sci.* 519:287–298.

Stensaas, L.J., Lauker, R.M., Monti-Bloch, L., Grosser, B. and Berliner, D.L. 1991. Ultrastructure of human VNO. *J.* of *Steroid Biochem. and Mol. Biol.* 39:553–560.

Swain, J.E., Lorberbaum, S.K. and Strathearn, L. 2007. Brain basis of early parent-infant interactions: psychology, physiology and *in vivo* functional neuroimaging studies. *J. of Child Psychol. and Psychiat.* 48:262–287.

Tabatadze, N., Huang, G., May, R.M., Jain, A. and Woolley, C.S. 2015. Sex differences in molecular signaling at inhibitory synapses in the hippocampus. *J. Neurosci.* 35:11252–11265.

Takagi, S. F. 1989. Human olfaction. V. Receptive mechanism for odors. Studies in the Electro-Olfactogram (EOG). University of Tokyo Press, Tokyo, pp. 147–233.

Tronick, E. and Reck, C. 2009. Infants of depressed mothers. *Harvard Rev. of Psychiat.* 17:147–156.

Valencia, A., Segovia, S. and Guillamón, A. 1986. Effects on sex steroids on the development of the accessory olfactory bulb mitral cells in the rat. *Dev. Brain Res.* 24:287–290.

Varendi, H., Porter, R.H. and Winberg, J. 1996. Attractiveness of amniotic fluid odor: evidence of prenatal olfactory learning. *Acta Paediatr.* 85:1223–1227.

Vinader-Caerols, C., Collado, P., Segovia, S. and Guillamón, A. 2000. Estradiol masculinizes the posteromedial cortical nucleus of the amygdala in the rat. *Brain Res. Bull.* 53(3):269–273.

Weinstock, M. 2008. The long term behavioral consequences of prenatal stress. *Neurosci. Neurobehav. Rev.* 32:1073–1086.

Weinstock, M. 2011. Sex-dependent changes induced by prenatal stress in cortical and hippocampal morphology and behavior in rats: an update. *Stress.* 14:604–613.

Weinstock, M., Mattina, E., Maor, G.I., Rosen, J.H. and McEwen, B.S. 1992. Prenatal stress selectively alters the reactivity of the hypothalamic-pituitary-adrenal system in the female rat. *Brain Res.* 65:427–451.

Wysocki, C.J. 1979. Neuro behavioral evidence for the involvement of the vomeronasal system in mammalian reproduction. *Neurosci. Biobehav. Rev.* 3:301–341.

Wysocki, C.J. 1981. Stimuli for male mouse (Mus musculus) ultrasonic courtship vocalizations: presence of female chemosignals and/or absence of male chemosignals. *J. Comp. Physiol. Psychol.* 95:623–629.

Wysocki, C.J. and Lepri, J.J. 1991. Consequences of removing the vomeronasal organ. *J. Steroid. Biochem. Mol. Biol.* 39:661–669.

Wysocki, C.J. and Meredith, M. 1987. The vomeronasal system. In *Neurobiology of Taste and Smell,* ed. Finger, T.L. and Silver, W.L. New York: Wiley, 125–150.

Zou, J.N., Hofman, M.A. Goren, L.J.G. and Swaab, D.F. 1995. A sex difference in the human brain and its relation to transexuality. *Nature.* 378:68–70.

13 Ubiquitous Modulators of Brain Activity

GABA and Carlos Beyer-Flores, PhD

Margaret M. McCarthy and Rae Silver

CONTENTS

13.1 Introduction 231
13.2 Paradoxical GABA Effects 233
13.3 GABA and the Suprachiasmatic Nucleus 233
13.4 Modulation of GABA Action Occurs at Multiple Levels 234
13.5 Neurosteroids and GABA 237
13.6 GABA is an Excitatory Neurotransmitter During Development 237
13.7 GABA Action Directs Brain Development 239
13.8 Hormonal Modulation of GABA is Central to Reproduction 241
13.9 Conclusion 243
References 244

13.1 INTRODUCTION

This review is our tribute to Carlos Beyer Flores PhD, a pioneering investigator in the field of neuroendocrinology who had an outsized influence both locally in his laboratories at the National Autonomous University of Mexico (Institute for Biomedical Research), at the Mexican Institute for Social Security, at the Autonomous Metropolitan University (in Mexico City); in Tlaxcala, where he established a laboratory involving a collaboration between the Center for Research and Advanced Studies of the National Polytechnic Institute and the Autonomous University of Tlaxcala and in the US at the Institute of Animal Behavior, Rutgers University, Newark, N.J. His influence extended even more broadly via his inspiring talks and courses around the world. We are privileged to count ourselves among his many students and colleagues and hold an unredeemable debt to him for the gift of training in scientific rigor, the art of experimental design and the joy of discovery. He was a pervasive and ubiquitous modulatory influence, much like one of his favorite molecules, the amino acid transmitter gamma-aminobutyric acid (GABA).

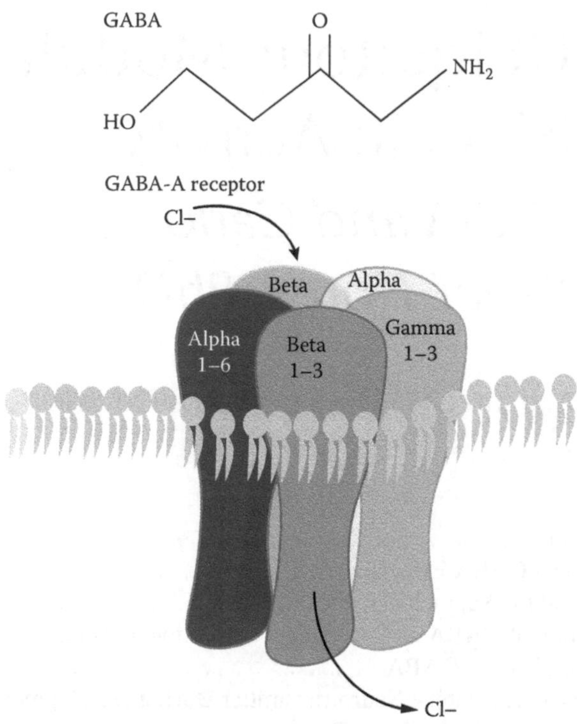

FIGURE 13.1 Structure of GABA and GABA-A receptors. Gamma-aminobutyric acid (GABA) is a simple amino acid with a big impact. It is distinct from glutamate in that it is not used as an essential amino acid building block for proteins but instead functions only via its receptors, GABA-A and GABA-B. The GABA-A receptor is a heteromericpentamer that always contains at least two alpha and two beta subunits with a fifth that may be a gamma or, if located extrasynaptically, delta (not shown). GABA-A receptors are permeable to chloride (Cl^-) and the direction of chloride flow is a function of the transmembrane concentration gradient and the electrical driving force.

GABA is unique among amino acids for its singular use as a neurotransmitter. Different from other amino acids, it is not incorporated into proteins, it has a ubiquitous distribution, and a vital role in controlling the neural excitation that would quickly "overheat" and kill the brain from excitotoxicity. The simple structure of this amino acid (Figure 13.1) belies its role as a critical molecule that maintains the brain on the narrow knife edge of consciousness. Too much GABA and you have catatonia, coma, and ultimately death. Too little GABA and you have seizures, convulsions, and ultimately death. In the narrow zone between, GABA modulates every aspect of neuronal functioning, from simple reflexes to higher cognition. Two receptors mediate GABA action, the GABA-A and GABA-B which are ionotropic and metabotropic, respectively. The GABA-A receptor is a heteromeric pentamer consisting of at least two alpha and two beta subunits combined with additional gamma or delta subunits (Figure 13.1) (Bowery 1993; Burt and Kamatchi 1991). The universal feature of GABA-A receptors is their permeability to chloride ions, which impact membrane potential and thereby the probability of achieving an action potential (Burt 1994). The GABA-B receptor comes in two isoforms, one of which is presynaptic and the other postsynaptic (Kasten and Boehm 2015). The presynaptic GABA-B plays a modulatory role, impacting release of various neurotransmitters (Bowery 1993), whereas the postsynaptic receptors impact neuronal excitability by modulating membrane potential via potassium flux (Loose et al. 1991).

The most frequently used GABA-A agonist, muscimol, is derived from the deadly mushroom *Aminita muscaria* (Johnston 2014) and the basis of most anesthetics is modulation of GABAergic tone (Antkowiak 2015). GABA receptors are so central to anesthesia that multiple components are independently modulated by different receptor subtypes, and studies by Carlos Beyer revealed that the analgesic portion of barbiturate-induced sedation is mediated by the GABA-A receptor (McCarthy et al. 1989). Thus, GABA has two faces, one life threatening and the other life preserving.

13.2 PARADOXICAL GABA EFFECTS

One of the most striking features of GABA action is the completely opposite nature of its effects in the developing versus mature brain and its changing effects over time of day in the mature brain. Early in development, GABA is the dominant excitatory transmitter, due to an increased accumulation of intracellular chloride compared to extracellular (Cherubini et al. 1991; McCarthy et al. 2002). As neurons mature, the transmembrane chloride gradient reverses such that extracellular chloride is higher and opening of the GABA-A channel results in chloride influx and membrane hyperpolarization. This, combined with a reduction in membrane resistance due to the opening of membrane pores (i.e., the receptor), so-called shunting inhibition, is the basis for central nervous system (CNS)-wide inhibition. Thus GABA is both excitatory and inhibitory depending on the maturational state of the neuron. This diametric shift in effect is also seen in certain unique circumstances, such as across the circadian rhythm (discussed below), in newly born neurons of the adult brain (Sernagor et al. 2010), and on the GnRH neurons controlling the release of gonadotropins from the anterior pituitary (DeFazio et al. 2002; Han et al. 2002).

13.3 GABA AND THE SUPRACHIASMATIC NUCLEUS

GABA also plays a major but complex role in the brain's master circadian clock, located in the suprachiasmatic nucleus (SCN) of the anterior hypothalamus. The SCN is made up of approximately 20,000 neurons that somehow produce a coherent signal to indicate time of day to the rest of the brain and body. These SCN timing signals set the phase of behavior patterns, such as sleep–wake cycles, and they are part of the mechanism that resets body clocks in response to "jet lag"—the phase shift that occurs with long distance travel across time zones. Thus the SCN can be viewed as a multiscale neuronal system that determines the timing of daily events. Much work has shown that the SCN is made up of several different peptidergic cell types. The SCN peptides, vasoactive intestinal peptide (VIP) and arginine vasopressin (AVP) are important in producing synchronization of the individual cells that contribute to SCN oscillation. A great deal of research has revealed the importance of interneuronal communication, and of the peptides of neuronal origin in achieving synchronization of SCN rhythmicity. The role of GABA is less well understood.

Though not well understood, compared to the wealth of data on peptides of neuronal origin, it has long been evident, based on immunocytochemistry and in situ hybridization studies, that GABA occurs in *all* neurons of the SCN [reviewed in Moore (2013)]. Paradoxically, GABA appears to both synchronize and desynchronize, to both excite and inhibit, "clock" neurons in the adult SCN. GABA is also thought to exert excitatory signals in fetal and neonatal brains when chloride equilibrium potential is relatively high (Rivera et al. 1999). In the SCN, GABA is reported to act as an inhibitory neurotransmitter throughout the day, as in other CNS areas (Aton et al. 2006). But in the SCN, GABA also acts as an excitatory transmitter depending on time of day (Albus et al. 2005; Choi et al. 2008), and on season of the year (Farajnia et al. 2014). GABA seems to transmit coupling signals between subpopulations of neurons in the SCN, notably between dorsal (shell) and ventral (core) regions (Albus et al. 2005; Myung et al. 2015) and in long photoperiods (Evans et al. 2013), GABA destabilizes networks at a millisecond order of transmission, counteracting the stabilizing role of VIP (Freeman et al. 2013). In summary, the evidence from careful studies indicates that GABA can act to synchronize, desynchronize, inhibit, or excite SCN neurons. The balance of stabilizing and destabilizing signals appears to be important for robust and resilient circadian oscillation.

Despite substantial evidence of an important presence of GABA and its uptake in the regulation of SCN rhythmicity, the localization and expression of GABA transporters (GATs) in the SCN was assumed but only recently investigated (Moldavan et al. 2015). A surprising finding, based on immunohistochemistry and electron microscopy, demonstrated the presence of GABA transporter 1 (GAT1) and GAT3 in glial processes surrounding unlabeled neuronal perikarya (Figure 13.2), axons, dendrites, and enveloped symmetric and asymmetric axodendritic synapses in adult rats. GAT1 and GAT3 were not expressed in the perikarya of AVP- or VIP-immunoreactive (-ir) neurons, nor in the neuronal processes labeled with the neurofilament heavy chain. These data demonstrate that understanding the SCN network and its production of a coherent circadian rhythm requires consideration of synapses that include glia. Astrocytes that regulate GAT may be key to understanding the paradoxical effects of GABA.

13.4 MODULATION OF GABA ACTION OCCURS AT MULTIPLE LEVELS

Neurotransmission is subject to regulation at multiple key nodes. Beginning with transmitter synthesis, the enzymes that either limit the rate of production or degradation can profoundly impact ligand levels and therefore receptor binding and

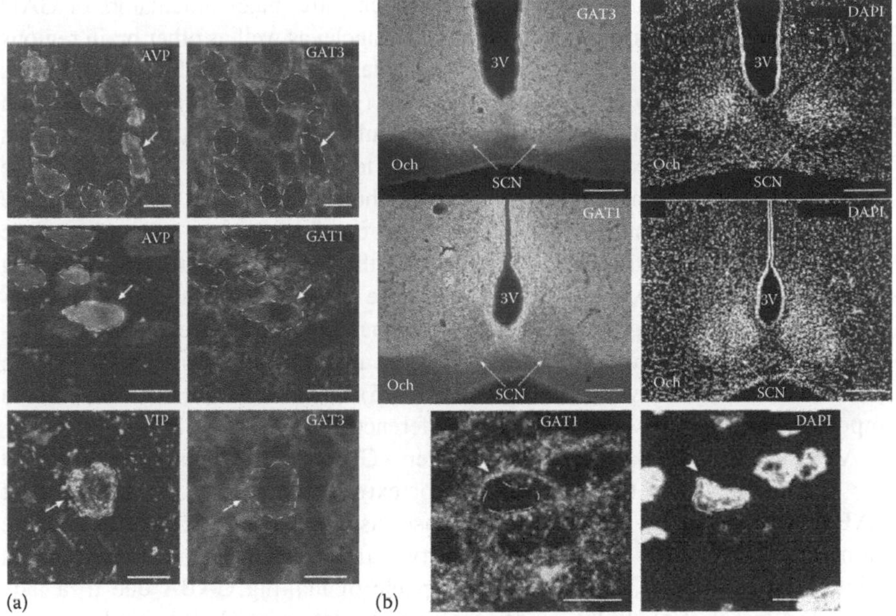

FIGURE 13.2 GABA is central to the SCN and circadian rhythms. (a) GAT1 and GAT3 were not expressed in neurons immunoreactive for arginine vasopressin (AVP) or vasoactive intestinal peptide (VIP). Double labeling with antibodies to GAT1 or GAT3 and AVP or VIP. Arrows show corresponding points on adjacent panels. The dashed line was drawn around AVP- or VIP-immunoreactive cells viewed in the original color image. Scale bar 10 μm. Rats, housed in a 12:12 light:dark cycle, were perfused for immunohistochemistry at 6 hours after lights out. (b) Coronal sections of the hypothalamus including the SCN, demonstrate GAT3 (upper panel) and GAT1 (middle panel) expression in hypothalamus. GAT3 is highly expressed around 3rd ventricles and in SCNs, while GAT1 is highly expressed in the region of the periventricular hypothalamic nuclei and the region between the lobes of the SCN. Scale bar 200 μm. Lower panel shows higher magnification image with GAT1-ir puncta (arrowhead) surrounding a DAPI stained cell body (dashed line) in the SCN. Scale bar 10 μm.. SCN, hypothalamic suprachiasmatic nucleus; 3V, third ventricle; Och, optic chiasm. Rats, housed in a 12:12 light:dark cycle, were perfused for immunohistochemistry at 6 hours after lights out. (Reprinted with permission from Moldavan, M. et al., *Eur J Neurosci*, 42, 3018–3032, 2015.)

activation. GABA is synthesized from glutamate by the enzyme glutamic acid decarboxylase (GAD), which comes in two isoforms, GAD-65 and GAD-67, and are distinguished by their molecular weight. There is some variation in the cellular distribution of the two GAD isoforms but it is not absolute (Fukuda et al. 1997; Martin and Barke 1998). More important is that GABA is synthesized and released both synaptically and extrasynaptically, with each being regulated differently. This means there is a constant baseline of GABAergic tone that maintains inhibition throughout the brain, but also changes in local inhibition that are critical to dynamic neural activity. Steroid hormones, in particular estrogens, are major modulators of GAD levels by altering transcription in hypothalamic nuclei as well as other brain regions (McCarthy et al. 1995a; Searles et al. 2000; Weiland 1992). Physiologically, steroids are associated with two dominant states: stress (glucocorticoids) and reproduction (estrogens, androgens, and progestins). By modulating the GABA system, steroids transduce changes in excitatory and inhibitory tone in the brain. These modulate both internal motivational drives, for example, hunger, fear, drug seeking, and the valence of external stimuli such as pheromones, conditioned cues and novelty. One example was demonstrated by Carlos Beyer with the discovery that the pain-reducing abilities of the GABA-A agonist, muscimol, were potently modulated by prior treatment with estradiol, progesterone or their combination (McCarthy et al. 1990b). This work presaged by 30+ years the now dawning awareness that drugs act differently in the CNS of men versus women (Klein et al. 2015), and that steroid hormones are an important contributing variable to those differences.

As in the case of many neurotransmitters, GABA action is also regulated by how long it lingers in the synaptic cleft or extrasynaptic space. Degradation of GABA is achieved by GABA-transaminase, as a part of the GABA-glutamate shunt that involves the shuttling of these two amino acids between neurons and neighboring astrocytes, which are not capable of making GABA due to a lack of GAD (Hassal et al. 1998). Understanding regulation of GABA synthesis thus requires the context of this two-cell system. The existence of changes in GABA in various nuclei of the hypothalamus across development and in adulthood provides indirect evidence of changes in the GABA-glutamate shunt, and in some regions the rate of change is different in males and females, also suggesting hormonal modulation (Davis 1999; Luine et al. 1997). This is more directly demonstrated in castration and hormone replacement experiments, in which the rate of GABA turnover in the arcuate nucleus is found to be directly related to androgen levels (Grattan and Selmanoff 1994).

Transmitter release is the second mode of influence on GABA action but as noted above, for this particular transmitter system it is not straightforward. Up to 50% of GABA release is calcium-independent and therefore does not involve vesicles (Belhage et al. 1993). GABA can be quantified in microdialysates but any conclusions reached are tempered by the component of mass action. This turns attention to the postsynaptic receptor. Changes in GABA binding can be addressed from several perspectives. Receptor autoradiography provides regional anatomical resolution and has generated useful insights into how steroids can increase or decrease binding (Davis 2000a; McCarthy et al. 1992). Changes in the amount of particular subunits suggest that the composition of receptors also changes and this has implications for channel open time and other aspects of receptor kinetics (Benke et al. 1996; Davis 2000b; Fenelon and Herbison 1996; Fritschy and Mohler 1995; Gao et al. 1995; Mohler 2006). Different subunits confer specific properties on the receptor so that gamma-2 containing receptors are sensitive to benzodiazepines while those lacking it are not (Benke et al. 1996). The delta subunit is associated with extrasynaptic GABA-A receptors, which are critical to baseline maintenance of inhibition and are likely fully occupied most of the time by the nonvesicularly released GABA, as discussed above.

13.5 NEUROSTEROIDS AND GABA

Perhaps the most important modulation of the GABA receptor outside of ligand binding is that achieved by neurosteroids, a class of endogenous steroids synthesized in the brain from cholesterol (Mellon and Vaudry 2001). All steroids are derived from cholesterol, but the term neurosteroids is usually reserved for those that do not have an associated canonical nuclear receptor such as allopregnanolone and tetrahydrocorticosterone. These steroids are produced in high concentrations locally in the brain and act as positive allosteric modulators of the GABA-A receptor. That is, they increase GABA action by increasing the probability of channel opening and the duration of open time (Majewska 1992; Robel and Baulieu 1995). Indeed, the initial high profile publication of neurosteroid modulation of the GABA-A receptor (Majewska et al. 1986) was brought to the attention of one of us (MMM) by Carlos Beyer, launching a career-long love affair with this amino acid and its actions in the brain.

Neurosteroids and their actions in the brain remain a topic of active investigation. An exhaustive and decade-long search for a binding pocket for neurosteroids was eventually abandoned and replaced with the notion of "intercalation" of the neurosteroid into the lipid bilayer surrounding the channel and thereby modifying its conformation. However more recently, it was found that delta subunit-containing GABA-A receptors, which are extrasynaptic, are highly sensitive to neurosteroid modulation (Stell et al. 2003), revealing how these endogenous signaling molecules can broadly enhance inhibition. But steroids can also directly impact the binding of the GABA-A receptor. As shown by Carlos Beyer, progesterone is a potent enhancer of the binding capacity of muscimol (Lopez-Colome et al. 1990).

13.6 GABA IS AN EXCITATORY NEUROTRANSMITTER DURING DEVELOPMENT

While the receptor subunit composition of the GABA-A receptor is important to its function, ultimately the transduction of information to the postsynaptic cell is via chloride flux. It is not the chloride per se that signals (as in the case for calcium for instance), but the negative charge on the chloride ion, which impacts the polarization of the membrane. The GABA-A channel is freely permeable to chloride and thus the direction of flow is a function of the driving force, which is a combination of ionic and electrical concentration gradients. As noted above, early in development, chloride accumulates inside the cell so that when GABA-A receptor channels open the chloride leaves the cell and depolarizes the membrane (McCarthy et al. 2002; Stein and Nicoll 2003). In some instances the depolarization may be sufficient to trigger an action potential (Ben-Ari et al. 1989), but more often, and perhaps more importantly, the depolarization is sufficient to open voltage-gated-calcium channels of the L-type (Perrot-Sinal et al. 2003). This results in calcium rushing into the cell and initiating any number of signal transduction pathways that are responsive to calcium, many of which are related to growth and differentiation. Indeed it appears that the influx of calcium can lead to maturation of the neuron and a reversal of the chloride gradient so that GABA takes on its adult inhibitory role (Ganguly et al. 2001). Additional modulation of the maturational process comes from the GABA-B receptor which when activated along with the GABA-A can inhibit the influx of calcium, thereby delaying the maturational process (Obrietan and van den Pol 1998).

The status of the transmembrane chloride gradient is fundamental to GABA action, but is controlled independently of the amino acid. Chloride, along with the cations sodium and potassium, must be actively transported across the membrane by chloride–cation–cotransporters (CCC). NKCC1 transports one sodium, one potassium, and two chlorides into the cell in an electroneutral fashion, while KCC2 essentially does the opposite, only it transports equal amounts of potassium and chloride out of the cell (Payne et al. 2003). Together, these two CCCs regulate the transmembrane chloride gradient. Their respective contributions shift across development, with NKCC1 being higher in young neurons and KCC2 being dominant in mature neurons (Stein et al. 2004). As a result, opening of the GABA-A channel in young neurons is depolarizing, or excitatory, while opening of the same channel in mature neurons is hyperpolarizing or inhibitory.

Identifying and decoding the variables that shift GABA action from excitatory to inhibitory is central to the challenge of understanding this ubiquitous amino acid. Steroids are potent modulators of the excitatory/inhibitory shift in the immature brain during the period for sexual differentiation which occurs perinatally in rodents. Estrogen is higher in the developing male brain compared to females (Amateau et al. 2004; Konkle and McCarthy 2011) because it is synthesized in the brain from testicular androgens that only fetal males produce (Lephart 1996). Among the many actions of this neuronally derived estrogen is an enhancement of the excitatory action of GABA, such that the threshold for depolarization is lower, more calcium enters the cell with each depolarization and the developmental time course over which GABA is excitatory is extended (Nunez and McCarthy 2008; Perrot-Sinal 2001; Perrot-Sinal et al. 2003) (Figure 13.3). The impact on the developing brains of males versus females is likely profound; that possibility is strongly supported by the observation that the ubiquitous signaling molecule, camp response element-binding protein (CREB), is induced in neonatal male but reduced in neonatal female hypothalamic nuclei after administration of the GABA-A agonist, muscimol (Auger et al. 2001). Ongoing work seeks to further explore the importance of this divergence in sex-specific brain development, but the fact that depolarizing GABA is central to hypothalamic development is firmly established (Chen et al. 1996)

Importantly, the cellular mechanisms by which estradiol enhances depolarizing GABA are quite removed from the primary response. The steroid has no impact on the permeability of the receptor to chloride but there is an estradiol-induced increase in the transporter that ships chloride into the cell, thereby increasing the transmembrane gradient (Perrot-Sinal et al. 2007). However, the amount the transporter, NKCC1, that is increased in males is very modest and not in keeping with the dramatic sex difference in depolarizing GABA. Further study revealed that the activity of the transporter is modulated by phosphorylation, and that this is induced by a highly specific kinase called odd-skipped related transcription factor 1 (OSR1) and its close relative, Ste20p-related Proline Alanine-rich Kinase (SPAK). Estradiol

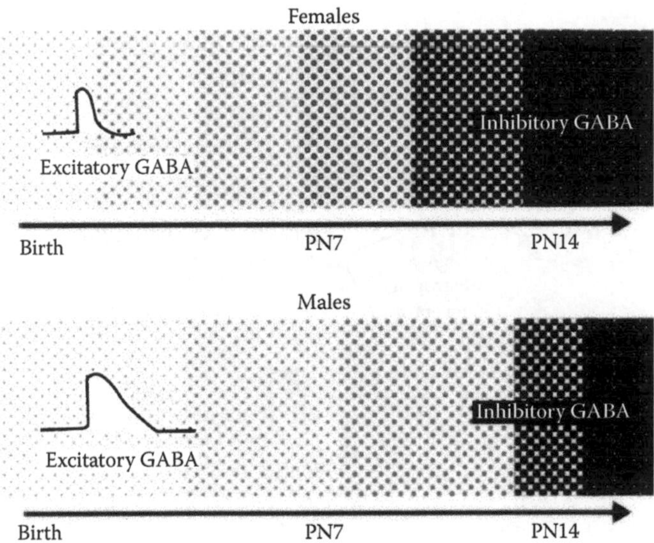

FIGURE 13.3 Divergence in excitatory and inhibitory effects of GABA in males and females during brain development. GABA is the dominant source of excitation in the developing brain by opening calcium influx into neurons via voltage-gated calcium channels following membrane depolarization. Gradually over development this excitatory action is replaced with membrane hyperpolarization and inhibitory actions of GABA. In females this developmental switch occurs earlier, and during the period when GABA is excitatory it is less so in females, with a smaller calcium transient (inset). This sex differences is mediated by estradiol, which is higher in the neonatal male brain following aromatization from testicular androgens.

up-regulates these kinases in a classic genomic fashion. This requires transcription followed by translation, a process that in this case takes at least 24 hours. Then these kinases phosphorylate NKCC1 to increase its activity and elevate intracellular chloride (Nugent et al. 2012) (Figure 13.4). This process of steroid modulation of a kinase is consistent with a larger body of evidence of hormonally induced changes in the amount or activity of enzymes and kinases, as opposed to the amount of neurotransmitter receptors or even the ligand themselves. [see for review McCarthy, (2008)]

13.7 GABA ACTION DIRECTS BRAIN DEVELOPMENT

The appropriate differentiation, migration, and integration of GABAergic neurons in the brain are foundational to all aspects of neural functioning, as ultimately it is a balance of excitation and inhibition that must be achieved. Dysregulation of GABA cells is common to multiple disorders with origins in development, including autism spectrum disorders, early onset schizophrenia, and attention deficit hyperactivity disorders (Nestler and Hyman 2010). Each of these disorders also displays a sex bias, in being more frequent or more severe in males (Abel et al. 2010; Bangasser and

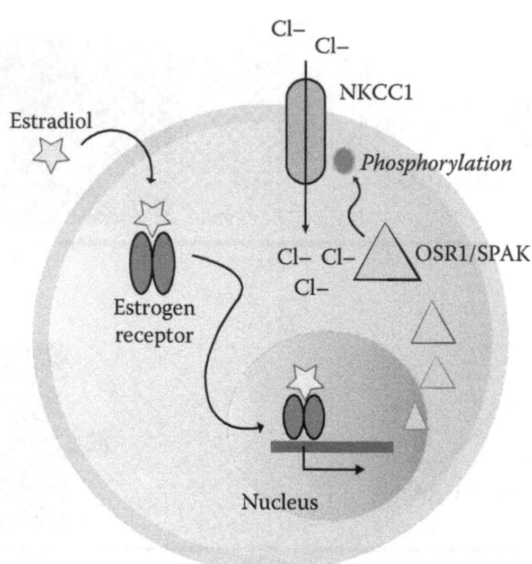

FIGURE 13.4 Estrogen regulation of kinases that phosphorylate NKCC1 and chloride gradient. The excitatory actions of GABA are a function of the transmembrane chloride concentration gradient. Early in development intracellular chloride is high because of the activity of chloride co-transporters such as NKCC1. The activity of NKCC1 is regulated by phosphorylation which is mediated by highly specific and related kinases SPAK and OSR1. Estradiol enhances depolarizing GABA actions by regulating the activity of NKCC1 to increase intracellular chloride. The regulation of NKCC1 activity is indirect via increased transcription of SPAK and OSR1. Thus the actions of estradiol are genomic and take at least 24 hours to manifest.

Valentino 2014; Goldstein et al. 2014; Werling and Geschwind 2013). Identifying biological origins of sex differences in the brain in the context of GABAergic neurotransmission offers the potential for novel insights into the etiology of these complex disorders.

Most sex differences identified in the brain to-date can be traced to the differential exposure of males and females to steroid hormones during a perinatal sensitive window (McCarthy et al. 2009, 2015). In males, the embryonic testes produce copious quantities of testosterone, which gains access to the brain and is converted into estradiol by the aromatase enzyme. In rodents such as rats and mice, the elevated estradiol of the male initiates cellular processes that lead to masculinization of brain and ultimately adult behavior (McCarthy 2008). Estradiol has many diverse actions in multiple brain regions. In the hippocampus and hypothalamus, this includes enhancing the excitatory actions of GABA (Perrot-Sinal 2001). Thus, if a GABA agonist is applied to male versus female neurons, either in vivo or in vitro, the male neurons exhibit a more robust depolarization with greater calcium influx for a longer duration compared to the female neurons (McCarthy et al. 2000). Excitation is the currency for promoting proliferation, axonal growth, and synaptogenesis. Moreover, many more neurons will be born and many more synapses will be made than are necessary, and will therefore undergo a process of pruning. The selective elimination of cells and synapses is often determined by excitation, either too much or too little. For cells, too much excitation leads to excitotoxicity and cell death. In neonatal males, over excitation is more quickly achieved following GABA-A receptor activation than in females, leading to greater cell death (Nunez et al. 2003a,b). This may contribute to the greater vulnerability of males to developmental neuropsychiatric disorders as well as the worse outcome experienced by males following neonatal brain injury (Nunez and McCarthy 2003).

Excitatory actions of GABA are not limited to neurons, as astrocytes also respond with depolarization (MacVicar et al. 1989; Nilsson 1993). In the arcuate nucleus of the hypothalamus, depolarizing GABA action on astrocytes increases the length and branching frequency of processes and this is correlated with a decrease in local excitatory synapses on the neurons (Mong and McCarthy 1999, 2002). Recall that the GABA originated in the neuron, but acts on the astrocytes (which cannot make GABA) which, in turn, modify the neurons, revealing a cell-to-cell feedback loop.

13.8 HORMONAL MODULATION OF GABA IS CENTRAL TO REPRODUCTION

Exploring the neural control of reproductive physiology and behavior offered an excellent opportunity to connect specific neuroanatomical substrates and neurotransmitters systems with clearly defined and easily quantifiable endpoints. In the case of physiology this took the form of the control of luteinizing hormone (LH) release from the anterior pituitary, which is determined by the firing rate of gonadotropin-releasing hormone (GnRH) neurons. GABAergic input onto those neurons is an essential driver. This is true both for the regulation of pulsatile, versus surge, release in adulthood in males versus females, respectively, and in the reawakening of the hypothalamic–pituitary–gonadal axis at puberty (Kasuya et al. 1999; Sagrillo et al. 1996). It is noteworthy that there are no estrogen or androgen receptors in GnRH neurons (Pfaff et al. 1994b), but there is a high concordance with GABA neurons in the essential brain regions (Herbison et al. 1993). Consequently, GABA serves the role of transducer of peripheral steroid physiology to the CNS.

Further coordination is essential between the periphery and the CNS in the context of ensuring that reproductive receptivity is in synchrony with the capacity for fertilization. That is, females in most species mate only when they have recently ovulated or are about to ovulate. Males, on the other hand, maintain a state of constant readiness with the continuous production of gametes and essentially continuous interest in mating. For females, there is a challenge in ensuring that behavior and physiology are coordinated. Steroid hormone modulation of GABAergic transmission provides the mechanism by which the necessary coordination can be achieved. For females to be sexually receptive, a combination is required of removal of a tonic inhibitory network and excitation of a facilitative neural network. These two nodes reside in the preoptic area and ventromedial nucleus, respectively, with additional integration of excitatory input in the midbrain central gray (Pfaff et al. 1994a). GABA provides modulation at each node, but the direction of the effect is complex (Figure 13.5). Increased GABA activity in the preoptic

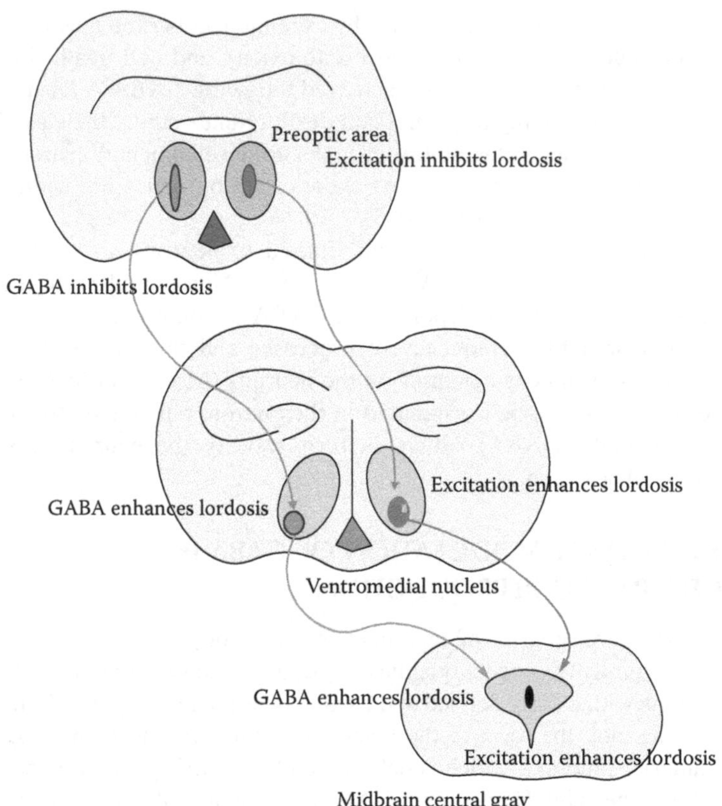

FIGURE 13.5 The neural circuitry of lordosis is affected by GABA at three critical nodes. The neural circuitry controlling female sexual receptivity, as evidenced by the lordosis response, has been clearly established. Three critical nodes for steroid hormone action are the preoptic area (POA), the ventromedial nucleus (VMN), and the midbrain central gray (MCG). In general, excitation of the POA inhibits lordosis responding, but GABA has the opposite effect as predicted and instead when agonists are infused into this brain region it reduces lordosis responding. Conversely, excitation of the VMN and MCG are essential to the lordosis response but again, GABA has the opposite effect to that predicted and facilitates rather than inhibits lordosis. The reasons for this physiological conundrum remain unknown.

area inhibits female sexual responding, yet this brain region is generally inhibitory to receptivity and so the opposite would be predicted. Likewise, increased GABA in the hypothalamus increases receptivity (McCarthy et al. 1990a), and yet increased excitation in this brain region is associated with increased receptivity. The same is true for the midbrain central gray (McCarthy et al. 1995b). The disconnect between behavioral response and neuronal excitation could be due to a complex network of interneurons and disinhibition following GABA-A activation, as was presumed at the time. Alternatively, with the hindsight that GABA action can be excitatory under varying conditions, such as the changes across the circadian cycle, it is possible that GABA becomes an excitatory neurotransmitter in response to the hormonal conditions that promote sexual receptivity. Unfortunately the answer is likely to remain unknown for the foreseeable future.

The role of GABA as a coordinator of physiology and behavior begins with the sculpting of the developing brain. Because there was considerable evidence of steroid hormone modulation of the GABA system at the level of the synthetic enzymes, GAD65 and GAD67 (Davis et al. 1996; Grattan et al. 1996; Luine et al. 1997; Weiland 1992), it was natural to speculate that this might play a role in the sexual differentiation of the brain to control adult reproductive behavior. The drugs available for inhibiting GAD are largely toxic, inducing epilepsy or seizures (Salazar and Tapia 2015), and receptor mimetics or antagonists are much too short-acting to be used for a developmental investigation. Antisense oligonucleotides preceded the more modern method of silencing RNAs; they are equally effective at down-regulating the translation of targeted mRNAs, and their effect can endure for days (McCarthy 1994). Infusion of antisense oligos against both forms of GAD into the three key nodes for controlling female receptivity noted above (preoptic area, ventromedial hypothalamus, and midbrain central gray), reduced GABA levels and modulated behavior in a manner wholly consistent with that seen using GABA-A agonists and antagonists (McCarthy et al. 1994). Thus, this was an effective approach for reducing GABA in the developing brain during the early postnatal sensitive period for sexual differentiation. Davis and colleagues found that reducing both GADs during early life significantly impaired both male and female sexual behavior in adulthood (Davis et al. 2000). This was somewhat surprising, as both GAD and GABA levels had been found to be significantly higher in the relevant regions of the male brain during the same time points (Davis 1999; Davis et al. 1996), leading to the prediction that only masculinization, not feminization, would be impaired by a reduction in GAD and GABA levels. In retrospect, however, the effectiveness of reducing GAD in both sexes makes sense in light of the realization that GABA acts profoundly differently, even oppositely, in developing male and female hypothalamus. Opening of GABA-A receptors in neonatal male brains leads to membrane depolarization and calcium influx and activation of CREB with many downstream activational effects. Conversely, opening of GABA-A receptors in neonatal female brains leads to membrane hyperpolarization and neuronal inhibition. Thus there is a dramatic divergence in the development of the neural circuits controlling sex behavior in males and females that is the result of GABA action in both cases, but via profoundly different mechanisms (McCarthy et al. 2002).

13.9 CONCLUSION

In summary, the amino acid transmitter GABA is arguably the most wide-ranging and impactful chemical in the brain. It mediates life and death, alertness and somnolence. Circadian rhythms both direct and affect GABA. GABA action is an excitatory driver of the developing brain and an essential inhibitor of activity in the mature brain. Steroids are potent modulators of the GABAergic system throughout life, generating divergent effects in the brains of developing males and females and mediating the divergent physiological and behavioral processes inherent to adult reproduction. It is a molecule worthy of attention. Carlos Beyer-Flores, PhD, was a creative neuroendocrinologist and an enthusiastic pioneer in GABA research, with students throughout Mexico and the U.S. As in the case of GABA, he was a ubiquitous modulator of both excitation and inhibition, providing guidance, encouragement and enthusiasm for science. Carlos' influence lives on in his many students and colleagues around the world, who continue in his tradition of conducting sound basic science research that asks and answers fundamental questions about how the brain controls complex behavior.

REFERENCES

Abel, K.M., Drake, R., Goldstein, J.M., 2010. Sex differences in schizophrenia. *Int Rev Psychiatry*. 22: 417–428.

Albus, H., Vansteensel, M.J., Michel, S., Block, G.D., Meijer, J.H., 2005. A GABAergic mechanism is necessary for coupling dissociable ventral and dorsal regional oscillators within the circadian clock. *Curr Biol*. 15: 886–893.

Amateau, S.K. et al. 2004. Brain estradiol content in newborn rats: sex differences, regional heterogeneity, and possible de novo synthesis by the female telencephalon. *Endocrinology*. 145: 2906–2917.

Antkowiak, B. 2015. Closing the gap between the molecular and systemic actions of anesthetic agents. *Adv Pharmacol*. 72: 229–262.

Aton, S.J. et al. 2006. GABA and Gi/o differentially control circadian rhythms and synchrony in clock neurons. *Proc Natl Acad Sci U S A*. 103: 19188–19193.

Auger, A.P., Perrot-Sinal, T.S., McCarthy, M.M., 2001. Excitatory versus inhibitory GABA as a divergence point in steroid-mediated sexual differentiation of the brain. *Proc Natl Acad Sci U S A*. 98: 8059–8064.

Bangasser, D.A., Valentino, R.J., 2014. Sex differences in stress-related psychiatric disorders: neurobiological perspectives. *Front Neuroendocrinol*. 35(3): 303–319.

Belhage, B., Hansen, G.H., Schousboe, A., 1993. Depolarization by K+ and glutamate activates different neurotransmitter release mechanisms in GABAergic neurons: vesicular versus non-vesicular release of GABA. *Neuroscience*. 54: 1019–1034.

Ben-Ari, Y. et al. 1989. Giant synaptic potentials in immature rat CA3 hippocampal neurones. *J Physiol*. 416: 303–25.

Benke, D. et al. 1996. GABA-A receptor subtypes differentiated by their gamma-subunit variants: prevalence, pharmacology, and subunit architecture. *Neuropharmacology*. 35: 1413–1422.

Bowery, N.G. 1993. GABA-B receptor pharmacology. *Ann Rev Pharm Tox*. 33: 109–147.

Burt, D.R. 1994. GABA-A receptor activated chloride channels. In *Current Topics in Membranes*, ed. W. Guggino, vol. 42, 215–263. New York: Academic Press.

Burt, D.R. Kamatchi, G.L., 1991. GABA-A receptor subtypes: from pharmacology to molecular biology. *FASEB J*. 5: 2916–2923.

Chen, G., Trombley, P.Q., van den Pol, A.N., 1996. Excitatory actions of GABA in developing rat hypothalamic neurons. *J Physiol*. 494: 451–464.

Cherubini, E., Gaiarsa, J.L., Ben-Ari, Y., 1991. GABA: an excitatory transmitter in early postnatal life. *Trend Neurosci*. 14: 515–519.

Choi, H.J. et al. 2008. Excitatory actions of GABA in the suprachiasmatic nucleus. *J Neurosci*. 28: 5450–5459.

Davis, A.M., Grattan, D.R., McCarthy, M.M., 2000. Decreasing GAD neonatally attenuates steroid-induced sexual differentiation of the rat brain. *Behav Neurosci*. 114: 923–933.

Davis, A.M., McCarthy, M.M., 2000a. Developmental increase in [3H]muscimol binding to the GABAA receptor in the rat excludes the ventromedial nucleus of the hypothalamus. *Neurosci Lett*. 288: 223–227.

Davis, A.M., Penshuck, S., Fritschy, J-M., McCarthy, M.M., 2000b. Developmental switch in the expression of GABAA receptor subunits a1 and a2 in the hypothalamus and limbic system of the rat. *Dev Brain Res*. 119: 127–138.

Davis, A.M., Ward, S.C., Selmanoff, M., Herbison, A.E., McCarthy, M.M., 1999. Developmental sex differences in amino acid neurotransmitter levels in hypothalamic and limbic areas of rat brain. *Neuroscience*. 90: 1471–1482.

Davis, A.M., Grattan, D.R., Selmanoff, M., McCarthy, M.M., 1996. Sex differences in glutamic acid decarboxylase mRNA in neonatal rat brain: implications for sexual differentiation. *Horm Behav*. 30: 538–552.

DeFazio, R.A., Heger, S., Ojeda, S.R., Moenter, S.M., 2002. Activation of A-tType gamma-aminobutyric acid receptors excites gonadotropin-releasing hormone neurons. *Mol Endocrinology*. 16: 2872–2891.

Evans, J.A., Leise, T.L., Castanon-Cervantes, O., Davidson, A.J., 2013. Dynamic interactions mediated by nonredundant signaling mechanisms couple circadian clock neurons. *Neuron*. 80: 973–983.

Farajnia, S., van Westering, T.L., Meijer, J.H., Michel, S., 2014. Seasonal induction of GABAergic excitation in the central mammalian clock. *Proc Natl Acad Sci U S A*. 111: 9627–9632.

Fenelon, V.S., Herbison, A.E., 1996. Plasticity in GABA-A receptor subunit mRNA expression by hypothalamic magnocellular neurons in the adult rat. *J Neurosci*. 16: 4872–4880.

Freeman, G.M. Jr., Krock, R.M., Aton, S.J., Thaben, P., Herzog, E.D., 2013. GABA networks destabilize genetic oscillations in the circadian pacemaker. *Neuron*. 78: 799–806.

Fritschy, J.-M., Mohler, H., 1995. GABA-A receptor heterogeneity in the adult rat brain: differential regional and cellular distribution of seven major subunits. *J Comp Neurol*. 359: 154–194.

Fukuda, T., Heizmann, C.W., Kosaka., 1997. Quantitative analysis of GAD65 and GAD67 immunoreactivities in somata of GABAergic neurons in the mouse hippocampus proper (CA1 and CA3 regions), with special reference to parvalbumin-containing neurons. *Brain Res*. 764: 237–243.

Ganguly, K., Schinder, A.F., Wong, S.T., Poo, M., 2001. GABA itself promotes the developmental switch of neuronal GABAergic responses from excitation to inhibition. *Cell*. 105: 521–532.

Gao, B., Fritschy, J.-M., Moore, R.Y., 1995. GABA-A receptor subunit composition in the circadian timing system. *Brain Res*. 700: 142–156.

Goldstein, J.M., Holsen, L., Handa, R., Tobet, S., 2014. Fetal hormonal programming of sex differences in depression: linking women's mental health with sex differences in the brain across the lifespan. *Front Neurosci*. 8: 247.

Grattan, D.R., Selmanoff, M., 1994. Castration-induced decrease in the activity of medial preoptic and tuberoinfundibular GABAergic neurons is prevented by testosterone. *Neuroendocrinology*. 60: 141–149.

Grattan, D.R., Rocca, M.S., Strauss, K.I., Sagrillo, C.A., Selmanoff, M., McCarthy, M.M., 1996. GABAergic neuronal activity and mRNA levels for both forms of glutamic acid decarboxylase (GAD65 and GAD67) are reduced in the diagonal band of Broca during the afternoon of proestrus. *Brain Res*. 733: 46–55.

Han, S.K., Abraham, I.M., Herbison, A.E., 2002. Effect of GABA on GnRH neurons switches from depolarization to hyperpolarization at puberty in the female mouse. *Endocrinology*. 143: 1459–1466.

Hassal, B., Johannessen, C.U., Sonnewald, U., Fonnum, F., 1998. Quantification of the GABA shunt and the importance of the GABA shunt versus the 2-oxoglutarate dehydrogenase pathway in GABAergic neurons. *J Neurochem*. 71: 1511–1518.

Herbison, A.E., Robinson, J.F., Skinner, D.C., 1993. Distribution of estrogen receptor-immunoreactive cells in the preoptic area of the ewe: co-localization with glutamic acid decarboxylase but not luteinizing hormone releasing hormone. *Neuroendocrinology*. 57: 751–759.

Johnston, G.A. 2014. Muscimol as an ionotropic GABA receptor agonist. *Neurochem Res.* 39: 1942–1947.

Kasten, C.R., Boehm, S.L., 2nd, 2015. Identifying the role of pre- and postsynaptic GABA(B) receptors in behavior. *Neurosci Biobehav Rev.* 57: 70–87.

Kasuya, E., Nyberg, C.L., Mogi, K., Terasawa, E., 1999. A role of gamma-amino butyric acid (GABA) and glutamate in control of puberty in female rhesus monkeys: effect of an antisense oligodeoxynucleotide for GAD67 messenger ribonucleic acid and MK801 on luteinizing hormone-releasing hormone release. *Endocrinology*. 140: 705–712.

Klein, S.L. et al. 2015. Opinion: sex inclusion in basic research drives discovery. *Proc Natl Acad Sci U S A*. 112: 5257–5258.

Konkle, A.T., McCarthy, M.M., 2011. Developmental time course of estradiol, testosterone, and dihydrotestosterone levels in discrete regions of male and female rat brain. *Endocrinology*. 152: 223–235.

Lephart, E.D. 1996. A review of brain aromatase cytochrome P450. *Brain Res Rev*. 22: 1–26.

Loose, M.D., Ronnekleiv, O.K., Kelly, M.J., 1991. Neurons in the rat arcuate nucleus are hyperpolarized by GABA-B and u-opioid receptor agonists: evidence for convergence at a ligand-gated potassium conductance. *Neuroendocrinology*. 54: 537–544.

Lopez-Colome, A.M., McCarthy, M.M., Beyer, C., 1990. Enhancement of 3H-muscimol binding to brain synaptic membranes by progesterone and related pregnanes. *Euro J Pharm.* 176: 297–304.

Luine, V.N., Grattan, D.R., Selmanoff, M., 1997. Gonadal hormones alter hypothalamic GABA and glutamate levels. *Brain Res*. 747: 165–168.

MacVicar, B.A. et al. 1989. GABA-activated Cl- channels in astrocytes of hippocampal slices. *J Neurosci.* 9: 3577–3583.

Majewska, M.D. 1992. Neurosteroids: endogenous bimodal modulators of the GABA-A receptor, mechanism of action and physiological significance. *Prog Neurobiol.* 38: 379–395.

Majewska, M.D., Tse, F.W., Crichton, S.A., Kettenmann, H., 1986. Steroid hormone metabolites are barbiturate-like modulators of the GABA receptor. *Science*. 232: 1004–1007.

Martin, D.L., Barke, K.E., 1998. Are GAD65 and GAD67 associated with specific pools of GABA in brain? *Perspect Dev Neurobiol.* 5: 119–129.

McCarthy, M.M., 1994. Use of antisense oligonucleotides to block gene expression in the central nervous system. In *Methods in Neurosciences: Neurobiology of Steroids*, ed. E.R.d.K.a.W. Sutano, vol. 22, 342–356. New York: Academic Press.

McCarthy, M.M. 2008. Estradiol and the developing brain. *Physiol Rev.* 88: 91–124.

McCarthy, M.M., Auger, A.P., Perrot-Sinal, T.S., 2002. Getting excited about GABA and sex differences in the brain. *TINS* 25: 307–312.

McCarthy, M.M., Beyer, C., Komisaruk, B.R., 1989. Barbiturate-induced analgesia: permissive role of a GABA A agonist. *Pharm Biochem Behav*. 32: 897–900.

McCarthy, M.M., Malik, K., Feder, H.H., 1990a. Increased GABAergic neurotransmission in medial hypothalamus facilitates lordosis but has the opposite effects in preoptic area. *Brain Res*. 507: 40–44.

McCarthy, M.M., Wright, C.L., Schwarz, J.M., 2009. New tricks by an old dogma: mechanisms of the Organizational/Activational Hypothesis of steroid-mediated sexual differentiation of brain and behavior. *Horm Behav*. 55: 655–665.

McCarthy, M.M. et al. 1990b. Modulation by estrogen and progesterone of the effect of muscimol on nociception in the spinal cord. *Pharmacol Biochem Behav*. 37: 123–128.

McCarthy, M.M., Coirini, H., Schumacher, M., Johnson, A.E., Pfaff, D.W., Schwartz-Giblin, S., McEwen B.S., 1992. Steroid regulation and sex differences in [3H]muscimol binding in hippocampus, hypothalamus and midbrain in rats. *J Neuroendocrinol*. 4(4): 393–399.

McCarthy, M.M., Masters, D.B., Rimvall, K., Schwartz-Giblin, S., Pfaff, D.W., 1994. Intracerebral administration of antisense oligodeoxynucleotides to GAD65 and GAD67 mRNAs modulate reproductive behavior in the female rat. *Brain Res*. 636: 209–220.

McCarthy, M.M., Kaufman, L.C., Brooks, P.J., Pfaff, D.W., Schwartz-Giblin, S., 1995a. Estrogen modulation of mRNA levels for the two forms of glutamic acid decarboxylase (GAD) in the female rat brain. *J Compar Neurol.* 360: 685–697.

McCarthy, M.M., Felzenberg, E., Robbins, A., Pfaff, D.W., Schwartz-Giblin, S., 1995b. Infusions of diazepam and allopregnanolone into the midbrain central gray facilitate open-field and reproductive behavior in female rats. *Horm Behav.* 29: 279–295.

McCarthy, M.M., Pickett, L.A., VanRyzin, J.W., Kight, K.E., 2000. Excitatory GABA as a mediator of steroid-induced brain sexual differentiation. In *Neuroplasticity, Development, and Steroid Hormone Action* ed. R. J. Handa, E. Terasawa, S. Hayashi, and M. Kawata, 323–345. Boca Raton, FL: CRC Press.

McCarthy, M.M., Pickett, L.A., VanRyzin, J.W., Kight, K.E., 2015. Surprising origins of sex differences in the brain. *Horm Behav.* 76: 3–10.

Mellon, S.H. Vaudry, H., 2001. Biosynthesis of neurosteroids and regulation of their synthesis. *Int Rev Neurobiol.* 46: 33–78.

Mohler, H. 2006. GABA(A) receptor diversity and pharmacology. *Cell Tissue Res.* 326: 505–516.

Moldavan, M., Cravetchi, O., Williams, M., Irwin, R.P., Aicher, S.A., Allen, C.N., 2015. Localization and expression of GABA transporters in the suprachiasmatic nucleus. *Eur J Neurosci.* 42: 3018–3032.

Mong, J.A., McCarthy, M.M., 1999. Steroid-induced developmental plasticity in hypothalamic astrocytes: implications for synaptic patterning. *J Neurobiol.* 40: 602–619.

Mong, J.A., McCarthy, M.M., 2002. Ontogeny of sexually dimorphic astrocytes in the neonatal rat arcuate. *Dev Brain Res.* 139: 151–158.

Moore, R.Y., 2013. The suprachiasmatic nucleus and the circadian timing system. *Prog Mol Biol Transl Sci.* 119: 1–28.

Myung, J., Hong, S., DeWoskin, D., De Schutter, E., Forger, D.B., Takumi, T., 2015. GABA-mediated repulsive coupling between circadian clock neurons in the SCN encodes seasonal time. *Proc Natl Acad Sci U S A.* 112: E3920–E3929.

Nestler, E.J., Hyman, S.E., 2010. Animal models of neuropsychiatric disorders. *Nat Neurosci.* 13: 1161–1169.

Nilsson, M., Eriksson, P.S., Ronnback, L., Hansson, E., 1993. GABA induces Ca2+ transients in astrocytes. *Neuroscience.* 54: 605–614.

Nugent, B.M. et al. 2012. Kinases SPAK and OSR1 are upregulated by estradiol and activate NKCC1 in the developing hypothalamus. *J Neurosci.* 32: 593–598.

Nunez, J.L., Alt, J., McCarthy, M.M., 2003a. A new model for prenatal brain damage: I. GABAA receptor activation induces cell death in developing rat hippocampus. *Exp. Neurol.* 181: 258–269.

Nunez, J.L., Alt, J., McCarthy, M.M., 2003b. A new model for prenatal brain damage: II. Long-term deficits in hippocampal cell number and hippocampal dependent behavior following neonatal GABAA receptor activation. *Exp Neurol.* 181: 270–280.

Nunez, J.L., McCarthy, M.M., 2003. Estradiol exacerbates hippocampal damage in a model of preterm brain injury. *Endocrinology.* 144: 2350–2359.

Nunez, J.L., McCarthy, M.M., 2008. Resting intracellular calcium concentration, depolarizing GABA and possible role of local estradiol synthesis in the developing male and female hippocampus. *Neuroscience.* 158: 623–634.

Obrietan, K., van den Pol, A.N., 1998. GABA-B receptor mediated inhibition of GABA-A receptor calcium elevations in developing hypothalamic neurons. *J Neurophysiol.* 79: 1360–1370.

Payne, J.A., Rivera, C., Voipio, J., Kaila, K., 2003. Cation-chloride co-transporters in neuronal communication, development and trauma. *Trends Neurosci.* 26: 199–206.

Perrot-Sinal, T.S., Auger, A.P., McCarthy, M.M., 2003. Excitatory GABA-induced pCREB in developing brain is mediated by L-type Ca+2 channels and dependent on age, sex and brain region. *Neuroscience.* 116: 995–1003.

Perrot-Sinal, T.S., Davis, A.M., Gregerson, K.A., Kao, J.P.Y., McCarthy, M.M., 2001. Estradiol enhances excitatory gamma-aminobutyric acid-mediated calcium signaling in neonatal hypothalamic neurons. *Endocrinology*. 143: 2238–2243.

Perrot-Sinal, T.S. et al. 2007. Sex difference in the chloride cotransporters NKCC1 and KCC2, in the developing hypothalamus. *J Neuroendo*. 19: 1–7.

Pfaff, D.W., Lewis, C., 1994a. Cellular and molecular mechanisms of female reproductive behaviors. In *Physiology of Reproduction*, ed. E. Knobil and J. D. Neill, vol. 2, 107–220. New York: Raven Press.

Pfaff, D.W., Schwanzel-Fukuda, M., Parhar, I.S., Lauber, A.H., McCarthy, L.M., Kow, L.M., 1994b. GnRH neurons and other cellular and molecular mechanisms for simple mammalian reproductive behaviors. In *Recent Progress in Hormone Research,* ed. C. W. Bardin, vol. 49, 1–25. New York: Academic Press.

Rivera, C. et al. 1999. The K+/Cl– co-transporter KCC2 renders GABA hyperpolarizing during neuronal maturation [see comments]. *Nature*. 397: 251–255.

Robel, P., Baulieu, E.E., 1995. Neurosteroids: biosynthesis and function. *Critic Rev Neurobiol*. 9: 383–394.

Sagrillo, C.A., Grattan, D.R., McCarthy, M.M., Selmanoff, M., 1996. Hormonal and neurotransmitter regulation of GnRH gene expression and related reproductive behaviors. *Behav Genet*. 26: 241–277.

Salazar, P., Tapia, R., 2015. Epilepsy and hippocampal neurodegeneration induced by glutamate decarboxylase inhibitors in awake rats. *Epilepsy Res*. 116: 27–33.

Searles, R.V., Yoo, M.J., He, J.R., Shen, W.B., Selmanoff, M., 2000. Sex differences in GABA turnover and glutamic acid decarboxylase (GAD(65) and GAD(67)) mRNA in the rat hypothalamus. *Brain Res*. 878: 11–19.

Sernagor, E., Chabrol, F., Bony, G., Cancedda, L., 2010. GABAergic control of neurite outgrowth and remodeling during development and adult neurogenesis: general rules and differences in diverse systems. *Front Cell Neurosci*. 4: 11.

Stein, V., Nicoll, R., 2003. GABA generates excitement. *Neuron*. 37: 375–378.

Stein, V., Hermans-Borgmeyer, I., Jentsch, T.J., Hübner, C.A., 2004. Expression of the KCl cotransporter KCC2 parallels neuronal maturation and the emergence of low intracellular chloride. *J Comp Neurol*. 468: 57–64.

Stell, B.M., Brickley, S.G., Tang, C.Y., Farrant, M., Mody, I., 2003. Neuroactive steroids reduce neuronal excitability by selectively enhancing tonic inhibition mediated by delta subunit-containing GABAA receptors. *Proc Natl Acad Sci U S A*. 100: 14439–14444.

Weiland, N.G. 1992. Glutamic acid decarboxylase messenger ribonucleic acid is regulated by estradiol and progesterone in the hippocampus. *Endocrinology.* 131: 2697–2702.

Werling, D.M., Geschwind, D.H., 2013. Sex differences in autism spectrum disorders. *Curr Opin Neurol*. 26: 146–153.

Section III

Neuroendocrine Insights toward Development of Therapeutic Agents

14 From Reproductive Neuroendocrinology and Lactation to Vasoinhibins and Angiogenesis

Carmen Clapp and
Gonzalo Martínez de la Escalera

CONTENTS

14.1 Introduction 251
14.2 Regulation of Prolactin Secretion 253
14.3 Reproductive Neuroendocrine Axis 253
14.4 Prolactin Structure–Function Relationship 254
14.5 Angiogenesis 254
14.6 Ocular Effects 255
14.7 Vasoinhibins 255
14.8 Physiological Implications 255
Epilogue 256
Acknowledgment 256
References 256

14.1 INTRODUCTION

We met Carlos Beyer the very same day that we met each other, in September of 1974. It was the inauguration day of the Iztapalapa campus of the Autonomus Metropolitan University (Universidad Autónoma Metropolitana or UAM-I) in Mexico City. Carlos Beyer was head of the Division of Health Sciences, and head of the Department of Reproductive Biology. As such, he played a seminal role in the academic organization of this brand new university. By showing us the excitement of scientific research he instantly became a trusted mentor and was instrumental in our decision to commit to a career in reproductive biology. Carlos Beyer's research group at that time was extremely large, extending from a laboratory that he still headed at the Mexican Institute of Social Security-Medical Center to numerous laboratories established by his senior former students in the Department of Reproductive Biology at UAM-I. On top of that he was extremely busy with the administrative work at UAM-I, so he distributed the freshmen students to the care of various colleagues. In our case, we started training with Pablo Pacheco, a neurophysiologist from the Physiology

Department of the Biomedical Research Institute at the National Autonomous University of Mexico (UNAM), and a superb teacher, who introduced us to the remarkable complexities of the nervous system and taught us how to study them. Eventually, we learned about the research projects going on in the laboratory of Flavio Mena, Carlos Beyer's first student and colleague. Beyer and Mena published a paper in 1961 reporting the letdown of milk in response to stimulation of the cingulate gyrus in the female rabbit (Beyer et al. 1961). This was probably the first paper published in the field of neuroendocrinology from a laboratory in Mexico. They pursued a collaborative effort for almost a decade, studying the pathways involved in the neuroendocrine control of lactation, particularly the tel- and di-encephalic structures responsible for the secretion of anterior and posterior pituitary hormones and their effects on endocrine targets such as the mammary gland and the uterus (Beyer et al. 1962; Beyer and Mena 1965a,b,c, 1970; Mena and Beyer 1963, 1968a,b; Anguiano et al. 1970). Additionally, they explored the neuroendocrine regulation of sexual behavior, a line of research that emerged from their pioneer observations in female rabbits with lesions in the temporal lobe (Beyer et al. 1964; Yaschine et al. 1967).

In contrast to Carlos Beyer's seductive personality, Flavio was seen as a bit gruff/intimidating (which he could be and, we think, took some pleasure in it), so we had no competition to join his group as undergraduate students in 1977 to work on the regulation of milk ejection and milk secretion. During the following 7 years we completed our undergraduate studies and received PhD degrees in Physiology from UNAM. The subject of our dissertations focused on the two neuroendocrine branches of lactation, one leading to milk ejection and another one to milk secretion.

We learned how to mimic the afferent impulse of the suckling stimulus under controlled conditions by electrically stimulating a mammary nerve. By simultaneously monitoring the secretion of oxytocin, prolactin, and catecholamines, we were able to study the relative contribution of the autonomic branch of the nervous system to the physiological regulation of milk secretion and milk ejection (Mena et al. 1978, 1979; Clapp et al. 1985).

We were also able to identify and dissect the role played by factors responsible for the initiation and maintenance of lactation, such as exteroceptive stimuli from the pups and length of the intervals between suckling episodes in rats and rabbits (Mena et al. 1981a,b, 1982b, 1990a,b, 1991). From these studies emerged the concept of "galactolysis" that described the mechanisms of the last stage of lactation characterized by the active decrease in milk production, and which complemented the classical concepts of "lactogenesis" (i.e., milk production) and "galactopoiesis" (i.e., maintenance of milk production).

In another series of experiments, we studied the dynamic relation between the processes of prolactin synthesis, storage and release within the pituitary gland, shedding light on the molecular mechanisms responsible for the depletion of the hormone prior to its secretion. At the time it was reported that suckling or exteroceptive stimuli from the pups induced the massive and rapid depletion of the hormone from the anterior pituitary that was temporally and quantitatively dissociated from its release into the circulation. We observed that the pituitary depletion of prolactin was caused by a transformation into an insoluble form (Mena et al. 1982a) dependent on the age of the hormone after its synthesis (Mena et al. 1984), and the result of the aggregation of prolactin molecules by a thiol-disulfide interchange mechanism (Martínez de la Escalera et al. 1986; Mena et al. 1986). This process was functionally linked to the degradation of prolactin by lysosomal enzymes (Mena et al. 1987) and to hypothalamic regulatory factors such as dopamine and TRH (Mena et al. 1989a, 1989b, 1992).

The above studies provided some answers and also opened a large number of new questions. Among them, two related to prolactin were particularly fascinating to us. This hormone was shown to exert over 300 separate biological effects on many tissues (Bern and Nicoll 1968) and to be under unique regulation by the hypothalamus. This brain region exerts a predominantly inhibitory control (Martin et al. 1984), operating as a gate to the stimulatory factors that were the central players in the field of neuroendocrinology beginning with its formalization by Ernst and Berta Scharrer (Scharrer and Scharrer 1945) and by Geoffrey Harris (Green and Harris 1947), and continuing until the isolation and characterization of hypophysiotropic hormones that merited the Nobel Prize in Physiology and Medicine to Roger Guillemin and Andrew Schally in 1977. As postdoctoral fellows in the laboratories of Charles (Karl) Nicoll at UC Berkeley (Carmen) and Richard (Dick) Weiner at UC San Francisco (Gonzalo), we addressed issues related to the structure–function relationship of prolactin and the role of dopamine tone as a gatekeeper for prolactin secretion. We continued to pursue these questions after our return to UNAM, initially at the Biomedical Research Institute in Mexico City, and later at the Neurobiology Institute in Juriquilla, Querétaro.

14.2 REGULATION OF PROLACTIN SECRETION

To address the issue of dopamine regulation of prolactin secretion, we designed a superfusion system of primary cultured lactotrophs to measure very rapid changes in prolactin release (Martínez de la Escalera et al. 1989a). This system enabled us to establish a hierarchical chain of control, in which dopamine regulates the secretion of prolactin both by occupying, as well as by dissociating from, specific D2 dopamine receptors. The escape from dopamine results in a rapid activation of cyclic adenosine monophosphate (cAMP)-dependent protein kinase A, phospholipase C and protein kinase C, and the sustained potentiation of the prolactin-releasing action of TRH via a Ca^{2+}/protein kinase C-dependent pathway, but not of the prolactin-releasing action of vasoactive intestinal peptide (VIP) (Martínez de la Escalera et al. 1987, 1988, 1989b; Martínez de la Escalera and Weiner 1988a,b, 1992a,b). At least two different mechanisms seem to be directly responsible for the long-term potentiation of exocytosis: the cAMP-dependent phosphorylation of voltage-dependent Ca^{2+} channels (Hernández et al. 1994, 1999a,b), and the protein kinase C-dependent phosphorylation of proteins such as MAP-80 and DARPP-32 (Martínez de la Escalera and Weiner 1992b).

14.3 REPRODUCTIVE NEUROENDOCRINE AXIS

The development and the experience that we had with the superfusion system allowed us to address a fundamental question in the physiology of reproduction, one that is related to the nature of the oscillator responsible for the physiologically relevant pulsatile secretion of gonadotropin-releasing hormone (GnRH), as well as to the nature of central and peripheral signals that directly modulate the function of the GnRHergic neurons. Using an immortalized cell line of the GnRH lineage (GT1-1 cells), we were able to show that the basal secretion of GnRH occurs in a pulsatile manner (Martínez de la Escalera et al. 1992b), with an interpulse interval almost identical to that reported for LH secretion *in vivo*, showing that the central pattern generator is an intrinsic property of the network of GnRH neurons. In addition, since a similar pattern of secretion was observed using a configuration of chambers containing two physically separated GT1-1 cell cultures, we concluded that one or more diffusible factors must coordinate nonadjacent cells to produce synchronized bursts. Furthermore, we characterized the nature of various receptors and the proximal and distal signal transduction mechanisms involved in the actions of multiple afferents, including adrenergic, dopaminergic, GABAergic, histaminergic and kisspeptinergic, as well as of some trophic factors, including FGF, EGF, TGF-alpha and IGF-1 (Martínez de la Escalera et al. 1992a,c, 1994, 1995; Noris et al. 1995; Trueta et al. 1996; Ochoa et al. 1997; Beltrán-Parrazal et al. 2001; Martínez de la Escalera and Clapp 2001; Jacobi et al. 2007; Martín et al. 2007).

14.4 PROLACTIN STRUCTURE–FUNCTION RELATIONSHIP

The question of how prolactin exerts so many functions presented two possible explanations. First, what is referred to as prolactin is, in fact, not a single molecular entity but several molecular forms. Second, diversity of action may arise from the molecular heterogeneity of the prolactin receptor, each isoform coupled to intracellular events mediating only certain effects of the hormone, but not others. Thus, our initial efforts were directed to the chemical isolation and biological characterization of specific isoforms, and to the identification of putative specific receptors (Clapp 1987; Clapp et al. 1988, 1989).

14.5 ANGIOGENESIS

The second step was to look for specific effects linked to a particular isoform of prolactin. Through an unexpected observation, it was possible to assign to an amino-terminal 16 kDa fragment of prolactin, a potent inhibitory action on the formation of new capillary blood vessels that is not triggered by the full-length 23-kDa prolactin isoform. Moreover, endothelial cells were found to contain what appeared to be a unique receptor for these prolactin fragments that differed structurally and functionally from the classic prolactin receptor. Not only did these findings support the concept that the molecular heterogeneity of prolactin and its receptor could account for its diverse biological actions, but they also disclosed a previously unknown field for the actions of the prolactin family, that is, the regulation of the formation of new capillary blood vessels, a process termed angiogenesis or neovascularization (Ferrara et al. 1991; Clapp and Weiner 1992; Clapp et al. 1993).

Discovery of this novel action of prolactin fragments stimulated the search for the endogenous source of these factors. In addition to the anterior pituitary, other sites were quickly described: the hypothalamo–neurohypophyseal system, the vascular endothelium and some elements of the connective tissue. We initially found that hypothalamic paraventricular (PVN) and supraoptic (SON) neurons express prolactin mRNA, contain a prolactin-like immunoreactive protein of 14 kDa that corresponds to the N-terminal part of the prolactin molecule, and display inhibitory actions on endothelial cell proliferation (Clapp et al. 1994; López-Gómez et al. 1995; Torner et al. 1995, 1999; Mejía et al. 1997, 2003; Vega et al. 2010).

We also found that human endothelial cells express prolactin mRNA and produce a 16 kDa protein that corresponds to the N-terminal part of the prolactin molecule and is capable of exerting autocrine effects (Clapp et al. 1998; Corbacho et al. 2000b).

Prolactin mRNA and a 16 kDa protein that corresponded to the N-terminal part of the prolactin molecule were also observed in rat pulmonary fibroblasts. Incubation of exogenous prolactin with a lysate of these fibroblasts resulted in the formation of a 16 kDa prolactin fragment (Corbacho et al. 2000a; Macotela et al. 2002; Corbacho et al. 2003).

14.6 OCULAR EFFECTS

Since ocular angiogenesis is a leading cause of blindness, an obvious next step was to look for angiogenesis inhibitors such as prolactin isoforms in the eye. Indeed, a 16 kDa prolactin-immunoreactive protein was found in corneal homogenates (Dueñas et al. 1999). Furthermore, antiprolactin antibodies specifically stimulated the outgrowth of blood vessels in the rat cornea, suggesting that endogenous, antiangiogenic prolactin isoforms may be responsible for maintaining the avascularity of eye structures such as the cornea (Dueñas et al. 1999). These antiangiogenic prolactin fragments may be produced within the eye, as suggested by findings that prolactin mRNA is expressed both by endothelial cells from rat retinas (Ochoa et al. 2001) and by fibrovascular tissue within the vitreous compartment of patients with premature retinopathy (Dueñas et al. 2004).

14.7 VASOINHIBINS

By the year 2006, the body of evidence that had been collected over the previous 15 years on the antiangiogenic actions of peptides derived from prolactin, growth hormone, and placental lactogen, prompted us to propose to refer to them collectively as vasoinhibins (Clapp et al. 2006, 2009). The rationale behind this proposal was that they all share a common set of vascular effects, including inhibition of angiogenesis, vasodilatation, and vasopermeability, that are not shared with their precursors. Vasoinhibins appear to be involved in the various physiological processes and in the pathogenesis of a number of angiogenesis-dependent diseases.

14.8 PHYSIOLOGICAL IMPLICATIONS

Different enzymes, including matrix metalloproteinases and cathepsin D, have been shown to cleave prolactin into biologically active vasoinhibins (Macotela et al. 2006; Cruz-Soto et al. 2009). Vasoinhibins act directly on endothelial cells to inhibit the action of several vasoactive substances, via various signaling pathways that include nitric oxide and calcium (Gonzalez et al. 2004). In this way, they affect functions in a wide range of tissues, such as: (1) the adhesion of circulating cells to endothelial cells (Montes de Oca et al. 2005); (2) the regulation of follicular maturation in the ovary (Castilla et al. 2010); (3) vascular function in the mammary gland (Clapp et al. 2008); (4) maintaining the avascular status of various tissues in the eye (Aranda et al. 2005); (5) regeneration of the liver (Moreno-Carranza et al. 2013); (6) anxiety behaviors in response to stress (Zamorano et al. 2014).

Vasoinhibins also play a role in other angiogenesis-dependent pathologies, including autoimmune diseases such as lupus erythematosus (Cruz et al. 2001), rheumatoid arthritis (Zermeño et al. 2006; Adán et al. 2013, 2014), pituitary adenomas (Cosío et al. 2003, Méndez et al. 2010), pre-eclampsia (González et al. 2007), cardiomyopathy (Clapp et al. 2007), and diabetic retinopathy (García et al. 2008; Arnold et al. 2010, 2014; Triebel et al. 2011; Ramírez et al. 2011). In addition, by inhibiting angiogenesis, vasoinhibins can inhibit tumor development (Gutiérrez de la Barrera et al. 2006).

EPILOGUE

Carlos Beyer was our mentor and our friend. He had an encyclopedic knowledge of almost everything. We sought his advice often and benefitted greatly from his deep understanding of life and academia. Our graduate training was supervised by Flavio Mena, Carlos Beyer's first student and colleague; thus, both directly and indirectly Carlos Beyer shaped our academic interest in the field of reproductive neuroendocrinology (central regulation of the reproductive axis), lactation (physiology of milk secretion and ejection), and nonreproductive actions of prolactin and its metabolites (vasoinhibins). We will cherish his guidance and friendship for the rest of our life, and we will try to live up to his standards.

ACKNOWLEDGMENT

Supported by the National Council of Science and Technology of Mexico (CONACYT, grant 220574 to CC).

REFERENCES

Adán, N., J. Guzmán-Morales, G. Ledesma-Colunga et al. 2013. Prolactin promotes cartilage survival and attenuates inflammation in rheumatoidarthritis. *J ClinInvest* 123:3902–13.

Adán, N., M. G. Ledesma-Colunga, A. L. Reyes-López et al. 2014. Arthritis and prolactin: A phylogenetic viewpoint. *Gen Comp Endocrinol* 203:132–6.

Anguiano, G., F. Mena, and C. Beyer. 1970. Effect of electrical stimulation and sections of the neuraxis on uterine motility in the cat. *Bol Inst Estud Med Biol Mex* 26:311–14.

Aranda, J., J. C. Rivera, H. Quiróz-Mercado et al. 2005. Prolactins are natural inhibitors of angiogenesis in the retina. *IOVS* 46:2947–53.

Arnold, E., J. C. Rivera, S. Thebault et al. 2010. High levels of serum prolactin protect against diabetic retinopathy by increasing ocular vasoinhibins. *Diabetes* 59:3192–7.

Arnold, E., S. Thebault, G. Baeza-Cruz et al. 2014. The hormone prolactin is a novel endogenous trophic factor able to regulate reactive glia and to limit retinal degeneration. *J Neurosci* 34:1868–78.

Beltrán-Parrazal, L., G. Noris, C. Clapp et al. 2001. GABA inhibition of immortalized GnRH neuronal excitability involves GABAA receptors negatively coupled to cyclic AMP formation. *Endocrine* 14:189–95.

Bern H. A. and C. S. Nicoll. 1968. The comparative endocrinology of prolactin. *Rec ProgrHorm Res* 24:681–720.

Beyer, C. and F. Mena. 1965a. Blockage of milk removal in the cat by periventricular diencephalic lesions. *Am J Physiol* 208:585–8.

Beyer, C. and F. Mena. 1965b. Effect of ovariectomy and barbiturate administration on lactation in the cat and the rabbit. *Bol Inst Estud Med Biol Mex* 23:89–99.

Beyer, C. and F. Mena. 1965c. Induction of milk secretion in the rabbit by removal of the telencephalon. *Am J Physiol* 208:289–92.

Beyer, C. and F. Mena. 1970. Parturition and lactogenesis in rabbits with high spinal cord transection. *Endocrinology* 87:195–7.

Beyer, C., F. Mena, P. Pacheco et al. 1962. Blockage of lactation by brain stem lesions in the cat. *Am J Physiol* 202:465–8.

Beyer, C., G. Anguiano, and F. Mena. 1961. Oxytocin release in response to stimulation of the cingulate gyrus. *Am J Physiol* 200:625–7.

Beyer, C., T. Yaschine, and F. Mena. 1964. Alterations in sexual behavior induced by temporal lobe lesions in female rabbits. *Bol Inst Estud Med Biol Mex* 22:379–86.

Castilla, A., C. García, M. Cruz-Soto et al. 2010. Prolactin in ovarian follicular fluid stimulates endothelial cell proliferation. *J Vascul Res* 47:45–53.

Clapp, C. 1987. Analysis of the proteolytic cleavage of prolactin by the mammary gland and liver of the rat. Characterization of the cleaved and 16K forms. *Endocrinology* 121:2055–64.

Clapp, C., F. López-Gómez, G. Nava et al. 1998. Expression of PRL mRNA and of prolactin-like proteins in endothelial cells. Evidence for autocrine effects. *J Endocrinol* 158:137–44.

Clapp, C., G. Martínez de la Escalera, M. T. Morales et al. 1985. Release of catecholamines follows suckling or electrical stimulation of mammary nerve in lactating rats. *Endocrinology* 117:2498–504.

Clapp, C., J. Aranda, C. González et al. 2006. Vasoinhibins: Endogenous regulators of angiogenesis and vascular function. *Trends Endocr Metab* 17:301–7.

Clapp, C., J. Martial, F. Rentier-Delrue et al. 1993. The 16 kDa N-terminal fragment of human prolactin is a potent inhibitor of angiogenesis. *Endocrinology* 133:1292–9.

Clapp, C., L. Torner, G. Gutiérrez-Ospina et al. 1994. The prolactin gene is expressed in the hypothalamic-neurohypophyseal system and the protein is processed into a 14 kDa fragment with 16K prolactin-like activity. *Proc Natl Acad Sci, USA*. 91:10384–8.

Clapp, C., P. S. Sears, and C. S. Nicoll. 1989. Binding studies with intact rat prolactin and a 16K fragment of the hormone. *Endocrinology* 125:1054–9.

Clapp, C., P. S. Sears, D. R. Russell et al. 1988. Biological and immunological characterization of cleaved and 16K forms of rat prolactin. *Endocrinology* 122:2892–9.

Clapp, C. and R. I. Weiner. 1992. A specific, high affinity, saturable binding site for the 16 kd fragment of prolactin on capillary endothelial cells. *Endocrinology* 130:1380–6.

Clapp, C., S. Thebault, and G. Martínez de la Escalera. 2007. Hormones and postpartum cardiomyopathy. *Trend Endocr Metab* 18:329–30.

Clapp, C., S. Thebault, and G. Martínez de la Escalera. 2008. Role of Prolactin and vasoinhibins in the regulation of vascular function in mammary gland. *J Mamm Gland Biol Neopl* 13:55–67.

Corbacho, A. M., G. Nava, J. P. Eiserich et al. 2000a. Proteolytic cleavage confers nitric oxide synthase inducting activity upon prolactin. *J Biol Chem* 275:13183–6.

Corbacho, A. M., Y. Macotela, G. Nava et al. 2000b. Human umbilical vein endothelial cells express multiple prolactin-like isoforms. *J Endocrinol* 166:53–62.

Corbacho A. M., Y. Macotela, G. Nava et al. 2003. Cytokine induction of prolactin receptors mediates prolactin inhibition of nitric oxide synthesis in pulmonary fibroblasts. *FEBS Lett* 544:171–5.

Cosío, G., M. C. Jeziorski, F. López-Barrera et al. 2003. Hypoxia inhibits expression of prolactin and secretion of cathepsin-D in the GH4C1 pituitary adenoma cell line. *Lab Invest* 84:1627–36.

Cruz, J., A. Aviña-Zubieta, G. Martínez de la Escalera et al. 2001. Molecular heterogeneity of prolactin in the plasma of patients with systemic lupus erythematosus. *Arthr Rheum* 44:1331–5.

Cruz-Soto, M., G. Cosío, M. C. Jeziorski et al. 2009. Cathepsin D is a primary protease for the generation of adeno hypophyseal vasoinhibins: Cleavage occurs within the prolactin secretory granules. *Endocrinology* 150:5446–54.

Dueñas, Z., J. C. Rivera, H. Quiróz-Mercado et al. 2004. Prolactin in eyes of patients with retinopathy of prematurity: implications for vascular regression. *IOVS* 45:2049–55.

Dueñas, Z., L. Torner, A. Corbacho et al. 1999. Inhibition of rat corneal angiogenesis by 16kDa prolactin and endogenous prolactin-like molecules. *IOVS* 40:2498–505.

Ferrara, N., C. Clapp, and R. I. Weiner. 1991. The 16K fragment of prolactin specifically inhibits basal or FGF stimulated growth of capillary endothelial cells. *Endocrinology* 129:896–900.

García, C., J. Aranda, E. Arnold et al. 2008. Vasoinhibins prevent retinal vasopermeability associated with diabetic retinopathy via protein phosphatase 2A-dependent eNOS inactivation. *J Clin Invest* 118:2291–300.

Green, J. D. and G. W. Harris. 1947. The neurovascular link between the neurohypophysis and adenohypophysis. *J Endocrinol* 5:136.

González, C., A. Parra, J. Ramírez-Pulido et al. 2007. Elevated vasoinhibins may contribute to endothelial cell dysfunction and low birth weight in preeclampsia. *Lab Invest* 87:1009–17.

González, C., A. M. Corbacho, J. P. Eiserich et al. 2004. 16K-prolactin inhibits activation of endothelial nitric oxide synthase, intracellular calcium mobilization and endothelium-dependent vasorelaxation. *Endocrinology* 145:5714–22.

Gutiérrez de la Barrera, M., B. Trejo, P. Luna-Pérez et al. 2006. Opposite association of serum prolactin and survival in patients with colon and rectal carcinomas: influence of preoperative radiotherapy. *Digest Dis Sci* 51:54–62.

Hernández, M. E., C. Clapp, and G. Martínez de la Escalera. 1994. Dopamine-escape potentiation of prolactin release involves the activation of calcium channels by protein kinases A and C. *Endocrine* 2:779–86.

Hernández, M. E., M. Díaz M., M. M. Hernández et al. 1999a. Potentiation of prolactin secretion induced by lactotrope-escape from dopamine action: II. Phosphorylation of the α1-subunit of voltage-dependent Ca^{2+} channels. *Neuroendocrinology* 70:31–42.

Hernández, M. M., R. E. García Ferreiro, D. E. García et al. 1999b. Potentiation of prolactin secretion following lactotrope escape from dopamine action: I. Dopamine withdrawal augments L-type Ca^{2+} channels. *Neuroendocrinology* 70:20–30.

Jacobi, J. S., C. Martin, G. Nava et al. 2007. 17β-estradiol directly regulates the expression of adrenergic and kisspeptin receptors in GT1-7 GnRH neurons. *Neuroendocrinology* 86:260–9.

López-Gómez, F. J., L. Torner, S. Mejía et al. 1995. Immunoreactive prolactins of the neurohypophyseal system display actions characteristic of prolactin and 16K prolactin. *Endocrine* 3:573–8.

Macotela Y., C. Mendoza, G. Cosío et al. 2002. 16K prolactin induces NF-kB activation in pulmonary fibroblasts. *J Endocrinol* 175:R13–18.

Macotela, Y., M. Aguilar, J. Guzmán-Morales et al. 2006. Matrix metalloproteases from chondrocytes generate antiangiogenic 16K-prolactin. *J Cell Sci* 119:1790–800.

Martin, C., J. S. Jacobi, G. Nava et al. 2007. GABA inhibition of cyclic AMP production in immortalized GnRH neurons is mediated by calcineurin-dependent dephosphorylation of adenylyl cyclase 9. *Neuroendocrinology* 85:257–66.

Martin, M. C., R. I. Weiner, S. E. Monroe et al. 1984. Prolactin-secreting adenomas in women. VII. Dopamine regulation of prolactin secretion. *J Clin Endocrinol Metab* 59:485–9.

Martínez de la Escalera, G., A. L. H. Choi, and R. I. Weiner. 1992a. ß1-adrenergic regulation of the GT1 GnRH neuronal cell lines: stimulation of GnRH release via receptors positively coupled to adenylate cyclase. *Endocrinology* 131:1397–403.

Martínez de la Escalera, G., A. L. H. Choi, and R. I. Weiner. 1992b. Generation and synchronization of GnRH pulses: intrinsic properties of the GT1-1 GnRH neuronal cell line. *Proc Natl Acad Sci USA* 89:1852–5.

Martínez de la Escalera, G., A. L. H. Choi, and R. I. Weiner. 1994. Biphasic GABAergic regulation of GnRH secretion in GT1 cell lines. *Neuroendocrinology* 59:420–5.

Martínez de la Escalera, G., A. L. H. Choi, and R. I. Weiner. 1995. Signalling pathways involved in the pulsatile release of GnRH from GT1-1 cells. *Neuroendocrinology* 61:310–7.

Martínez de la Escalera, G., B. W. Porter, T. F. J. Martin, and R. I. Weiner. 1989b. Dopamine withdrawal and addition of TRH stimulate membrane translocation of protein kinase C and phosphorylation of an endogenous 80-kDa substrate in enriched lactotrophs. *Endocrinology* 125:1168–73.

Martínez de la Escalera, G. and C. Clapp. 2001. Regulation of GnRH secretion: lessons from GT1 immortalized GnRH neurons. *Arch Med Res* 32:486–98.

Martínez de la Escalera, G., C. Clapp, M. T. Morales et al. 1986. Reversal by thiols of dopamine-, stalk-median eminence-, and zinc-induced inhibition of prolactin transformation in adenohypophyses of lactating rats. *Endocrinology* 118:1803–7.

Martínez de la Escalera, G., F. Gallo, A. L. H. Choi et al. 1992c. Dopaminergic regulation of the GT1 GnRH neuronal cell lines: stimulation of GnRH release via D1-receptors positively coupled to adenylate cyclase. *Endocrinology* 131:2965–71.

Martínez de la Escalera, G., J. Guthrie, and R. I. Weiner. 1988. Transient removal of dopamine potentiates the stimulation of PRL release by TRH but not VIP: stimulation via Ca^{2+}/ protein kinase C pathway. *Neuroendocrinology* 47:38–45.

Martínez de la Escalera, G., K. C. Swearingen, and R. I. Weiner. 1989a. Superfusion and static culture techniques for the measurement of rapid changes in prolactin secretion. *Meth Enzymol* 168:254–63.

Martínez de la Escalera, G. and R. I. Weiner. 1988a. Effect of DA withdrawal on activation of adenylate cylase and phospholipase C in enriched lactotrophs. *Endocrinology* 123:1682–7.

Martínez de la Escalera, G. and R. I. Weiner. 1988b. Mechanism(s) by which the transient removal of dopamine regulation potentiates the PRL releasing action of TRH. *Neuroendocrinology* 47:186–93.

Martínez de la Escalera, G. and R. I. Weiner. 1992a. Dissociation of dopamine from its receptor as a signal in the pleiotropic hypothalamic regulation of prolactin secretion. *Endocrine Reviews* 13:241–55.

Martínez de la Escalera, G. and R. I. Weiner. 1992b. Hypothalamic regulation of microtubule-associated protein phosphorylation in lactotrophs. *Neuroendocrinology* 55:327–35.

Martínez de la Escalera, G., T. F. J. Martin, and R. I. Weiner. 1987. Phosphoinositide hydrolysis in response to the withdrawal of dopamine inhibition in enriched lactotrophs in culture. *Neuroendocrinology* 46:545–8.

Mejía, S., L. M. Torner, M. C. Jeziorski et al. 2003. Prolactin and 16k prolactin stimulate the release of vasopressin by a direct effect on the hypothalamo-neurohypophyseal system. *Endocrine* 20:155–61.

Mejía, S., M. A. Morales, M. E. Zetina et al. 1997. Immunoreactive prolactins co-localize with vasopressin in neurons of the hypothalamic paraventricular and supraoptic nuclei. *Neuroendocrinology* 66:151–9.

Mena, F. and C. Beyer. 1963. Effect of high spinal section on established lactation in the rabbit. *Am J Physiol* 205:313–16.

Mena, F. and C. Beyer. 1968a. Effect of spinal cord lesions on milk ejection in the rabbit. *Endocrinology* 83:615–17.

Mena, F. and C. Beyer. 1968b. Induction of milk secretion in the rabbit by lesions in the temporal lobe. *Endocrinology* 83:618–20.

Mena, F., C. Clapp, D. Aguayo et al. 1986. Thiol regulation of depletion-transformation and release of prolactin by the pituitary of the lactating rat. *Endocrinology* 118:1795–802.

Mena, F., C. Clapp, D. Aguayo et al. 1989a. Differential effects of thyrotropin-releasing hormone on *in vitro* release of *in vivo* or *in vitro* newly synthesized and mature prolactin by lactating rat adenohypophyses; further evidence for a sequential pattern of hormone release. *Neuroendocrinology* 49:207–14.

Mena, F., C. Clapp, D. Aguayo et al. 1989b. Regulation of prolactin secretion by dopamine and thyrotropin-releasing hormone in lactating rat adenohypophyses: influence of intracellular age of the hormone. *Endocrinology* 125:1814–20.

Mena, F., C. Clapp, D. Aguayo et al. 1990a. Prolactin and propranolol prevent the suckling-induced inhibition of lactation in rabbits. *Physiol & Behav* 48:311–15.

Mena, F., C. Clapp, and G. Martínez de la Escalera. 1990b. Age related stimulatory and inhibitory effects of suckling regulate lactation in rabbits. *Physiol & Behav* 48:307–10.

Mena, F., D. Aguayo, G. Martínez de la Escalera et al. 1981a. Effect of short-term food deprivation and prolactin upon milk yield in the lactating rabbit. *Physiol & Behav* 27:529–32.

Mena, F., G. Hummelt, D. Aguayo et al. 1992. Changes in molecular variants during *in vitro* transformation and release of prolactin by the pituitary gland of the lactating rat. *Endocrinology* 130:3365–77.

Mena, F., G. Martínez de la Escalera, C. Clapp et al. 1981b. Effect of acute increases in suckling frequency upon food intake and milk secretion in the rabbit. *Proc Soc Exp Biol Med* 163:373–7.

Mena, F., G. Martínez de la Escalera, C. Clapp et al. 1982a. A solubility shift occurs during depletion-transformation of prolactin within the lactating rat pituitary. *Endocrinology* 111:1086–91.

Mena, F., G. Martínez de la Escalera, C. Clapp et al. 1984. *In vivo* and *in vitro* secretion of prolactin by lactating rat adenohypophyses as a function of intracellular age. *J Endocrinol* 101:27–32.

Mena, F., G. Martínez de la Escalera, D. Aguayo et al. 1982b. Latency and duration of bromocriptine and prolactin on milk secretion in lactating rabbits. *J Endocrinol* 94:389–95.

Mena, F., G. Martínez de la Escalera, C. Clapp et al. 1987. Investigation into the role of dopamine and lysosomes in the impairment of prolactin transformation and release imposed by long periods of non-suckling in the rat. *Acta Endocrinologica* (Copenh) 114:371–8.

Mena, F., P. Pacheco, D. Aguayo et al. 1978. A rise in intramammary pressure follows electrical stimulation of a mammary nerve. *Endocrinology* 103:1929–36.

Mena, F., P. Pacheco, D. Aguayo et al. 1979. Reflex regulation of autonomic influences upon the oxytocin-induced contractile response in the mammary gland in the anesthetized rat. *Endocrinology* 104:751–6.

Méndez, I., C. Vega, M. Zamorano et al. 2010. Vasoinhibins and the pituitary gland. *Front Horm Res* 38:184–9.

Montes de Oca, P., Y. Macotela, G. Nava et al. 2005. Prolactin stimulates integrin-mediated adhesion of circulating mononuclear cells to endothelial cells. *Lab Invest* 85:633–42.

Moreno-Carranza, B., M. Goya-Arce, C. Vega et al. 2013. Prolactin promotes normal liver growth, survival, and regeneration in rodents: effects on heptic IL-6, suppressor of cytokine signaling-3, and angiogenesis. *Am J Physiol: Regulatory, Integrative and Comparative Physiology* 305:R720–6.

Noris, G. F., D. Hol, C. Clapp et al. 1995. Histamine directly stimulates gonadotropin releasing hormone secretion from GT1-1 cells via H1 receptors coupled to phosphoinositide hydrolysis. *Endocrinology* 136:2967–74.

Ochoa, A., C. Domenzáin, C. Clapp et al. 1997. Differential effects of bFGF, EGF, TGF-Alpha and IGF-I on a hypothalamic GnRH neuronal cell line. *J Neurosci Res* 49:739–49.

Ochoa, A., P. Montes de Oca, J. C. Rivera et al. 2001. Expression of prolactin gene and secretion of prolactin by rat retinal capillary endothelial cells. *IOVS* 42:1639–45.

Ramírez, M., Z. Wu, B. Moreno-Carranza et al. 2011. Vasoinhibin gene transfer by Adeno-Associated Virus type 2 protects against VEGF- and diabetes-induced retinal vasopermeability. *IOVS* 52:8944–50.

Scharrer, E. and B. Scharrer. 1945. Neurosecretion. *Physiological Reviews* 25:171–81.

Torner, L., G. Nava, Z. Dueñas et al. 1999. Changes in the expression of neurohypophyseal prolactins during the estrus cycle and after estrogen treatment. *J Endocrinol* 161:423–32.

Torner, L., S. Mejía, F. J. López-Gómez et al. 1995. A 14 kDa prolactin-like fragment is secreted from the hypothalamo-neurohypophyseal system of the rat. *Endocrinology* 136:5454–60.

Triebel, J., Y. Macotela, G. Martínez de la Escalera et al. 2011. Prolactin and Vasoinhibins: endogenous players in diabetic retinopathy. *IUBMB Life* 63:806–10.

Trueta, C., Z. Salgado, C. Clapp et al. 1996. The catecholaminergic stimulation of gonadotropin-releasing hormone release by GT1-1 cells does not involve phosphoinositide hydrolysis. *Life Sci* 58:1453–9.

Vega, C., B. Moreno-Carranza, M. Zamorano et al. 2010. Prolactin promotes oxytocin and vasopressin release by activating neuronal nitric oxide synthase in the supraoptic and paraventricular nuclei. *Am J Physiol—Regulatory, Integrative and Comparative Physiology* 299:R1701–8.

Yaschine, T., F. Mena, and C. Beyer. 1967. Gonadal hormones and mounting behavior in the female rabbit. *Am J Physiol* 213:867–72.

Zamorano, M., M. G. Ledesma-Colunga, N. Adán et al. 2014. Vasoinhibin increases anxiety- and depression-related behaviors. *Psychoneuroendocrinology* 44:123–32.

Zermeño, C., J. Guzmán-Morales, Y. Macotela et al. 2006. Prolactin inhibits the apoptosis of chondrocytes induced by serum-starvation. *J Endocrinol* 189:R1–8.

15 Neuroendocrine and Molecular Aspects of the Physiology and Pathology of the Prostate

Maria Elena Hernández, Gonzalo E. Aranda-Abreu, and Fausto Rojas-Durán

CONTENTS

15.1 Introduction 263
15.2 Overview 264
15.2.1 Prostate Anatomy 264
15.2.2 Histology of the Prostate 265
15.3 Function of the Prostate Gland 266
15.3.1 Hormonal Regulation 266
15.3.2 Regulation of the Prostate by the Peripheral Nervous System 267
15.4 Prostatic Diseases Induced by Prolactin and the Autonomic Nervous System 268
15.4.1 Prolactin 268
15.4.2 Autonomic Nervous System 270
15.5 Conclusions 272
Acknowledgments 272
References 272

15.1 INTRODUCTION

Currently, prostate cancer is the number one cause of mortality in men in Mexico, and number two globally (Meneses-Garcia et al. 2012; Li et al. 2016). Thus, there is an imperative to attend to the causes and treatment of this disease. Although there are therapies that inhibit progression of the disease, they fail to eradicate it. Relapse after controlling the progression can occur, accompanied by more aggressive cell growth that is often fatal. Because the prostate is an accessory sex gland, the development and maintenance of which depends on androgen, therapies designed to prevent the progression of prostate disease are directed at inhibiting the action of testosterone or its metabolite, dihydrotestosterone (DHT) (Hoque et al. 2015). Prolactin (PRL) also plays a role in the regulation of the prostate, and in some reports, PRL blood levels are elevated in cases of prostatic disease (Harvey et al. 2008). The innervation of the prostate may also play a significant role in the control of the prostate. In the present chapter, we address the relative roles and interaction of androgens, PRL, and peripheral innervation in prostate disease. At the cellular level, testosterone can activate signaling pathways responsive to PRL, both hormones sharing activation of these pathways in their effects on the prostate. We also address evidence of morphological changes in the major pelvic ganglion in males related to sexual experience, and how this may affect the prostate.

15.2 OVERVIEW

15.2.1 Prostate Anatomy

The prostate is an accessory sex gland that is located in the pelvic area posterior (caudal) to the bladder. In humans, this gland is divided into four zones: (a) the inner transition zone surrounds the urethra and is a common site of origin of benign prostatic hyperplasia (BPH), (b) the central zone is located superior to the transition area, (c) the peripheral zone is the outermost part of the gland, located superior to the central zone; these last two areas are the regions where the incidence of prostate cancer is most prevalent, 90% of cancer developing in the peripheral zone, and (d) the fibrous zone. In the rat, the prostate is mainly divided into two ventral lobes the same size as the empty bladder, two lateral lobes, and one dorsal lobe. In the rat, the prostate is also located caudal to the bladder, but in the rat, prostatic ducts only converge on the urethra (Ahmed et al. 1997; Kindblom et al. 2001; Rojas-Duran 2005) (Figure 15.1).

The literature emphasizes that the prostate is a male sexual gland; however, there is evidence that in female humans and rodents, the Skene's paraurethral gland has the same tissue characteristics as the prostate in the male. The existence

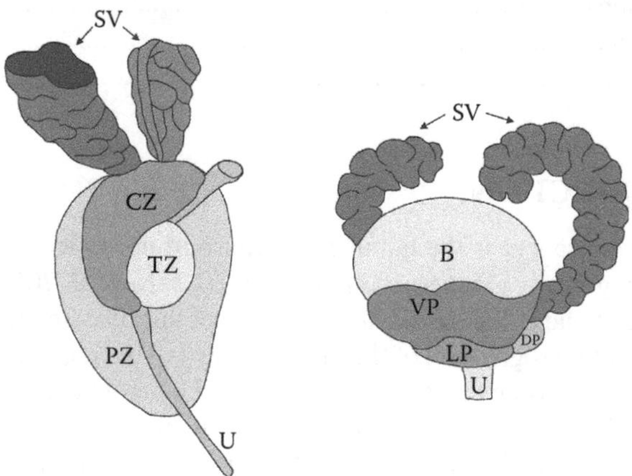

FIGURE 15.1 Drawing of the human (left) and rat prostate (right). TZ = transition zone; CZ = central zone; PZ = peripheral zone; SV = seminal vesicles; B = bladder; VP = ventral prostate; LP = lateral prostate; DP = dorsal prostate; U = urethra.

of such gland in females was first reported by Reiner de Graaf in 1672, but not named until 1880 by Alexander J.C. Skene as the paraurethral gland. Hence, the gland is now termed either the Skene's gland or the paraurethral gland (Custodio et al. 2008). This female prostate comprises clusters of alveoli and ducts embedded in a fibromuscular stroma, although compared to the male prostate, it is poorly developed.

15.2.2 Histology of the Prostate

In their gross anatomy, the human and rat prostate are different from each other, but histologically, they are comparable. Simple epithelial cells constitute the prostate alveoli and they produce the glandular fluid. A second type of epithelium forms the prostatic ducts, which contain smooth muscle, the contraction of which expels the prostatic fluid into the prostatic urethra. A feature of the epithelial cells comprising the alveoli is their internal organization: they show a polarity in the cytoplasm, in which the nucleus is typically located in the basal region close to the prostatic stroma, while the endoplasmic reticulum and Golgi apparatus are situated close to the lumina (Maslova et al. 2015). This organization enables components of the prostatic fluid to be evacuated into the alveolar luminae. The prostate contributes about 30% of the ejaculated seminal fluid. The prostatic fluid contains carnitine, phospholipids, citric acid, calcium ions, sodium, zinc, high levels of glucose (Kuosa et al. 1977) and fructose (Grayhack 1965), the latter of which provides energy to the sperm, alkaline phosphatase, prostatic-specific antigen (PSA; also known as "seminin"), and cathepsin D. The latter two promote the release of sperm from the coagulated ejaculate, which enables fertilization (Golan et al. 1983; Ando et al. 1989; Wilson et al. 1991; Cooper et al. 1990; Schieferstein 1999; Hernandez and Wilson 2012; Rojas-Duran et al. 2015).

In the female, the epithelium of the alveolus contains distinct basal proliferating, intermediate, and secretory cells. Secreting cells are the most abundant and are responsible for the production of secretory products such as PSA, acid phosphatase (Zaviacic et al. 1993, 2000), and citric acid, and respond to the addition of testosterone (Biancardi et al. 2010), while the proliferating basal cells are responsible for the renewal of the cells giving rise to the secretory cells.

More detailed studies report that the secretory glands of the female prostate have merocrine secretion (secretory vacuoles and granules) and apocrine secretion (apical "buds" that enter the lumen) (Zaviacic et al. 2000). The composition of this secretion is similar to that found in the male prostate; for that reason, it is speculated that it promotes egg fertilization (Biancardi et al. 2010). Also reported are an abundant stroma, and high levels of collagen and elastin fibers associated with smooth muscle cells and fibroblasts. Despite being a small organ, physiologically, the female prostate seems to be active because the secretory products are discharged into the ducts in response to sexual events (Zaviacic et al. 2000; dos Santos et al. 2003). Perhaps the prostate gland is rudimentary in females as a result of the absence of testosterone during embryonic and neonatal periods (Biancardi et al. 2010; Dwyer 2012), because addition of this steroid promotes its development, in which case it closely resembles the male prostate. Under these conditions, a developed female prostate is observed having two ventral lobes similar of the ventral lobes of the male prostate, and histologically, it contains columnar epithelial cells, which are more active in the presence of testosterone (Biancardi et al. 2010).

15.3 FUNCTION OF THE PROSTATE GLAND

15.3.1 Hormonal Regulation

Approximately 90% of testosterone in the prostate is metabolized to DHT by the enzyme 5α-reductase (5α-R). This metabolite is 5 to 10 times more active than testosterone itself because it forms a more stable complex with the androgen receptor (Ekman 2000). Pharmacological antagonism of DHT action produces involution and a decrease in prostate weight, which results in atrophy of the epithelial cells in the alveoli. When the hormone is given to castrated animals, the parameters return to levels similar to those observed in normal tissue (Wright et al. 1996; Rojas-Duran 2005). Loss of DHT affects the ventral lobe more than other lobes. This may be due to the high content of androgen receptors in this region. It is also possible that the relatively low content of zinc in the ventral lobe makes it more susceptible to the loss of DHT, as zinc acts as an inhibitor of apoptosis (Banerjee et al. 1995). Androgens also regulate growth, differentiation, and programmed cell death of the prostate, and there is evidence that DHT influences the development of hypertrophy and prostatic hyperplasia by promoting transcription of growth factors in the epithelium and stroma (Ullmann and Ross 1967). Clearly, the prostate gland is dependent on the presence of androgens to maintain its weight and adequate functioning, but with the inherent risk of hormonal dysfunction, which can lead to prostate disease.

The combination of PRL and androgens induces the synthesis of PSA; PRL itself regulates the synthesis and/or accumulation of zinc and citrate, and activation of Bcl2, an antiapoptotic protein. Like testosterone, PRL has tissue-specific effects, inducing in the ventral prostate the expression of C3 subunit of prostatein, and in the dorsolateral prostate, the production of the secretory protein of the seminal vesicle SVS-II or RWB and probasin expression (a prostate-specific gene) (Costello and Franklin 1994; Reiter et al. 1995, 1999; Kindblom et al 2001; Xu et al. 2003; Zhang et al. 2004; Harper et al. 2009). Nevertheless, at present, the exact function of these proteins is not known. However, as they have a high level of expression they are used as markers of activity of the different lobes of the prostate (Reiter et al. 1995).

At present, it is not known how PRL and androgens regulate the production of prostatic fluid in relation to sexual activity. Testosterone increases and remains elevated over the entire period of sexual activity in men and male rats, whereas PRL increases acutely during sexual activity and then declines to basal levels after ejaculation (Hernandez et al. 2006, 2007). PRL, secreted by the anterior pituitary gland, is under inhibitory control by dopamine, which is produced in neurons of the arcuate nucleus of the hypothalamus (dopamine is "the" Prolactin Inhibitory Factor, PIF) (Hernandez et al. 1999). When levels of dopamine in the hypothalamo–hypophyseal portal system decrease as a result of sexual behavior, the secretion of PRL is disinhibited and its blood levels increase. PRL secretion is again inhibited when the portal vessel levels of dopamine increase, returning PRL blood levels to the basal state.

This phasic release of PRL may prevent its possible adverse effects on the prostate (Hernandez et al. 2006; Galdiero et al. 2012; Herrera-Covarrubias et al. 2015). By contrast, testosterone levels in blood persist in two forms: free, which can be captured by the target organ, and bound to Sex Hormone Binding Globulin (SHBG), which delays its release.

The actions of testosterone and PRL are triggered upon binding to their receptors located in the prostate. The prostate has only one type of receptor for both testosterone and DHT through which they act. However, the subsequent molecular mechanisms of regulation activated by testosterone or DHT are different beyond their binding to the receptor (Askew et al. 2007; Brinkmann 2011). There are two types of membrane receptors for PRL in the prostate—the long and the short types, differing in the length of the intracellular domain. However, the extracellular domains of both receptors seem to have the same affinity for PRL (Barash and Madar 1992). The two types of receptors differ in the signaling pathways triggered and the regulatory mechanisms involved. The long receptor uses mainly the STAT pathway, the functions of which are related to cellular metabolism, whereas the short receptor uses the MAPK pathway, activation of which is related to the regulation of cell proliferation (Hernandez and Wilson 2012). Sexual activity not only increases the serum PRL and androgen serum levels, but also affects the expression of these receptors. That is, sexual activity promotes a significant moderate response in the long PRL messenger and a trend toward an increase in the short PRL messenger. This response is consistent with the long PRL receptor being involved in cellular metabolism and production of prostatic fluid, and the short PRL receptor being associated with cellular proliferation (Silva-Morales 2009), although the possibility that the latter also participates in the production of prostatic fluid cannot be excluded.

The regulation of PRL secretion and expression of its receptors and signaling pathways is affected by prolonged sexual activity. PRL blood levels of rats that periodically engage in sexual behavior remain elevated for the entire duration of sexual stimulation. In that case, there is an increase of PRL messenger RNA for both receptors, with corresponding activation of STAT and MAPK signaling pathways. This process provides a higher production rate and semen quality that promotes both the survival of sperm and egg fertilization (Sofikitis and Miyagawa 1993; Soto-Cid et al. 2006; Rojas-Duran et al. 2015).

15.3.2 Regulation of the Prostate by the Peripheral Nervous System

The function of the prostate is also controlled by the autonomic nervous system (ANS). The main nerves known to regulate the prostate are the Pelvic nerve (PvN) and the Hypogastric Nerve (HgN), which emerge from the thoracic, lumbar, and sacral segments of the spinal cord. The PvN originates from the L6-S1 segments forming a trunk; the HgN fibers originate from the T12 and L6 segments, converging in the mesenteric ganglion and emerging as the HgN. Both nerves converge in the major pelvic ganglion, and then distribute to the pelvic area including the prostate gland (Wang et al. 1991; McVary et al. 1998; Diaz et al. 2010). It is not known whether the ANS normally regulates the content of prostatic fluid, but experimental stimulation of the HgN induces the secretion of prostatic fluid, which is not observed when the PvN is stimulated (Vaalasti et al. 1986; Watanabe et al. 1988).

There are also structural changes in the pelvic ganglion that are related to sexual experience (White et al. 2013). The pelvic ganglion contains large monopolar ganglion cells, smaller SIF (Small Intensely Fluorescent) cells, which are labeled by retrograde markers, and satellite cells that surround the ganglion cells (Dail and Evan 1978; Dail and Dziurzynski 1985; Hanani 2010). This organization of the pelvic ganglion changes depending on sexual experience. The ganglion from male rats that experienced sexual activity is more highly organized and has a greater number of satellite cells that surround, or are near, the ganglion cells, than the ganglion of sexually inactive males, that despite being in the presence of a receptive female showed low, or no, interest in courtship or copulation. Perhaps these effects are due to low testosterone production, the decline in androgen receptor expression in the ganglion or low synthesis of the survival neurotrophic factor, neurturin (Yan and Keast 2008; Yan et al. 2012).

The organization and development of the ganglion depends on the presence of androgens (Melvin and Hamill 1989; Keast 1999), and the HgN and PvN are involved in the secretory functions of the prostatic fluid by inducing the contraction of the smooth muscle that surrounds epithelial cells and ducts that converge in the prostatic urethra (Bruschini et al. 1978; McVary et al. 1998; Ventura et al. 2002). Apparently, the nerve-mediated effects on the prostate are not only mediated by the classical neurotransmitters norepinephrine and acetylcholine, through receptors such as the alpha and beta adrenergic (sympathetic), and M1 and M2 muscarinic (parasympathetic), but also by neuromodulators such as vasoactive intestinal peptide (VIP), neuropeptide Y (NPY), somatostatin, bombesin, neurotensin, calcitonin gene-related peptide (CGRP), nitric oxide, and even PRL (Rodriguez et al. 2005; Pozuelo et al. 2010; Hernandez and Wilson 2012; Morgat et al. 2014). It is important to recognize that neurotransmitters and neuromodulators are also related to other functions, such as cell proliferation, repair of nerve tissue, neural network "training" with neurotrophic factors, and regulation of synaptic transmission; all of these impact the functioning of the pelvic ganglion and thereby the prostate gland (Calenda et al. 2012).

15.4 PROSTATIC DISEASES INDUCED BY PROLACTIN AND THE AUTONOMIC NERVOUS SYSTEM

15.4.1 Prolactin

The etiology of chronically elevated levels of PRL (hyperPRL) is diverse and includes the micro or macroadenomas and administration of certain medications. HyperPRL is classified according to the levels in blood as mild, medium, and severe. Mild hyperPRL does not exceed 100 ng/ml, medium up to 500 ng/ml, and severe may exceed 1000 ng/ml (Kooy et al. 1988; Hernandez and Wilson 2012). Behaviorally, the effects of hyperPRL are controversial, some authors reporting that this elevation either does or does not affect sexual activity. The discrepancy may be due to different experimental paradigms used to produce high systemic levels of PRL (Shrenker and Bartke 1987; Drago and Lissandrello 2000; Krüger et al. 2003; Galdiero et al. 2012; Roke et al. 2012; Montalvo et al. 2013).

When the PRL level in blood is higher than 100 ng/ml, male rats only perform mounts and intromissions, and sometimes can ejaculate but with a long latency; these effects are not observed in males with mild hyperPRL (levels below 100 ng/ml), no parameter being affected by this condition (Hernandez et al. 2006). A possible basis for this difference is the lack of control in the medial preoptic area where this behavior pattern is regulated. However, at the prostatic level the elevated PRL levels disrupt the morphological characteristics of the prostate. After 15 days of treatment with PRL, there was an increase in alveolar area and epithelial height, suggesting early prostate pathology even though sexual behavior appeared normal (Hernandez et al. 2006). After three months, elevated PRL levels induce severe prostatic tissue changes in the form of pseudostratification and increased cell proliferation, leading to an increase in the formation of buds—the invasion of the alveolar lumen by epithelial tissue. As this pathology progresses, it becomes difficult to identify individual epithelial cells, leading to a phase of metaplasia, and over time, these changes develop into prostatic dysplasia (Herrera-Covarrubias et al. 2015). Histologically, metaplasia corresponds to a mature epithelium that resembles the normal tissue but with a characteristic mitotically active area, and can be considered as the origin for the generation of dysplastic cells with the potential of malignancy. This dysplasia can occur after the first month of mild hyperPRL and is characterized by atypical proliferation and alterations in the size, shape, and organization of the cells, but its growth is restricted to the epithelial layer without invasion of other tissue. This is the early stage of development that can be transformed, over time, into neoplasia. For this reason, dysplasia is considered a preneoplastic or precancerous stage (Herrera-Covarrubias et al. 2015). While the neoplasia remains in situ, it is considered benign, but when it spreads outside the limits of the tissue, it is considered malignant (Van Cleef et al. 2014). Thus, there are multiple factors affecting the cellular morphology of the prostate: changes in the systemic levels of PRL and/or testosterone, variations in the expression of the PRL receptor, and activation of the intracellular mechanisms that regulate both cell function and the morphological characteristics of the epithelial cells. In a preliminary study, we found that systemic elevation of PRL (mild hyperPRL) in subjects with sexual experience up-regulates short PRL receptors and down-regulates large PRL receptors (Figure 15.2). Stat5a/Stat3, MAPK, and PI3K/Akt isoforms are the main mediators that promote the prostatic histological changes (Thompson et al. 2002; Abdulghani et al. 2008; Park et al. 2014; Siveen et al. 2014). However, activation of these pathways can be as varied as the expressed pathology in the prostate, possibly as the NFkB is also involved in promoting cancer. Other molecular factors in prostate cancer are (1) histone trimethylation of the p53 gene (a tumor suppressor gene), induced by the constitutive activation of the PRL receptor (Tan et al. 2013, 2014) and (2) the reduced activation of the inhibitor protein Raf kinase that eventually is closely related to the PRL-activated Stat3 pathway (Saha et al. 2014; Yousuf et al. 2014). While the roles of each of the signaling pathways and the significance of trimethylation of the p53 gene are not well understood, they all converge on two processes—increased cell proliferation and decreased apoptosis. MicroRNAs regulate all these proteins, regardless of their mechanism of action, and Stat3 can accelerate their cancer-producing effects via miR124 and miR106a (Wei et al. 2013; Zhang et al. 2013; Siveen et al. 2014).

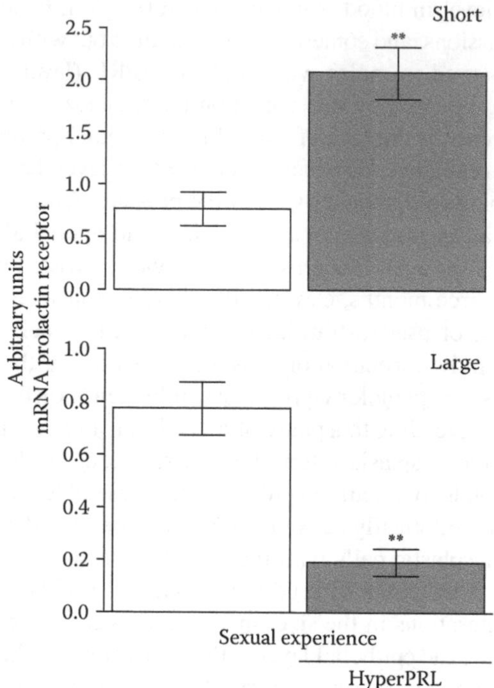

FIGURE 15.2 Expression levels of the transcript for the PRL receptor in sexually active subjects treated with prolactin. HyperPRL = hyperprolactinemia.

Thus, pathologies of the prostate are at least the result of the participation of several signaling pathways in response to hormonal activation.

The female prostate is also susceptible to disease, for example, ductal infections and abscesses that are prevalent in childhood and puberty (Nickles et al. 2009). Prostatic disease is also associated with age in women and female rodents (Pongtippan et al. 2004; Biancardi et al. 2010). Because few studies have been conducted with the female prostate, at present, the etiology and possible role of hormones in female prostatic disease is not known. However, it is known that circulating levels of estradiol, testosterone, and dehydroepiandrosterone (DEHA) are higher in adulthood than in youth and senility, and a reduction in estradiol and DHEA in senility was reported (Custodio et al. 2008). While these findings are suggestive of a correlation between hormonal levels and female prostatic pathology, further research is required.

15.4.2 Autonomic Nervous System

Interruption of the connection between the ANS and the prostate gland by axotomy produces effects that are similar to those related to elevated PRL levels; that is, both promote invasion of epithelial cells (forming "papillae") within

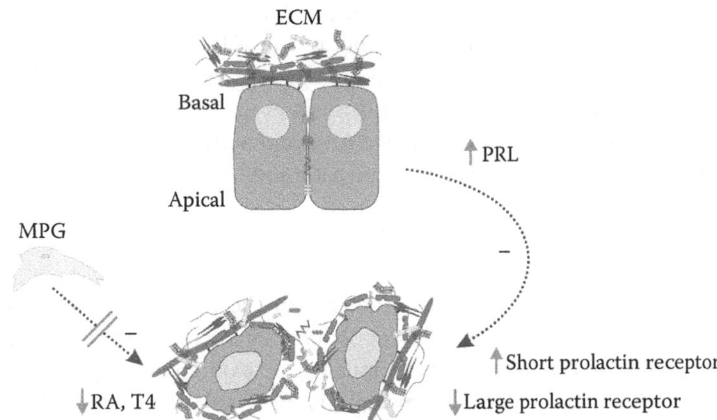

FIGURE 15.3 Effects of axotomy and hyperPRL on the morphological characteristics of epithelial cells in the prostate gland. Left arrow shows that axotomy produces a reduction in RA and T4 levels that, in turn, affect the morphology of the cells. The same situation is observed in the right arrow showing that the elevation of systemic PRL respectively increases and decreases the short and large prolactin receptor. MPG = major pelvic ganglion; ECM = extracellular matrix; PRL = prolactin; RA = androgen receptor; T4 = testosterone.

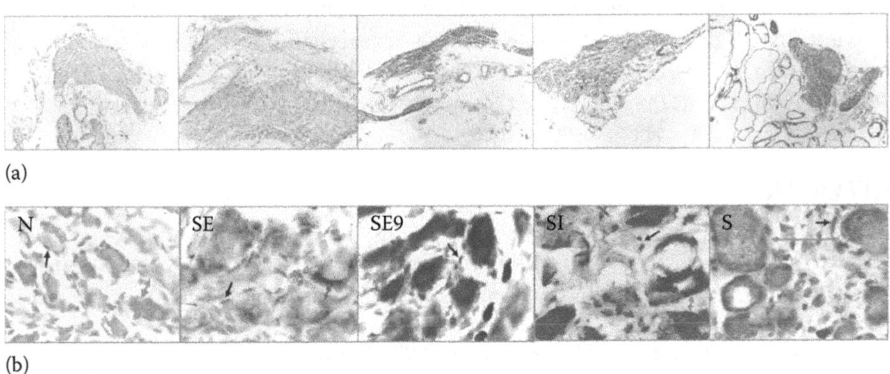

FIGURE 15.4 Nissl stained tissue sections at 4X (a) and 100X (b) of the major pelvic ganglion from the naive (N), sexually expert (SE), sexually experienced for 9 months (SE9), sexually inactive (SI), and sluggish (S) subjects. (a) Shows the changes in the size of the ganglion depending on the sexual characteristics of the subjects; the one from the sexually expert subject is the larger ganglion and that from the sluggish subject is the smaller ganglion. (b) Shows the hypertrophy of ganglion cells of all subjects placed in the presence of a receptive female; a better organization and proximity of ganglion cells (gray arrow) is observed in SE and SE9 subjects. Satellite cells (black arrow) have closer proximity to the ganglion cells in the SE subject.

the alveoli, as a result of accelerated cell proliferation, and reduction in the glandular epithelium (Figure 15.3).

While the pelvic ganglia of sexually experienced subjects show a reorganization and increase in the size of ganglion cells (Figure 15.4), after denervation, we found a higher rate of cell proliferation, resembling cellular hyperplasia (unpublished data).

NpY, one of the neuromodulators synthesized in the pelvic ganglion, regulates cell proliferation in the prostate by binding to the Y1 receptor and activating the Mapk pathway, while adrenergic and cholinergic actions use cAMP (Prieto et al. 1990; Arciszewski and Wąsowicz 2006; Ruscica et al. 2006, 2007). Evidently, both neural and hormonal mechanisms respond to sexual activity and can affect the prostate, even to the extent of inducing prostate pathology.

15.5 CONCLUSIONS

Sexual behavior maintains the healthy function of the prostate by inducing trophic effects on the major pelvic ganglion. It is possible that alteration of behavioral events, and the hormonal and autonomic concomitants, or axotomy, can produce prostatic pathology, for example, hyperplasia, metaplasia, or dysplasia via their effects on the structure of the pelvic ganglion and/or prostatic hormone receptors. Further research is required to ascertain the relative contributions of specific neural and endocrine factors to the healthy physiology and pathology of the prostate in females as well as in males.

ACKNOWLEDGMENTS

Although none of the authors was a direct student of Dr. Carlos Beyer, his wisdom in the use of simple methodologies to answer complicated scientific questions had a major impact on us as students and now as researchers. This work was supported by CONACYT Project No. 106531 (MEH), PROMEP UV-PTC-716 (RDF), and by PRODEP support to the Academic Group of Neurochemistry UV-CA 304.

REFERENCES

Abdulghani, J., L. Gu, A. Dagvadorj, J. Lutz, B. Leiby, G. Bonuccelli, M.P. Lisanti, T. Zellweger, K. Alanen, T. Mirtti, T. Visakorpi, L. Bubendorf, and M.T. Nevalainen. 2008. "Stat3 promotes metastatic progression of prostate cancer." *The American Journal of Pathology* 172(6):1717–28.

Ahmed, M.M., C.T. Lee, and J.E. Oesterling. 1997. "Current trends in the epidemiology of prostate diseases: benign hyperplasia and adenocarcinoma." In *Prostate: Basic and Clinical Aspects*, edited by K. N. Rajesh. New York: CRC Press. 1997; pp. 3–25.

Ando, S., A. Carpino, M. Buffone, M. Maggiolini, and D. Sisci. 1989. "The evaluation of free-L-carnitine, zinc and fructose in the seminal plasma of patients with varicocele and normozoospermia." *Andrologia* 21(2):155–60.

Arciszewski, M.B. and Wąsowicz, K. 2006. "Noradrenergic and cholinergic innervation of the accessory sexual glands in male sheep." *Acta Veterinaria Hungarica* 54(1):71–83.

Askew, E.B., R.T. Gampe, T.B. Stanley, J.L. Faggart, and E.M. Wilson. 2007. "Modulation of androgen receptor activation function 2 by testosterone and dihydrotestosterone." *The Journal of Biological Chemistry* 282(35):25801–16.

Banerjee, P.P., S. Banerjee, K.I. Tilly, J.L. Tilly, T.R. Brown, and B.R. Zirkin. 1995. "Lobe-specific apoptotic cell death in rat prostate after androgen ablation by castration." *Endocrinology* 136(10):4368–76.

Barash, I., Z. Madar, and A. Gertler. 1992. "Short-term in vivo regulation of prolactin receptors in the liver, testes, kidneys, and mammary gland of rats." *Receptor* 2(1):39–44.

Biancardi, M.F., F.C. Santos, L. Madi-Ravazzi, R.M. Goes, P.S. Vilamaior, S.L. Felisbino, and S.R. Taboga. 2010. "Testosterone promotes an anabolic increase in the rat female prostate (Skene's paraurethral gland) which acquires a male ventral prostate phenotype." *Anatomical Record* 293(12):2163–75.

Brinkmann, A.O., 2011. "Molecular mechanisms of androgen action—a historical perspective." *Methods Mol Biol* 776:3–24.

Bruschini, H., R.A. Schmidt, and E.A. Tanagho. 1978. "Neurologic control of prostatic secretion in the dog." *Investigative Urology* 15(4):288–90.

Calenda, G., T.D. Strong, C.P. Pavlovich, E.M. Schaeffer, A.L. Burnett, W. Yu, K.P. Davies, and T.J. Bivalacqua. 2012. "Whole genome microarray of the major pelvic ganglion after cavernous nerve injury: New insights into molecular profile changes after nerve injury." *BJU International* 109(10):1552–64.

Cooper, T.G., W. Weidner, and E. Nieschlag. 1990. "The influence of inflammation of the human male genital tract on secretion of the seminal markers alpha-glucosidase, glycerophosphocholine, carnitine, fructose and citric acid." *Int J Androl* 13(5):329–36.

Costello, L.C. and R.B. Franklin. 1994. "Effect of prolactin on the prostate." *The Prostate* 24:162–6.

Custodio, A.M., F.C. Santos, S.G. Campos, P.S. Vilamaior, R.M. Goes, and S.R. Taboga. 2008. "Aging effects on the mongolian gerbil female prostate (Skene's paraurethral glands): Structural, ultrastructural, quantitative, and hormonal evaluations." *Anatomical Record* 291(4):463–74.

Dail, W.G. and R. Dziurzynski. 1985. "Substance P immunoreactivity in the major pelvic ganglion of the rat." *Anatomical Record* 212(1):103–9.

Dail, W.G. and A.P. Evan. 1978. "Effects of chronic deafferentation on adrenergic ganglion cells and small intensely fluorescent cells." *J Neurocytol.* 7 (1):25–37.

Diaz, R., L.I. Garcia, J. Locia, M. Silva, S. Rodriguez, C.A. Perez, G.E. Aranda-Abreu, J. Manzo, R. Toledo, and M.E. Hernandez. 2010. "Histological modifications of the rat prostate following transection of somatic and autonomic nerves." *Anais da Academia Brasileira de Ciencias* 82(2):397–404.

dos Santos, F.C., H.F. Carvalho, R.M. Goes, and S.R. Taboga. 2003. "Structure, histochemistry and ultrastructure of the epithelium and stroma in the gerbil (Meriones unguiculatus) female prostate." *Tissue and Cell* 35(6):447–57.

Drago, F. and C.O. Lissandrello. 2000. "The 'low-dose' concept and the paradoxical effects of prolactin on grooming and sexual behavior." *European Journal of Pharmacology* 405(1–3):131–137.

Dwyer, P.L., 2012. "Skene's gland revisited: Function, dysfunction and the G spot." *International Urogynecology Journal and Pelvic Floor Dysfunction* 23(2):135–7.

Ekman, P., 2000. "The prostate as an endocrine organ: Androgens and estrogens". *Prostate* Suppl 10:14–18.

Galdiero, M., R. Pivonello, L.F. Grasso, A. Cozzolino, and A. Colao. 2012. "Growth hormone, prolactin, and sexuality." *Journal of Endocrinological Investigation* 35(8):782–94.

Golan, R., Y. Soffer, S. Katz, R. Weissenberg, O. Wasserzug, and L.M. Lewin. 1983. "Carnitine and short-chain acylcarnitines in the lumen of the human male reproductive tract." *International Journal of Andrology* 6(4):349–57.

Grayhack, J.T., 1965. "Effect of testosterone-estradiol administration on citric acid and fructose content of the rat prostate." *Endocrinology* 77(6):1068–74.

Hanani, M., 2010. "Satellite glial cells in sympathetic and parasympathetic ganglia: In search of function." *Brain Research Reviews* 64(2):304–27.

Harper, C.E., B.B. Patel, L.M. Cook, J. Wang, T. Shirai, I.A. Eltoum, and C.A. Lamartiniere. 2009. "Characterization of SV-40 Tag rats as a model to study prostate cancer." *BMC Cancer* 9:30.

Harvey, P.W., D.J Everett, and C.J. Springall. 2008. "Adverse effects of prolactin in rodents and humans: breast and prostate cancer." *Journal of Psychopharmacology (Oxford, England)* 22(2 Suppl):20–27.

Hernandez, M.E., A. Soto-Cid, F. Rojas, L.I. Pascual, G.E. Aranda-Abreu, R. Toledo, L.I. Garcia, A. Quintanar-Stephano, and J. Manzo.2006. "Prostate response to prolactin in sexually active male rats." *Reproductive Biology and Endocrinology* 4:28–39.

Hernandez, M.E., A. Soto-Cid, G.E. Aranda-Abreu, R. Dias, F. Rojas, L.I. Garcia, R. Toledo, and J. Manzo. 2007. "A study of the prostate, androgens and sexual activity of male rats." *Reproductive Biology and Endocrinology* 5:11.

Hernández M.E., M. del Mar Hernández, M. Díaz-Muñoz, C. Clapp, G. Martínez de la Escalera. 1999. "Potentiation of prolactin secretion following lactotrope escape from dopamine action. II. Phosphorylation of the alpha(1) subunit of L-type, voltage-dependent calcium channels." *Neuroendocrinology* 70(1):31–42.

Hernandez, M.E. and M.J. Wilson. 2012. "The role of prolactin in the evolution of prostate cancer." *Open Journal of Urology* 2(23):188–97.

Herrera-Covarrubias, D., G.A. Coria-Avila, P. Chavarria-Xicotencatl, C. Fernandez-Pomares, J. Manzo, G.E. Aranda-Abreu, and M.E Hernandez. 2015. "Long-term administration of prolactin or testosterone induced similar precancerous prostate lesions in rats." *Experimental Oncology* 37(1):13–18.

Hoque, A., S. Yao, C. Till, A.R. Kristal, P.J. Goodman, A.W. Hsing, C.M. Tangen, E.A. Platz, F.Z. Stanczyk, J.K. Reichardt, A. vanBokhoven, M.L. Neuhouser, R.M. Santella, W.D. Figg, D.K. Price, H.L. Parnes, S.M. Lippman, C.B. Ambrosone, and I.M. Thompson. 2015. "Effect of finasteride on serum androstenedione and risk of prostate cancer within the prostate cancer prevention trial: Differential effect on high- and low-grade disease." *Urology* 85(3):616–20.

Keast, J.R., 1999. "The autonomic nerve supply of male sex organs—An important target of circulating androgens." *Behavioural Brain Research* 105(1):81–92.

Kindblom, J., K. Dillner, J. Törnell, and H. Wennbo. 2001. "Actions of prolactin in the prostate gland." In *Prolactin*, edited by N. Horseman. Boston, MA: Kluwer Academic Publishers, pp. 233–246.

Kooy, A., R.F. Weber, M.P. Ooms, and J.T. Vreeburg. 1988. "Deterioration of male sexual behavior in rats by the new prolactin-secreting tumor 7315b." *Hormones and Behavior* 22(3):351–61.

Krüger, T.H., P. Haake, J. Haverkamp, M. Krämer, M.S. Exton, B. Saller, N. Leygraf, U. Hartmann, and M. Schedlowski. 2003. "Effects of acute prolactin manipulation on sexual drive and function in males." *Journal of Endocrinology* 179(3):357–65.

Kuosa, A., P. Härkönen, and R.S. Santti. 1977. "Studies on the inhibition of testosterone action by cycloheximide: Evidence for a protein activator of glucose metabolism in the ventral prostate of the rat." *Journal of Steroid Biochemistry* 8(4):259–67.

Li, Q., Y. Li, Y. Wang, Z. Cui, L. Gong, Z. Qu, Y. Zhong, J. Zhou, Y. Zhou, Y. Gao, and Y. Li. 2016. "Quantitative proteomic study of human prostate cancer cells with different metastatic potentials." *International Journal of Oncology* 48:1437–6.

Maslova, K., E. Kyriakakis, D. Pfaff, A. Frachet, A. Frismantiene, L. Bubendorf, C. Ruiz, T. Vlajnic, P. Erne, T.J. Resink, and M. Philippova. 2015. "EGFR and IGF-1R in regulation of prostate cancer cell phenotype and polarity: opposing functions and modulation by T-cadherin." *FASEB Journal* 29(2):495–507.

McVary, K.T., K.E. McKenna, and C. Lee. 1998. "Prostate innervation." *Prostate* 36(Suppl. 8):2–13.

Melvin, E. and W. Hamill. 1989. "Androgen-specific pelvic ganglion critical periods for the organization of the major." *Changes* 9(2):738–42.

Meneses-Garcia, A., L.M. Ruiz-Godoy, A. Beltran-Ortega, F. Sanchez-Cervantes, R. Tapia-Conyer, and A. Mohar. 2012. "Main malignant neoplasms in Mexico and their geographic distribution, 1993–2002." *Revista de Investigacion Clinica* 64(4):322–9.

Montalvo, I., L. Ortega, X. Lopez, M. Sole, R. Monseny, J. Franch, E. Vilella, and J. Labad. 2013. "Changes in prolactin levels and sexual function in young psychotic patients after switching from long-acting injectable risperidone to paliperidone palmitate." *International Clinical Psychopharmacology* 28(1):46–9.

Morgat, C., A.K. Mishra, R. Varshney, M. Allard, P. Fernandez, and E. Hindie. 2014. "Targeting neuropeptide receptors for cancer imaging and therapy: Perspectives with bombesin, neurotensin, and neuropeptide-Y receptors." *J Nucl Med* 55(10):1650–7.

Nickles, S.W., J.T. Burgis, S. Menon, and J.L. Bacon. 2009. "Prepubertal skene's abscess." *Journal of Pediatric and Adolescent Gynecology* 22(1):e21–2.

Park, J.S., J.K. Kwon, H.R. Kim, H.J. Kim, B.S. Kim, and J.Y. Jung. 2014. "Farnesol induces apoptosis of DU145 prostate cancer cells through the PI3K/Akt and MAPK pathways." *International Journal of Molecular Medicine* 33(5):1169–76.

Pongtippan A., A. Malpica, C. Levenback, M.T. Deavers, and E.G. Silva. 2004. "Skene's gland adenocarcinoma resembling prostatic adenocarcinoma." *Int J Gynecol Pathol.* 23(1):71–4.

Pozuelo, J.M., R. Rodriguez, R. Arriazu, I. Ingelmo, R. Martin, and L. Santamaria. 2010. "Changes in the number and volume of NPY and VIP neurons from periprostatic accessory vegetative ganglia in pre- and peripubertal rats. A stereological study." *Tissue and Cell* 42(1):1–8.

Prieto J.C., C. Hueso, and M.J. Carmena. 1990. "Modulation of the beta-adrenergic stimulation of cyclic AMP accumulation in rat prostatic epithelial cells by membrane fluidity." *Gen Pharmacol.* 21(6):931–3.

Reiter, E., S. Lardinois, M. Klug, B. Sente, B. Hennuy, M. Bruyninx, J. Closset, and G. Hennen. 1995. "Androgen-independent effects of prolactin on the different lobes of the immature rat prostate." *Molecular and Cellular Endocrinology* 112(1):113–22.

Reiter, E., B. Hennuy, M. Bruyninx, A. Cornet, M. Klug, M. McNamara, J. Closset, and G. Hennen. 1999. Effects of pituitary hormones on the prostate." *Prostate* 38(2):159–65.

Rodriguez, R., J.M. Pozuelo, R. Martin, R. Arriazu, and L. Santamaria. 2005. "Stereological quantification of nerve fibers immunoreactive to PGP 9.5, NPY, and VIP in rat prostate during postnatal development." *J Androl* 26(2):197–204.

Rojas-Duran, F., L.I. Pascual-Mathey, K. Serrano, G.E. Aranda-Abreu, J. Manzo, A.H. Soto-Cid, and M.E. Hernandez. 2015. "Correlation of prolactin levels and PRL-receptor expression with Stat and Mapk cell signaling in the prostate of long-term sexually active rats." *Physiology and Behavior* 138:188–92.

Rojas-Duran, F., 2005. *Influencia de la prolactina y las hormonas sexuales esteroideas sobre la morfología e histología prostática en la rata.*, Instituto de Neuroetología, Universidad Veracruzana, Xalapa, Veracruz, México.

Roke, Y., J.K. Buitelaar, A.M. Boot, D. Tenback, and P.N. van Harten. 2012. "Risk of hyperprolactinemia and sexual side effects in males 10-20 years old diagnosed with autism spectrum disorders or disruptive behavior disorder and treated with risperidone." *Journal of Child and Adolescent Psychopharmacology* 22(6):432–9.

Ruscica, M., E. Dozio, S. Boghossian, G. Bovo, V. Martos Riano, M. Motta, and P. Magni. 2006. "Activation of the Y1 receptor by neuropeptide Y regulates the growth of prostate cancer cells." *Endocrinology* 147(3):1466–73.

Ruscica, M., E. Dozio, M. Motta, and P. Magni. 2007. "Modulatory actions of neuropeptide Y on prostate cancer growth: Role of MAP kinase/ERK 1/2 activatio." *Adv Exp Med Biol* 604:96–100.

Saha, A., J. Blando, E. Silver, L. Beltran, J. Sessler, and J. DiGiovanni. 2014. "6-Shogaol from dried ginger inhibits growth of prostate cancer cells both in vitro and in vivo through inhibition of STAT3 and NF-kB signaling." *Cancer Prevention Research* 7(6):627–638.

Schieferstein, G., 1999. "Prostate-specific antigen (PSA) in human seminal plasma." *Arch Androl.* 42(3):193–7.

Shrenker, P. and A. Bartke. 1987. "Effects of hyperprolactinaemia on male sexual behaviour in the golden hamster and mouse." *Journal of Endocrinology* 112(2):221–8.

Silva-Morales, M., 2009. *Niveles de mRNA del receptor a prolactina en la próstata durante la conducta sexual de la rata.* Investigación, Instituto de Neuroetología, Universidad Veracruzana, Xalapa, Veracruz, México.

Siveen, K.S., S. Sikka, R. Surana, X. Dai, J. Zhang, A.P. Kumar, B.K. Tan, G. Sethi, and A. Bishayee. 2014. "Targeting the STAT3 signaling pathway in cancer: role of synthetic and natural inhibitors." *Biochimica et biophysica acta* 1845(2):136–54.

Sofikitis, N.V and I. Miyagawa. 1993. "Endocrinological, biophysical, and biochemical parameters of semen collected via masturbation versus sexual intercourse." *J Androl* 14(5):366–73.

Soto-Cid, A., L.C. Hernandez-Kelly, M.E. Hernandez, J. Manzo, M.E. Gonzalez-Mejia, R.C. Zepeda, and A. Ortega. 2006. "Signal transducers and activators of transcription 1 and 3 in prostate: Effect of sexual activity." *Life Sciences* 79(9):919–24.

Tan, D., P. Tang, J. Huang, J. Zhang, W. Zhou, and A.M. Walker. 2014. "Expression of a constitutively active prolactin receptor causes histone trimethylation of the p53 gene in breast cancer." *Chinese Medical Journal* 127(6):1077–83.

Tan, D., S. Tan, J. Zhang, P. Tang, J. Huang, W. Zhou, and S. Wu. 2013. "Histone trimethylation of the p53 gene by expression of a constitutively active prolactin receptor in prostate cancer cells." *Chin J Physiol* 56(5):282–90.

Thompson, C.J., N.N. Tam, J.M. Joyce, I. Leav, and S.M. Ho. 2002. "Gene expression profiling of testosterone and estradiol-17 beta-induced prostatic dysplasia in Noble rats and response to the antiestrogen ICI 182,780." *Endocrinology* 143(6):2093–105.

Ullmann A.S. and O.A. Ross. 1967. "Hyperplasia, atypism, and carcinoma in situ in prostatic periurethral glands." *Am J Clin Pathol* 47(4):497–504.

Vaalasti, A., A.M. Alho, H. Tainio, and A. Hervonen. 1986. "The effect of sympathetic denervation with 6-hydroxydopamine on the ventral prostate of the rat." *Acta Histochem* 79(1):49–54.

Van Cleef, A., S. Altintas, M. Huizing, K. Papadimitriou, P. Van Dam, and W. Tjalma. 2014. "Current view on ductal carcinoma in situ and importance of the margin thresholds: A review." *Facts, Views & Vision in Ob Gyn* 6(4):210–18.

Ventura, S., J. Pennefather, and F. Mitchelson. 2002. "Cholinergic innervation and function in the prostate gland." *Pharmacol Ther* 94(1/2):93–112.

Wang, J.M., K.E. McKenna, K.T. McVary, and C. Lee. 1991. "Requirement of innervation for maintenance of structural and functional integrity in the rat prostate." *Biol Reprod* 44(6):1171–6.

Watanabe H., M. Shima, M. Kojima, and H. Ohe. 1988. "Dynamic study of nervous control on prostatic contraction and fluid excretion in the dog." *J Urol* 140(6):1567–70.

Wei, J., F. Wang, L.Y. Kong, S. Xu, T. Doucette, S.D. Ferguson, Y. Yang, K. McEnery, K. Jethwa, O. Gjyshi, W. Qiao, N.B. Levine, F.F. Lang, G. Rao, G.N. Fuller, G.A. Calin, and A.B. Heimberger. 2013. "miR-124 inhibits STAT3 signaling to enhance T cell-mediated immune clearance of glioma." *Cancer Res* 73(13):3913–26.

White, C.W., J.H. Xie, and S. Ventura. 2013. "Age-related changes in the innervation of the prostate gland: implications for prostate cancer initiation and progression." *Organogenesis* 9(3):206–15.

Wilson, M.J., J.N. Whitaker, and A.A. Sinha. 1991. "Immunocytochemical localization of cathepsin D in rat ventral prostate: evidence for castration-induced expression of cathepsin D in basal cells." *Anat Rec* 229(3):321–33.

Wright, A.S., L.N. Thomas, R.C. Douglas, C.B. Lazier, and R.S. Rittmaster. 1996. "Relative potency of testosterone and dihydrotestosterone in preventing atrophy and apoptosis in the prostate of the castrated rat." *J Clin Invest* 98(11):2558–63.

Xu, X., W. Wu, V. Williams, A. Khong, Y.H. Chen, C. Deng, and A.M. Walker. 2003. "Opposite effects of unmodified prolactin and a molecular mimic of phosphorylated prolactin on morphology and the expression of prostate specific genes in the normal rat prostate." *Prostate* 54(1):25–33.

Yan, H. and J.R. Keast. 2008. "Neurturin regulates postnatal differentiation of parasympathetic pelvic ganglion neurons, initial axonal projections, and maintenance of terminal fields in male urogenital organs." *J Comp Neurol* 507(2):1169–83.

Yan, J., B. Xie, J.L. Capodice, and A.E. Katz. 2012. "Zyflamend inhibits the expression and function of androgen receptor and acts synergistically with bicalutimide to inhibit prostate cancer cell growth." *Prostate* 72(3):244–52.

Yousuf, S., M. Duan, E.L. Moen, S. Cross-Knorr, K. Brilliant, B. Bonavida, T. LaValle, K.C. Yeung, F. Al-Mulla, E. Chin, and D. Chatterjee. 2014. "Raf kinase inhibitor protein (RKIP) blocks signal transducer and activator of transcription 3 (STAT3) activation in breast and prostate cancer." *PLoS One* 9(3):e92478.

Zaviacic, M., J. Sidlo, and M. Borovsky. 1993. "Prostate specific antigen and prostate specific acid phosphatase in adenocarcinoma of Skene's paraurethral glands and ducts." *Virchows Arch A Pathol Anat Histopathol* 423(6):503–5.

Zaviacic, M., V. Jakubovska, M. Belosovic, and J. Breza. 2000. "Ultrastructure of the normal adult human female prostate gland (Skene's gland)." *Anat Embryol (Berl)* 201(1):51–61.

Zhang, J., N. Gao, S. Kasper, K. Reid, C. Nelson, and R.J. Matusik. 2004. "An androgen-dependent upstream enhancer is essential for high levels of probasin gene expression." *Endocrinology* 145(1):134–48.

Zhang, M., Y. Ye, J. Cong, D. Pu, J. Liu, G. Hu, and J. Wu. 2013. "Regulation of STAT3 by miR-106a is linked to cognitive impairment in ovariectomized mice." *Brain Res* 1503:43–52.

16 From Sexual Behavior to Analgesia to an Antinociceptive Agent *Glycinamide*

Porfirio Gómora-Arrati, Galicia-Aguas Y., Oscar González-Flores, and Barry R. Komisaruk

CONTENTS

16.1 Introduction 279
16.1.1 Sexual Behavior in Female Rats Produces Analgesia 279
16.1.2 Glycine Released by Vaginal Stimulation Produces Analgesia 279
16.1.3 Glycine Analogs Produce Analgesia 280
16.1.4 Glycinamide, a Precursor of Glycine, Counteracts Allodynia 280
16.2 Potential Therapeutic Implications of the Antinociceptive Effect of Glycinamide 283
16.3 Summary 284
Acknowledgments 284
References 284

16.1 INTRODUCTION

16.1.1 Sexual Behavior in Female Rats Produces Analgesia

Previous research in our laboratory showed that stimulation of the vagina in rats with a probe inhibits withdrawal responses to noxious stimulation (Komisaruk and Larsson, 1971) and inhibits thalamic sensory neuronal response to noxious, but not innocuous, stimulation, providing evidence that vaginal stimulation induces analgesia (Komisaruk and Wallman, 1977). Subsequently, we reported that natural mating behavior in female rats produces analgesia during intromissions, and especially ejaculation, that is more potent than a supra-analgesic dose of morphine (Gomora, Beyer, Gonzalez-Mariscal, and Komisaruk, 1994).

16.1.2 Glycine Released by Vaginal Stimulation Produces Analgesia

In order to ascertain which neurotransmitters may mediate this analgesic effect, we assayed neurotransmitters released into spinal cord superfusates in response to vaginal stimulation in rats (Masters, Jordan, and Komisaruk, 1989), one of which was glycine. Carlos Beyer had hypothesized that as glycine is a major inhibitory neurotransmitter in the central nervous system (CNS), it could be responsible, at least in part, for the analgesia produced by vaginal stimulation. We tested his hypothesis by administering the glycine receptor antagonist strychnine directly to the spinal cord via intrathecal (i.t.) catheter (Roberts, Beyer, and Komisaruk, 1986) and found that, indeed, the rats became hyper-responsive to the mildest tactile stimuli, by displaying distress vocalization, and scratching and biting the skin, indicative of nociceptive responses. Response to mild, innocuous stimulation as if it were noxious is termed "allodynia." On the basis of our observations of the allodynic effects of strychnine, we proposed that glycine normally maintains a tonic inhibitory effect on tactile afferent activity in the spinal cord (Beyer et al., 1985, 1988; Roberts et al., 1986). When this tonic glycine-induced inhibition is blocked by strychnine administered to the spinal cord, normally innocuous tactile stimuli become aversive. These findings have been confirmed and extended by multiple other investigators based on behavioral and/or neurophysiological evidence (Baba et al., 2003; Chen, Dai, and Zeng, 2007; Daniele and MacDermott, 2009; Dickenson, Chapman, and Green, 1997; Ishikawa et al., 2000; Loomis et al., 2001; Lim and Lee, 2010; Sherman and Loomis, 1994; Yaksh, 1989).

16.1.3 Glycine Analogs Produce Analgesia

Other amino acids, which are structurally related to glycine and which bind to the glycine receptor, that is, beta alanine, taurine, and betaine, also counteract the allodynic effect of strychnine (Beyer et al., 1988; Sherman and Loomis, 1994). Comparable effects are produced by administration of the glycine precursor Milacemide. Milacemide is converted to glycinamide, which in turn is converted to glycine in the brain (Christophe et al., 1983; Doheny et al., 1996; Janssens de Varebeke et al., 1988). Milacemide administration counteracts the allodynic effect of strychnine (Khandwala and Loomis, 1998). Recently, we reported that glycinamide, administered to otherwise untreated rats, increased vocalization thresholds to tail shock (Beyer et al., 2013). We now extend this line of investigation by testing whether glycinamide i.p., as a precursor of glycine, can antagonize the allodynic effect of strychnine i.t.

16.1.4 Glycinamide, a Precursor of Glycine, Counteracts Allodynia

Sprague-Dawley female rats (200–250 g) were used in this study. The rats were maintained in a dark-light cycle (14 h light: 10 h dark: lights off at 12:00). Four rats were housed per cage; they were fed with Purina rat pellets and water ad libitum. Rats were ovariectomized (ovx) under ether anesthesia and injected after surgery with i.m. penicillin (100 I. U.).

After one week of ovariectomy, animals were anesthetized for surgery with Ketamine (Bristol laboratories, 25 mg. I.P.) and Xylazine (Haver Lockhart, 1.2 mg, I.P.). Then, a 7.5 cm catheter (Clay Adams PE-10 tubing; Fisher Chemical) was inserted into the subarachnoid intrathecal space though an incision in the atlanto-occipital space, according to the technique of Yaksh and Rudy (1976). The catheter extended to the lumbar level of the spinal cord. After surgery, subjects were treated with 5 mg, i.m. of Terramycin.

One week after implantation of the intrathecal catheter and before intrathecal injection, rats were observed in a Plexiglas cylindrical arena (50 cm diameter) for 5 min. Rats having motor problems as a result of the catheter implantation were discarded. The response to cutaneous stimulation with a von Frey fiber (5.5 g force; Stoelting Co., Chicago, IL) was determined by counting the number of vocalizations elicited by 10 successive applications of the stimulus (applied at approximately 10 sec intervals) to the lower half of the dorsum and flanks.

After control observations, rats were injected with one of the following solutions: glycinamide (50, 200, or 800 mg/kg body weight) dissolved in saline solution (1 ml/kg body weight) and injected i.p.; strychnine (25 μg) was dissolved in 5 μl saline and delivered intrathecally with an additional 7 μl saline flushed from the catheter. I.t. injection duration was 1.5 to 2 min. Rostrocaudal diffusion following intrathecal injection at this volume is usually limited to the spinal cord within the first 30 min after injection (Yaksh and Rudy, 1976). The postinjection test began immediately after completion of the injection. Glycinamide and strychnine were obtained from Sigma Chemicals (St. Louis MO) and dissolved in a saline solution.

Rats were replaced in the cylindrical Plexiglas cage and their behavior was recorded and registered. Based on previous studies using i.t. strychnine administration (Beyer et al., 1985), the following behavioral parameters were measured on a minute-to-minute basis: (1) grooming, (2) spontaneous bouts of scratching, (3) spontaneous bouts of skin biting, and (4) spontaneous vocalizations. Additionally, the response to cutaneous stimulation with a von Frey fiber was determined at 5, 10, 15, 20, and 25 min after injection, as described above. Other types of motor or aversive reactions were recorded with video equipment (Sony Co., Nippon). We also recorded the number of fecal boluses and urinations.

As strychnine has been reported to lower the vocalization threshold (VT) in response to tail shock (Bristow et al., 1986), this test was used to establish a possible analgesic effect of glycinamide. Ovx rats were placed inside the cylindrical Plexiglas arena and allowed to adapt for 5 min. An electrode pair was attached to the tail and connected to a DC stimulator (Coulbourn model E 13-51) via a commutator, as described previously (González-Mariscal et al., 1992). Stimulus parameters: 1 msec pulses, 50 Hz, 300 msec train duration. VT was determined by increasing the current in steps of 100 μA until the rat vocalized and then decreasing the current in steps of 100 μA until vocalization did not occur. The process was repeated three times and the inflection points thus obtained were averaged to determine the VT. The vocalization threshold to tail shock was ascertained at 30, 45, 60, and 90 min after i.t. injection.

Kruskal–Wallis ANOVA was used to analyze vocalization threshold in all groups, followed by the Mann–Whitney U-test for paired comparisons, with $p < 0.05$ considered significant.

Figure 16.1 shows the von Frey fiber-induced vocalization response in the strychnine-treated rats. In the control, zero glycinamide group, the vocalizations ranged from 80% at 5 and 10 min, decreasing steadily to 40% at 25 min. Glycinamide at all 3 dose levels significantly reduced this high proportion of vocalization at points during the first 15 min after strychnine injection. It is noteworthy that the strongest inhibition of vocalization (down to about 10% or less) was obtained with the lowest dose level of glycinamide at each of the test periods up to 15 min.

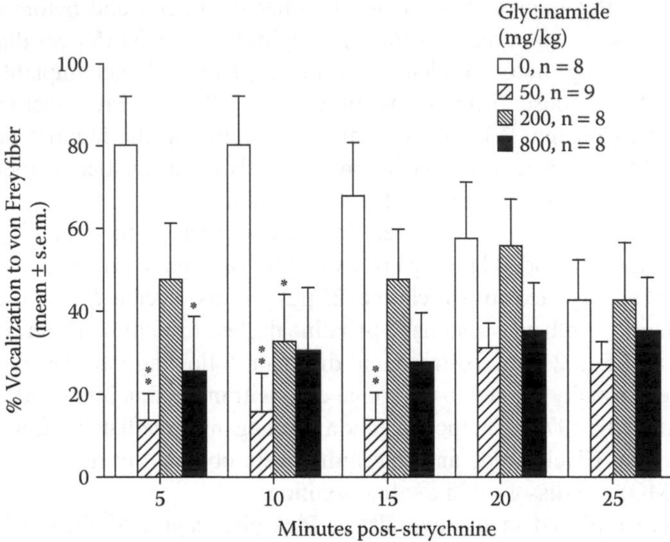

FIGURE 16.1 Effect of glycinamide on vocalization induced by von Frey fiber tactile stimulation. Glycinamide 0, 50, 200, or 800 mg/kg/ml was injected i.p. 30 min before strychnine i.t., 25 μg/5 μl. Each group was compared to strychnine-only; Mann-Whitney U test, $*p < 0.05$, $**p < 0.01$.

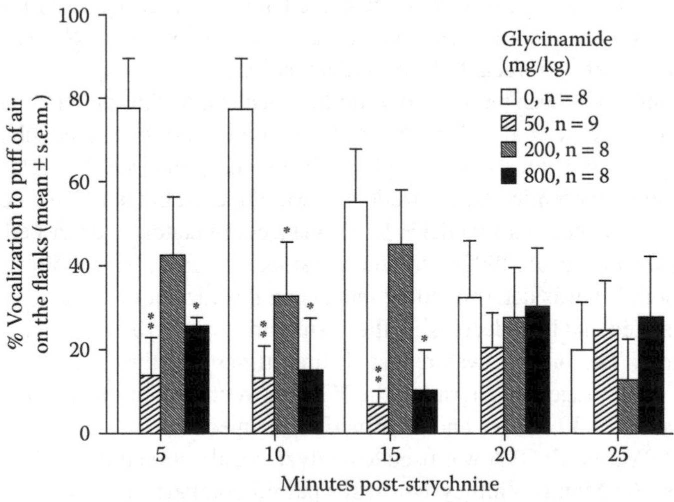

FIGURE 16.2 Effect of glycinamide on vocalization induced by hair deflection in response to a puff of air on the skin. Doses, times, and statistical analyses as in Figure 16.1.

Figure 16.2 shows a pattern of responses to a gentle puff of air to the flank fur that is comparable to the effects of von Frey fiber stimulation shown in Figure 16.1.

Figure 16.3 shows that the highest dose of glycinamide produced the greatest elevation of vocalization threshold, steadily increasing from 30 to 60 and 90 min after strychnine administration. The strychnine alone produced no significant change at any

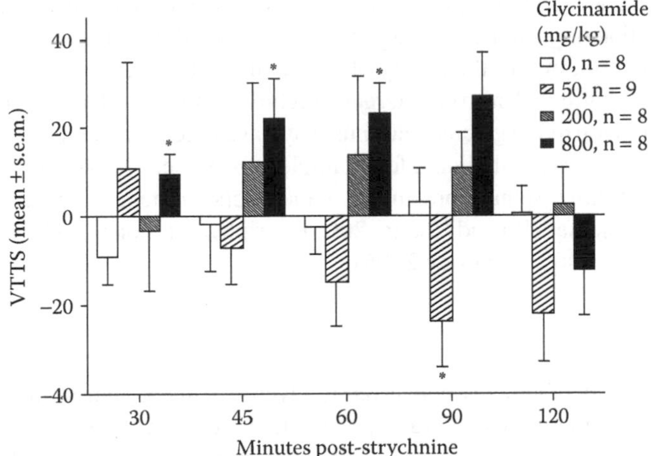

FIGURE 16.3 Effect of glycinamide on the vocalization threshold in response to tail shock (VTTS). Doses, times, and statistical analyses as in Figure 16.1.

point in time. It is curious that the lowest dose level of glycinamide actually tended to produce a decrease in vocalization threshold, reaching significance at 90 min.

The effect of strychnine injection i.t., which induced a high frequency of spontaneous scratching, vocalization, seizure, and hopping. When the animals received glycinamide 30 min i.p., prior to the strychnine, some of these spontaneous behavior patterns were actually potentiated. Thus, glycinamide 800 mg/kg potentiated scratching and seizures, and glycinamide 50 mg/kg potentiated hopping (data not shown).

16.2 POTENTIAL THERAPEUTIC IMPLICATIONS OF THE ANTINOCICEPTIVE EFFECT OF GLYCINAMIDE

The present findings provide evidence that glycinamide, a metabolic precursor of glycine, when injected intraperitoneally, can antagonize the allodynia-inducing, and vocalization threshold (to tail shock)-lowering, effect of strychnine, a glycine receptor antagonist. Allodynia is characterized by a painful response to otherwise innocuous or low-threshold tactile stimulation (Silvilotti and Woolf, 1994). The present findings are consistent with, and extend, previous findings that glycine, per se, can antagonize the allodynia-producing effect of strychnine (Beyer et al., 1985; Yaksh, 1989; Sherman and Loomis, 1994; Lim and Lee, 2010). In addition to glycine, other glycine receptor agonists, specifically, beta-alanine, taurine, betaine, and serine, also were reported to counteract strychnine-induced allodynia (Beyer et al., 1988).

At some doses, glycinamide paradoxically produced motor activity indicative of hyperalgesia (i.e., scratching and hopping behavior) and lowered the vocalization threshold to tail shock. This observation is consistent with the report that glycine administered intrathecally at certain doses also produced hyperalgesia (Beyer et al., 1985). This effect may be due to glycine at certain doses acting as a cotransmitter at the

N-methyl-D-aspartate (NMDA) receptor, thereby facilitating the excitatory hyperalgesia-producing action of glutamate and aspartate (see Caba, Komisaruk, Beyer, 1998).

The present findings of the ability of glycinamide administered intraperitoneally to rapidly (i.e., within 30 min) counteract strychnine-induced allodynia suggest that it is rapidly metabolized to glycine and thus may have potential therapeutic application in pathological cases of allodynia, for example, postherpetic or trigeminal neuralgia, diabetic neuropathy, spinal cord injury, cancer, chemotherapy, some degenerative neurological diseases (Ro and Chang, 2005), peripheral lesions, or CNS dysfunction (Finnerup et al., 2007; Hawksley 2006).

16.3 SUMMARY

Glycinamide is metabolically converted to glycine, a major neurotransmitter, in the CNS. Prior research in this laboratory found that administration of glycine directly to the spinal cord (i.e., intrathecally [i.t.]) counteracted the effect of strychnine, a glycine receptor antagonist, to induce allodynia, which is the aversive effect of otherwise innocuous sensory stimuli, for example, brushing the fur. In the present study, we analyzed the effect of graded doses of glycinamide on strychnine-induced allodynia and vocalization threshold to tail shock. Ovariectomized rats were used to assess the effect of glycinamide (0, 50, 200, or 800 mg/kg) i.p., against strychnine-induced allodynia. Thirty minutes after injection of glycinamide, rats received strychnine 25 μg i.t. We recorded behavioral responses to air puff on the fur and gentle von Frey fiber stimulation. The allodynic responses to strychnine were significantly reduced by all doses of glycinamide. Glycinamide (800 mg/kg) significantly elevated the vocalization threshold to tail shock against the vocalization threshold-lowering effect of strychnine. While some doses of glycinamide increased scratching, hopping, and seizures, and lowered the vocalization threshold to tail shock, overall, glycinamide exerted a dose–response counteraction of the allodynic effect of strychnine.

ACKNOWLEDGMENTS

We thank the support of CINVESTAV and Universidad Autónoma de Tlaxcala. We gratefully acknowledge the technical assistance of Guadalupe Dominguez-López. This work was supported by Consejo Nacional de Ciencia y Tecnología (CONACYT) Grant number 134291: Oscar Gonzalez Flores.

REFERENCES

Baba, H., Ji, R. R., Kohno, T., Moore, K. A., Ataka, T., Wakai, A., Okamoto, M., and Woolf, C. J., Removal of GABAergic inhibition facilitates polysynaptic A fiber-mediated excitatory transmission to the superficial spinal dorsal horn. *Mol Cell Neurosci.* 2003; 24:818–30.

Beyer, C., Banas, C., Gomora, P., and Komisaruk, B. R., Prevention of the convulsant and hyperalgesic action of strychnine by intrathecal glycine and related amino acids. *Pharmacol Biochem Behav.* 1988; 29:73–8.

Beyer, C., Komisaruk, B. K., González-Flores, O., and Gómora-Arrati, P., Glycinamide, a glycine pro-drug, induces antinociception by intraperitoneal or oral ingestion in ovariectomized rats. *Life Sci.* 2013; 92:576–81.

Beyer, C., Roberts, L. A., and Komisaruk, B. R., Hyperalgesia induced by altered glycinergic activity at the spinal cord. *Life Sci.* 1985; 37:875–82.

Bristow, D.R., Bowery, N.G., Woodruff, G.N., Light microscopic autoradiographic localisation of [3H]glycine and [3H]strychnine binding sites in rat brain. *Eur J Pharmacol.* 1986; 126(3):303–307.

Caba, M., Komisaruk, B. R., and Beyer, C., Analgesic synergism between AP5 (an NMDA receptor antagonist) and vaginocervical stimulation in the rat. *Pharmacol. Biochem. Behav.* 1998; 61:45–8.

Chen, Y., Dai, T. J., and Zeng, Y. M., Strychnine-sensitive glycine receptors mediate analgesia induced by emulsified inhalation anaesthetics in thermal nociception but not in chemical nociception. *Basic Clin Pharmacol Toxicol.* 2007; 100:165–9.

Christophe, J., Kutzner, R., Nguyen-Bui, N. D., Damien, C., Chatelain, P., and Gillet, L., Conversion of orally administered 2-n.pentylaminoacetamide into glycinamide and glycine in the rat brain. *Life Sci.* 1983; 33:533–41.

Daniele, C. A. and MacDermott A. B., Low-threshold primary afferent drive onto GABAergic interneurons in the superficial dorsal horn of the mouse. *J Neurosci.* 2009; 29:686–95.

Dickenson, A. H., Chapman, V., and Green G. M., The pharmacology of excitatory and inhibitory amino acid-mediated events in the transmission and modulation of pain in the spinal cord. *Gen Pharmacol.* 1997; 28:633–8.

Doheny, M. H., Nagaki, S., and Patsalos, P. N., A microdialysis study of glycinamide, glycine and other amino acid neurotransmitters in rat frontal cortex and hippocampus after the administration of milacemide, a glycine pro-drug. *Naunyn Schmiedebergs Arch Pharmacol.* 1996; 354:157–63.

Finnerup, N. B., Sørensen L., Biering-Sørensen F., Johannesen I. L., and Jensen T. S., Segmental hypersensitivity and spinothalamic function in spinal cord injury pain. *Exp Neurol.* 2007; 207:139–49.

Gomora, P., Beyer, C., Gonzalez-Mariscal, G., and Komisaruk, B. R., Momentary analgesia produced by copulation in female rats. *Brain Res.* 1994; 656:52–8.

González-Mariscal, G., Gómora, P., Caba, M., and Beyer, C., Copulatory analgesia in male rats ensues from arousal, motor activity, and genital stimulation: blockage by manipulation and restraint. *Physiol Behav.* 1992; 51:775–81.

Hawksley, H., Managing pain after shingles: a nursing perspective. *Br J Nurs.* 2006; 15(15):814–18.

Ishikawa, T., Marsala, M., Sakabe, T., and Yaksh, T. L., Characterization of spinal amino acid release and touch-evoked allodynia produced by spinal glycine or GABA (A) receptor antagonist. *Neuroscience* 2000; 95:781–6.

Janssens de Varebeke, P., Cavalier, R., David-Remacle, M., and Youdim, M. B., Formation of the neurotransmitter glycine from the anticonvulsant milacemide is mediated by brain monoamine oxidase B. *J Neurochem.* 1988; 50:1011–16.

Khandwala, H. and Loomis, C. W., Milacemide, a glycine pro-drug, inhibits strychnine-allodynia without affecting normal nociception in the rat. *Pain* 1998; 77:87–95.

Komisaruk, B. R. and Larsson, K., Suppression of a spinal and a cranial nerve reflex by vaginal or rectal probing in rats. *Brain Res.* 1971; 35:231–5.

Komisaruk, B. R. and Wallman, J., Antinociceptive effects of vaginal stimulation in rats: neurophysiological and behavioral studies. *Brain Res.* 1977; 137:85–107.

Lim, E. S. and Lee, I. O., Effect of intrathecal glycine and related amino acids on the allodynia and hyperalgesic action of strychnine or bicuculline in mice. *Korean J Anesthesiol.* 2010; 58:76–86.

Loomis, C. W., Khandwala, H., Osmond, G., and Hefferan, M. P., Coadministration of intrathecal strychnine and bicuculline effects synergistic allodynia in the rat: an isobolographic analysis. *J Pharmacol Exp Ther.* 2001; 296:756–61.

Masters, D. B., Jordan, F., and Komisaruk, B. R., Regional in vivo superfusion of the spinal cord and KCl-induced amino acid release. *Pharmacol Biochem Behav*. 1989; 34:107–12.

Ro, L. S. and Chang, K. H., Neuropathic pain: mechanisms and treatments. *Chang Gung Med J.* 2005; 28(9):597–605.

Roberts, L. A., Beyer, C., and Komisaruk, B. R., Strychnine antagonizes vaginal stimulation-produced analgesia at the spinal cord. *Life Sci.* 1986; 36:2017–23.

Sherman, S. E. and Loomis, C. W., Morphine insensitive allodynia is produced by intrathecal strychnine in the lightly anesthetized rat. *Pain* 1994; 56:17–29.

Sivilotti, L. and Woolf, C. J., The contribution of GABAA and glycine receptors to central sensitization: disinhibition and touch-evoked allodynia in the spinal cord. *J Neurophysiol.* 1994; 72:169–179.

Yaksh, T. L., Behavioral and autonomic correlates of the tactile evoked allodynia produced by spinal glycine inhibition: effects of modulatory receptor systems and excitatory amino acid antagonists. *Pain* 1989; 37:111–23.

Yaksh, T. L. and Rudy, T. A., Chronic catheterization of the spinal subarachnoid space. *Physiol Behav.* 1976; 17:1031–6.

Section IV

Epilogue

17 How Carlos Beyer Influenced Our Lives

Barry R. Komisaruk and Beverly Whipple

CONTENTS

17.1 My Remembrances of Carlos, by Barry 290
17.1.1 Getting My Research Credentials 290
17.1.1.1 Meeting Carlos at UCLA 291
17.1.1.2 Our Collaboration Develops into Friendship 292
17.1.1.3 Collaborating in Mexico and Newark 292
17.1.2 Carlos' Strange, Prescient, Idea about Testosterone as an Estrogen .. 292
17.1.2.1 Carlos at Newark 293
17.1.2.2 Deciding to Set Up a Mexico/Tlaxcala/Xalapa-Newark Exchange Program 294
17.1.2.3 One Night's Seminal Observations on Theta Rhythm, Sniffing, Lordosis, and Immobilization 294
17.1.2.4 Getting Arrested for "Reading or Writing in the School Cafeteria after Midnight" 295
17.1.2.5 Realizing that Vaginal Stimulation Blocks Pain Responses in Rats 296
17.1.3 Carlos Creates a Research Field 296
17.1.3.1 Tragic Loss of My Wife, Carrie, and Our Dear Friends 296
17.1.3.2 Carlos Takes My Sons and Me Under His Wing 297
17.1.3.3 Carlos Saves Our Lives 298
17.1.3.4 Reality Interferes: The Mexico City Earthquake of 1985 298
17.1.3.5 The Mexico Contingent Joins the U.S. Rat Sex Contingent, and Vice Versa 299
17.1.3.6 Fork in the Road. Theta/Sniffing or Vaginal Stimulation-Produced Analgesia? 299
17.1.4 Carlos Turns His Intellect Against Pain 300
17.1.4.1 Losing Yet Another Dear Friend … Where is Hugo? He's Never Late 301
17.1.4.2 Mating Produces Analgesia; Stronger than Morphine 301
17.1.5 Enter Beverly 302
17.1.5.1 We Interest Carlos in Human Sexuality Research; Writing Our Book 302

17.1.5.2 I Get to Realize My Childhood Intrigue with Baja California Sur 302
17.1.5.3 More Research Literature on Orgasm Than We Realized 303
17.1.6 Unfinished Work … Carlos' Life Cut Short 303
17.2 My Remembrances of Carlos, by Beverly 304
17.2.1 At the Institute of Animal Behavior 304
17.2.2 My Path to Research 304
17.2.2.1 Rediscovery and Naming of the Gräfenberg Spot 304
17.2.2.2 Female Ejaculation 305
17.2.3 My Path to Rutgers University 306
17.2.3.1 Analgesia in Women with Stimulation of the Area of the Gräfenberg Spot 306
17.2.4 Working with Carlos' Colleagues in Mexico 307
17.2.4.1 Time with Carlos and His Family 307
17.2.5 My New Human Physiology Laboratory at Rutgers University 308
17.2.5.1 Brain Imaging Research 309
17.2.6 Coauthoring Books with Carlos 309
References 310

17.1 MY REMEMBRANCES OF CARLOS, BY BARRY

17.1.1 Getting My Research Credentials

I first met Carlos when we both went to the Brain Research Institute at UCLA to work with Charles Sawyer in 1965. I had just finished my PhD with Danny Lehrman at the Institute of Animal Behavior (IAB) at Rutgers-Newark, in which I implanted progesterone crystals in ring dove brains. Danny and his graduate student, Rochelle Wortis, had previously shown that systemic injection of progesterone induced incubation behavior in male and female ring doves (Lehrman and Wortis, 1960). I repeated and confirmed their finding. In addition, I observed that systemically injected progesterone blocked courtship bow-cooing behavior in the male ring doves while also inducing their incubation behavior. These findings provided two objective endpoints to observe when I implanted progesterone crystals into the brain. Danny sent me to learn how to do stereotaxic implantation of crystalline hormones in the brain with Robert Lisk at Princeton University. This was exciting to me because it was one of Danny's lectures in the course, "Structure and Function of the Organism" (which he was teaching as a visiting professor at City College of New York, where I was an undergraduate), that Danny described Bob Lisk's recent research in which crystals of estrogen implanted into the brain of female rats induced their sexual behavior (Lisk, 1962). I was intrigued by that research and asked Danny more about it after his lecture. During that conversation, Danny asked me if I would like to do that kind of research as his graduate student. It was an offer I couldn't refuse! And so, in 1961, I started my doctoral studies as Danny's student at the IAB at Rutgers University in Newark, which Danny had founded in 1958, just a few years earlier. I had zero knowledge of brain anatomy, and especially that of the ring dove brain. So Danny then sent me to learn the identity and location of the nuclei of the brain in doves with Elizabeth Crosby, a delightful, diminutive, and frail, elderly lady who looked like she was in her late eighties (but she was born in 1888), at the University of Michigan in Ann Arbor. I only realized subsequently that she was the extant world's expert on comparative anatomy of the brain, having coauthored the three-volume timeless classic, *The Comparative Anatomy of the Nervous System of Vertebrates, Including Man*, by Ariëns Kappers, Huber, and Crosby, 1936. In my dissertation research, I found that the progesterone crystals inhibited male courtship behavior and induced incubation behavior in both sexes in *distributed* brain regions, rather than in brain "*centers*" (the latter of which was a popular, albeit naive, concept in the 1960s, which my data argued against; Komisaruk, 1967). I then wanted to know what the hormones do to neuronal activity to modify behavior. Charles Sawyer at UCLA was the only person in the United States studying the effects of hormones on brain activity. Danny introduced me to Sawyer at an Association for Research in Nervous and Mental Disease conference in New York City. I asked Sawyer if he was the same Charles H. Sawyer who had published a paper in 1947 on the cholinergic control of swimming movement ontogeny in larval salamanders, a study that I had read for an undergraduate term paper I wrote on the development of behavior of embryos. He was pleasantly surprised that I knew his earliest research. Sawyer told me that he would welcome me in his lab if I could get a postdoctoral fellowship. I did, and he did welcome me into his lab. When I arrived there in 1965, followed by a box of my essential books that I had mailed addressed to myself, everybody told me they were expecting that I would be Japanese… my immediate predecessors were Drs. Kawakami, Kanematsu, and Kawamura.

17.1.1.1 Meeting Carlos at UCLA

Carlos went to work with Sawyer for the second time in 1965 to also study the effects of hormones on neural activity. I was just a kid of 24 and Carlos was a grown-up of 31 when we both arrived in the lab. Carlos' demeanor, appearance, cheerfulness, and sense of humor reminded me so much of my father that I immediately took a liking to him. Especially, I was startled to meet "Dr. Carlos Beyer," the author of a paper that particularly impressed me as a graduate student, and which I was proud to have "found" and read … especially as it was in Spanish … published in the Boletín del Instituto de Estudios Médicos y Biológicos de la Universidad Nacional Autónoma de México in 1961, in which hypothalamic stimulation released oxytocin (Mena, Anguiano, and Beyer, 1961). I was intrigued by the connection between the brain and the endocrine system, and had reviewed all the papers I could find on the topic for my dissertation. Reciprocally, Carlos was very impressed with the fact that I knew his earlier research, in what he thought was a local specialized journal. So we hit it off shortly after meeting. We collaborated on a study of neuronal correlates of vaginal stimulation-induced pseudopregnancy, and of LHRH on hypothalamic neurons, with a third, more senior visitor to Sawyer's lab, Victor (Domingo) Ramirez, from Valdivia, Chile (Ramirez et al., 1967). Carlos and Domingo had worked with Sawyer previously, so I was the new kid on the block working with them. Also, I had become allergic to rats in Bob Lisk's lab when he taught me stereotaxic implantation methodology in rats, so I sneezed a lot. My Spanish was rudimentary; Domingo and Carlos appointed themselves as "El Capitan" and "El Jefe," respectively, and they named me "El Latoso." It was only years later, as my Spanish improved, that I realized that my name was the equivalent of the English-Yiddish, "hock mir nicht kein chinik." They both refer to noise—"banger on an empty can" in Spanish, and literally, "don't hit me no teapot," in Yiddish. Sawyer and everyone else told me that Carlos was a Renaissance man, with a brilliant, photographic mind, and an astonishing encyclopedic knowledge of the scientific literature. The more I got to know him, the truer I realized was their assessment.

17.1.1.2 Our Collaboration Develops into Friendship

Our friendship grew stronger. Carlos had a girlfriend in LA, Carol Percin, and with my wife, Carrie, we all became good friends and the four of us vacationed in Acapulco. Carlos knew all the obscure, unpretentious, delicious restaurants there and introduced us to the tremendously varied Mexican cuisine. The intriguing, varied, Mexican culture got to us then, and never left us. After our postdocs, Carlos returned to his lab at the Seguro Social in Mexico City, and I went to a position as assistant professor in Danny Lehrman's IAB at Rutgers-Newark, which he had offered to me prior to my postdoc with Sawyer. Carlos invited Carrie and me to visit him the next summer in Mexico City. We stayed in his home, which was steeped in ancient, traditional, and modern Mexican anthropology and culture, with pre-Columbian sculpture, antique masks, 1940s-American-style home furniture, modern paintings by upcoming Mexican artists, an enormous library of international art books and opera recordings, and literally thousands of ancient Mexican carvings and trinkets. We got to know Carlos' elegant mother and father. Carlos' father was a businessman, in excellent, trim physical condition, who impressed me with the fact that he always wore a suit with jacket, tie, and shoes even at home. The first thing Carlos took Carrie and me to see our very first day in Mexico City was the water purification plant! We often discussed that experience over the years, as Carlos' way of demonstrating to us his pride in Mexico's modernity. I must say that in traveling around Mexico City in those early years, I repeatedly commented to Carlos about the problems that would develop over the pollution caused by the flagrantly permissive belching of black smoke by the myriad buses and trucks in the city. It was only many years later, after the imposition of strict antipollution policies in Mexico City, that Carlos finally acknowledged that the pollution had, indeed, become a problem.

17.1.1.3 Collaborating in Mexico and Newark

Carlos invited me to work with him the summer of 1970, and arranged a beautiful apartment on Miguel Angel de Quevedo in Mexico City, with a housekeeper and her baby daughter, for us. Our first son, Adam, was 1½. The first thing Adam did when we took him for a walk along Insurgentes was to run down the block, turn around, crouch down, grin at us, rub his hand on the sidewalk, and put it in his mouth. He's had a great immune system ever since.

17.1.2 Carlos' Strange, Prescient, Idea about Testosterone as an Estrogen

That summer, Carlos told me about his idea, which seemed bizarre to me at the time, based on his reading of the literature in chemistry, that testosterone is aromatized to estrogen and may act on organs and behavior as an estrogen, rather than as an androgen. I didn't realize then that Carlos was about 20 years ahead of everyone else in the field of neuroendocrinology. It was my first exposure to Carlos' ability to create a field of science. We did a study with Gonzalez-Diddi at the Seguro Social, injecting testosterone, showing that it increased uterine weight, and that the estrogen receptor antagonist, MER-25, attenuated the testosterone effect on the uterus (Gonzalez-Diddi, Komisaruk, and Beyer, 1972). My contribution to that study was to find that just giving a very high dose of MER-25 by itself stimulated uterine development. I liked the idea that this estrogen antagonist could itself act as a very weak agonist, a kind of "dog in the manger." (The "dog in the manger" doesn't eat the straw that is stored there, but prevents the livestock from getting to the straw.) The MER-25 occupies the estrogen receptor, but has only a very weak estrogenic action, so it is an effective antiestrogen. More recently, I realized that methadone does the same kind of thing for the opiate system—it is a very weak opiate agonist.

17.1.2.1 Carlos at Newark

I reciprocated to Carlos by arranging to have him visit the IAB as a visiting scientist. We both liked the idea of olfactory input to the limbic system and thereby to the endocrine system, so we collaborated on a study in which we electrically stimulated the olfactory bulbs in rats and recorded responses of single neurons in the hypothalamus, mapping the distribution of responsive neurons (Komisaruk and Beyer, 1972). While it is generally accepted that the olfactory system is unique among the senses in that it projects directly to the cortex ("rhinencephalon") without first passing through the thalamus, Carlos and I found that the dorsomedial nucleus of the thalamus and also hypothalamic neurons were strongly and reliably activated by the olfactory bulb electrical stimulation. At that time, I had found a moment-to-moment correlation between individual theta waves and individual vibrissal and respiratory sniffing movements during exploratory behavior in rats, and I hypothesized that the functional significance is that the theta rhythm "gates" rhythmically the olfactory and tactile input to the brain, thus converting the environmental stimuli to sensory "quanta," which could simplify the rat brain task of processing environmental stimuli (Komisaruk, 1970). To test the idea of this temporal "gating" process, Carlos and I measured the number of hypothalamic neuronal spikes in response to single shocks of the olfactory bulb as a function of the phase of the theta rhythm that we were able to measure concurrently. We found that, indeed, the number of neuronal spikes in hypothalamic neurons that were triggered by the olfactory bulb shocks varied as a function of the phase of the theta wave during which each shock happened to be applied. This supported the concept of the "gating" function of the theta rhythm (Komisaruk and Beyer, 1972). Carlos and I presented preliminary results of this study at the first annual conference of the Society for Neuroscience in Washington, DC, in 1971, at which there were approximately 1,700 participants and only about 550 papers. In recent years, the number of participants at this conference typically varies between around 30,000 and 40,000! Carlos and I participated in that first conference as charter members of the Society for Neuroscience.

In a later study with my doctoral student Kazue Semba, we confirmed this "neuronal gating" observation with a different, behavioral, method. We trained rats to press a lever for a food pellet ad lib, during which we recorded their theta rhythm. Our null hypothesis was that the rats would press the lever in a random relation to the phase of the theta wave. However, we found that the rats pressed the lever significantly *non*randomly, that is, at a specific phase of the theta wave. The rats pressed the lever sporadically, so that many theta waves occurred between successive lever presses. However, whenever they pressed the lever, it turned out to be at a specific, limited phase of the theta wave, a significant nonrandom effect (Semba and Komisaruk, 1978). The concept of brain rhythms as representing a modulatory, gating, mechanism has subsequently been applied by others to concepts of learning and performance (e.g., Macrides, Eichenbaum, and Forbes, 1982).

17.1.2.2 Deciding to Set Up a Mexico/Tlaxcala/Xalapa-Newark Exchange Program

During the same year, 1971, Knut Larsson, the eminent behavioral endocrinologist/pharmacologist from Göteborg, Sweden, also came to the IAB as a visiting professor. Knut and Carlos had several mutual interests, and thus began a major collaboration, with Knut subsequently spending significant amounts of time in Mexico working with Carlos and his students and colleagues. There were many other colleagues Larsson interacted with in the lab at the Seguro Social. Knut became deeply absorbed in the culture of Mexico, which he came to adopt for himself for many years. Knut lived and worked mainly in Mexico City. Only after the lab was established in Tlaxcala did he visit there. While Carlos was at the IAB, we decided to initiate an exchange program so that students in the Mexican universities in which Carlos was involved could come to the IAB to collaborate in research, and our IAB students could do the same in universities in Mexico. So, in 1984, with Hugo Aréchiga, who was the chair of the Physiology Department at CINVESTAV (Centro de Investigación y Estudios Avanzados)—a leading research institution in Mexico City, we first organized a joint conference for our faculties and students. We then got funding from the Rutgers-Newark provost, Norman Samuels, to rent an apartment in the "Ironbound" section of Newark for the visiting students and faculty. Over the years, we published more than 20 joint peer-reviewed papers based on our collaborative research. Many, if not most, of the students from Mexican universities who participated in our exchange program have earned significant academic, research, and administrative positions in the leading universities of Mexico, which include CINVESTAV, UNAM, University of Tlaxcala, and Universidad Veracruzana (Xalapa campus). These students who participated in our program include Mario Caba, Porfirio Carrillo, Rosario Chirino, Rafael Cueva-Rolon (deceased), Ramon Eguíbar, Alonso Fernández-Guasti, Lisbet Gómez, Porfirio Gómora, Oscar González-Flores, Gabriela González-Mariscal, Helena Hernández, Rose-Angélica Lucio, Jorge Manzo, Margarita Martínez-Gómez, Angel Melo, Gabriela Moralí, and Maria Pacheco, who was a student of Rene Drucker-Colin.

17.1.2.3 One Night's Seminal Observations on Theta Rhythm, Sniffing, Lordosis, and Immobilization

During Knut Larsson's visit to the IAB, I demonstrated to him the observation I had made in James Olds' lab that probing against the uterine cervix immobilized rats and facilitated the lordosis response, regardless of the phase of their estrous cycle, and even if they were ovariectomized and hormonally untreated.

I had been in Olds' lab at the University of Michigan in Ann Arbor at his invitation. Olds had made the famous discovery that rats would press a lever vigorously to obtain a mild electrical stimulation of various regions of their "limbic system" to the extent that they would die of starvation with food available, preferring to apply brain stimulation. To characterize this effect, he coined the term "pleasure centers of the brain." Subsequently, Olds sought to identify the neuronal correlates of reinforcement during learning. In a visit to the IAB, he mentioned that he had developed the first methodology for recording the activity of single neurons in awake, behaving rats, and had fully automated the research protocol, but nobody was observing the rats' behavior. He invited me to see if I could identify correlations between the activity of single neurons and the rats' behavior, which was another offer I couldn't refuse. There was a seminal moment in my life and career one night in Olds' lab—I made two discoveries upon which I am still following up, and later the same night, I got arrested by the Ann Arbor police for "reading or writing in the school cafeteria after midnight." Let me explain. A movement artifact was evident in the brain microelectrodes that were correlated with the sniffing movements of the rats during exploratory behavior. I realized that the rate of the rhythmical sniffing movement artifact was the same as that of the theta rhythm—7 per second. Chewing and licking movements also occurred at the same rate. It was exciting to me that sniffing is olfactory, the theta rhythm is characteristic of the "limbic system" and rhinencephalon (the "olfactory brain"), so maybe there is a single pacemaker that drives all these fundamental behavior processes, each of which "plugs in" to the pacemaker selectively for sniffing, drinking, and eating. In attempting to stimulate the activity of single neurons whose "spike" (i.e., action potential) activity I was listening to and recording, I applied a variety of sensory stimuli, for example, tapping, lifting, pinching, chocolate, acetic acid, and vaginal stimulation, which in Sawyer's lab I saw produced strong effects on neuronal activity in the rats in his lab (Komisaruk et al., 1967). But in Sawyer's lab the rats were anesthetized. When I applied the vaginal stimulation in Olds' awake rats, they became immobilized and all showed lordosis, regardless of the phase of their estrous cycle, even those in diestrus (Komisaruk and Olds, 1968).

17.1.2.4 Getting Arrested for "Reading or Writing in the School Cafeteria after Midnight"

I was so excited with these observations that I suddenly realized I was in the lab past midnight. I was hungry and thirsty, so I went to the university cafeteria. It was open, large, brightly lit, and deserted. I bought a sandwich and drink from the machines and started writing notes and ideas to myself in my lab notebook. A night watchman walked over to me as I was munching and writing and said, "You see that sign over there?" I got up and walked over to the small sign hanging on the wall. It said, "No reading or writing in the school cafeteria after midnight." I mentioned to him that it was funny...actually, ridiculous!

I told him I'm minding my own business, nobody else is here, I'm not bothering anyone. He said, "Those are the rules," and plopped down directly across from me at my narrow table. Incredulous, I held my pen poised over my notebook. We stared at each other. Then I made a mark with my pen on the notebook. He slapped his hands on the table, stood up suddenly, said "OK, buddy," and went to a phone on the wall. In a few minutes, two city police officers arrived and told me I was under arrest. They jostled me as we walked to the police car, took me to the police station, held me there for 3 h, questioned me, and then told me to leave. I walked back to my apartment; it was after 4 a.m. The next morning I told Jim Olds what had happened. He told me I was crazy; don't mess with the Ann Arbor police; they could have shot me. I called the university administrative office and asked the purpose of the sign. I was told that if they allow ME to read or write in the school cafeteria after midnight, since the university library closes at midnight, then "EVERYONE" would leave the library and go to the school cafeteria after midnight and read or write there (???!!!).

17.1.2.5 Realizing that Vaginal Stimulation Blocks Pain Responses in Rats

At the IAB, Knut was impressed with the lordosis and immobilization effects of vaginal-cervical stimulation in the rats, and as we were leaving late one afternoon, discussing it in the elevator, our conversation turned to this: "Since the rats are immobilized by the cervical stimulation, what would happen if we pinched a foot?" So we hit the "up" button in the elevator, went back into the lab, got a rat out of its cage, probed the cervix, pinched its foot, and observed that it didn't respond—no leg withdrawal response, no squeak. In the next few days, we repeated the observations with many more rats, got the same response in every one, generated and photographed some leg-EMG data to provide objective evidence of the inhibitory effect, wrote it up on just one Saturday afternoon, and submitted it to *Brain Research* as a short communication; it was accepted without modification—the fastest, easiest, paper I ever published (Komisaruk and Larsson, 1971)! Knut subsequently worked with Carlos on the aromatization research, triggered by a collaboration with Peter McDonald from England, with whom Carlos and I became close friends and collaborators in Sawyer's lab.

17.1.3 Carlos Creates a Research Field

Carlos was a true intellectual catalyst, bringing virtually all with whom he came in contact into his research endeavors, using rats and rabbits. Starting in 1969 and in following years, Carlos, Knut, Knut's student, Per Sodersten, Pete McDonald, and Gabriela Moralí made the major discovery that only aromatizable testosterone, but not nonaromatizable 5-alpha-dihydrotestosterone, would induce male sexual behavior, and that adding estrogen to the latter would induce male sexual behavior (Beyer, Vidal, and McDonald, 1969; Larsson, Sodersten, and Beyer, 1973; Morali, Larsson, and Beyer, 1977). The concept developed through those studies was seminal in initiating virtually an entire research field involving many investigators over subsequent years.

17.1.3.1 Tragic Loss of My Wife, Carrie, and Our Dear Friends

In 1974, Pete and his wife, Wendy, were on a weekend vacation in Istanbul from their home in London, when the DC-10 on which they were returning crashed in Paris; there were no survivors among the 346 souls on board, one of the worst air disasters in history. Four years earlier, when I was working with Carlos in Mexico City, he entered the room, face ashen, to tell me that Shawn Shapiro and his wife, Lorraine, our dear friends also from Sawyer's lab, had just died in a plane crash in Cuzco, Peru, while on a tour to Macchu Pichu during a break from an international endocrinology conference in Lima. That was the worst air disaster in the history of Peru. The traumatic loss of these four dear friends in such a short interval never stopped haunting Carlos and me; we reminisced our fond, happy, humorous memories of them whenever we got together.

We shared other happy and traumatic events. In 1981, my older son, Adam, was Bar-Mitzvahed on the same day that my younger son, Kevin, at the age of 10 was singing the role of Amahl in Giancarlo Menotti's opera, "Amahl and the Night Visitors," in the New Jersey Opera Company. Carlos, his wife Josefina, and their daughters, María-Emilia and Gabriela, came to New Jersey to join us in the celebrations. My wife, Carrie, was severely debilitated with cancer on that exciting day. She died in December 1982.

17.1.3.2 Carlos Takes My Sons and Me Under His Wing

That summer, Carlos and Josefina invited Adam, Kevin, and me to vacation with them and María Emilia and Gaby in the jungle of Palenque (state of Tabasco, in southeast Mexico). It was a blessed respite from the sad adjustment we were struggling through. Some scenes from that memorable trip stick in my mind. One was waiting outside for a connecting train for 6 h, probably in Catemaco (Veracruz), during which one of the passengers took a nap and, as he was carrying a live chicken with him, tucked the chicken's legs under his belt, with the chicken quietly laying sideways on his belly for several hours while he slept. Another was trekking up the mountain in the afternoon to the settlement at Palenque, barely able to breathe in its insufferable heat and humidity. We all lay down on the grass until evening when it started to cool. Then we entered a tiny museum there that described the history of the settlement of Palenque and the construction of the remarkable stone temple, tower, and other buildings there in the midst of the jungle. I was intrigued and amused to see an etching of the moment of "discovery" of Palenque by Spanish explorers … about a dozen men lying prostrate on the ground or lying on the ground and propped up against some trees, depicted as totally exhausted. Our sentiments, exactly! How could the indigenous folks get up the energy to construct the enormous stone buildings of Palenque in such oppressive heat and humidity? The area was a clearing in the dense jungle, which enclosed it. In a brief walk through the jungle, I was astonished to realize that what looked like a hilly terrain was in fact pyramids that had been overgrown with vegetation! When I looked a little more closely, it was clear that the "hills" were made entirely of hand-cut stones, just covered by earth, brush, and trees. Carlos commented that there are literally hundreds of yet-unexplored pyramids throughout Mexico.

Months before, when Adam learned that we were going to the Palenque jungle, he became thrilled that his lifetime fantasy of hacking his way through the jungle with a machete would come true. We bought him a machete and calf-length leather boots to protect against the poisonous snakes in the jungle (and I brought along a poisonous snake antivenom kit). The day was oppressively hot and humid, but Adam was prepared! He stood defiantly on the hill in dense vegetation with all of us as witnesses, my camera was clicking...he started hacking. After about 2 min, he took off his shirt to cool off. About 2 min later he stopped hacking. It's too hot and this work is too strenuous! Adam came down off the hill, put his shirt back on, and sheathed his machete, and we returned to our hotel for lunch.

17.1.3.3 Carlos Saves Our Lives

On our return to Mexico City we stopped in Agua Azul (state of Chiapas), with its beautiful tropical pond at the base of a 4-m-high mountain waterfall, the water crashing onto the boulders in the pond, surrounded by lush foliage. It looked like paradise to me. We all walked up the mountain to picnic beside the rapidly running stream, about 30 m upstream of the waterfall. No guardrails, no warning signs. Adam waded out about 4 m into the stream. Then he called out that the current was strong, and the rocks were smooth and slippery, covered in slippery moss. Kevin sensed that Adam was a little distressed, so he waded out to help him. But then Kevin called out to me that they were both slipping. So I told Kevin to grab Adam's hand, and I immediately waded out to get them. But the current was so strong that I also began to lose my balance on the slippery rocks. I grabbed Kevin's hand and shouted to Carlos. Carlos prudently kept one foot on the bank and stretched into the river, extending his hand to mine, me holding Kevin's hand, and Kevin holding Adam's hand, and Carlos carefully dragged us all back to shore. I had visions of all of us swept down the river and over the waterfall to be dashed on the rocks below. It took a long time for us to pull ourselves together. Carlos said, "Let's walk up the mountain alongside the river to calm down." As we walked up the mountain, we saw one waterfall after another. In silent testimony, there were painted stone shrines with crosses, one after another, on the banks of the stream, all the way up the mountain. Carlos had saved our lives, truly! From that time on, I have been impressed and dismayed by the casual attitude toward physical safety in Mexico. In the United States, there would have been warning signs and guardrails physically blocking any access to the stream. In Mexico, we are on our own.

17.1.3.4 Reality Interferes: The Mexico City Earthquake of 1985

In September, 1985, Carlos was working at the IAB. At the end of the day I dropped him off at his apartment, and as I was driving home, there was breaking news on the radio of a major earthquake in Mexico City. As soon as I walked in my door, I ordered a ticket home for Carlos on the next flight out, figuring that plane tickets would suddenly get scarce. They did. Carlos left on the next flight to Mexico. He told me a year later that phone service was out when he arrived at the airport in Mexico City. He feared that he didn't know who and what, if anything, he would find when he reached his home. For more than 10 days, it was impossible to make any telephone connection between the United States and Mexico. On the radio news there was mention that a "Ham" amateur radio operator in New Jersey had made contact with his counterpart in Mexico City. I contacted him and asked him if he could somehow make contact with Carlos. Several days later he phoned me to say that he sent his contact to Carlos' home to ask their condition. Everyone there was "sano y salvo"—safe and sound! Many major office buildings including hospitals were demolished by the earthquake. After that fateful day in September, demolition of building remnants and construction of new replacement buildings proceeded throughout Mexico City for many years. Carlos said that at least 10,000 souls lost their lives in the catastrophe, one of the worst in the history of Mexico.

Years later, in 1997, as Kevin had become an accomplished organist, Carlos asked him to play the grand organ at the Cathedral of Puebla for the wedding of Gabriela and Juan. Kevin practiced long and hard, and played beautifully.

17.1.3.5 The Mexico Contingent Joins the U.S. Rat Sex Contingent, and Vice Versa

When Carlos and I were in Sawyer's lab in 1965, we heard about, and participated in, the small West Coast Sex Conference that was organized by Frank Beach and his students and postdocs at UC Berkeley. It was so stimulating that when the group of us participants landed positions in various universities around the country, we stayed in touch with each other, comparing notes. Then, in 1969, Lyn Clemens suggested that we all get together with our students and have a meeting to present our early research as new faculty members. We met in East Lansing, Michigan, and called the meeting the Eastern Regional Conference on Reproductive Behavior. Don Pfaff, Bruce McEwen, Harvey Feder, myself, our students, and a few others came to that meeting. Soon afterward, we joined with the West Coast Sex Conference folks to become the Annual Conference on Reproductive Behavior. Then we convinced Carlos to join in the festivities of the conference, which he did, and then became a regular, bringing his students, first of whom was Gaby González-Mariscal. We rotated hosting the conference each year, and in 1987, while Carlos and I were working together at the IAB in Newark, we organized the 19th Annual Conference on Reproductive Behavior in Tlaxcala at the hotel of the Seguro Social, La Trinidad. Carlos, an avid fan of bullfights in Mexico and Spain, who was the main host of the conference, had a great plan to have the conference banquet at an hacienda that raised brave bulls. He said that the conference participants would enjoy playing at being toreadors with the baby bulls there. Some of the conference participants, agitated by this idea, consulted with the NIH, and came back with a veto of the plan because the NIH considered that, as NIH grant travel funds would be used for some of the American conference participants, and as the treatment of the bulls would violate NIH policy on ethical treatment of animals, funds for travel to the conference would not be allowable. So that quashed our plans for the hacienda banquet. We ended up having our conference banquet at Albergue de la Loma, with an enthusiastic young folk band and pollo estilo Tocatlán (con epazote)!

17.1.3.6 Fork in the Road. Theta/Sniffing or Vaginal Stimulation-Produced Analgesia?

In my research career, confronted with, and torn between, two different lines of research—the rhythmical behavior synchronized with theta rhythm and the apparent blockage of pain by vaginal-cervical stimulation—I made a decisive choice one day. I was at Carrie's bedside in the hospital; she was suffering terribly from intractable pain of cancer. I felt desperate to take away her pain. I said to myself then and there, "If you think you're so smart, let's see you do something useful." At that moment,

I dedicated myself to study how to block pain. Studying theta rhythm would be interesting conceptually. But I preferred to try to do something against pain.

17.1.4 Carlos Turns His Intellect Against Pain

The next time Carlos came to the lab, he said, "Why don't we try to see if the main inhibitory neurotransmitters, gamma-aminobutyric acid (GABA) and glycine, will inhibit pain?" It was such a clear, direct, "elegantly simple" idea that we got right to testing it. Carlos immersed himself in the literature—again, as always—and came up with the plan to administer bicuculline or strychnine directly to the spinal cord (i.e., "intrathecally") in order to block GABA or glycine receptors, respectively. Together with our doctoral student Lowell Roberts, we tested miniscule doses of the two drugs and were amazed at their potency in blocking the pain response-blocking effect of vaginal-cervical stimulation. We made another observation of the effects of those two drugs: the rats became hypersensitive to the mildest of stimuli, such as gently blowing on their fur, to which they responded with prolonged stress vocalization, as if they were in pain. This is an example of allodynia, in which normally innocuous stimuli become noxious. We realized that it also implied that there must normally be tonic inhibitory activity of the GABA and glycine systems on the tactile system, for if the inhibition is blocked, tactile stimulation becomes noxious (Beyer, Roberts, and Komisaruk, 1985; Roberts, Beyer, and Komisaruk, 1985, 1986). Our report of that mechanism underlying allodynia has received significant citation in the literature, another example of Carlos' scientific "Midas touch"—he had an extraordinary ability to generate and open up new areas of scientific research. That ability was based on his voracious seeking, *understanding*, and learning the literature of chemical and biochemical mechanisms—the more molecular level below pharmacology and endocrinology, that is, chemistry, biochemistry, and molecular biology—and based on his acquiring that knowledge, made educated guesses as to the mechanisms underlying neurotransmitter and hormonal action mechanisms of behavior. Carlos had a rare ability...remarkably, his educated guesses were virtually always correct! Carlos' generosity in teaching us what he understood enlightened us, and was so compelling and convincing that we happily worked with him to test his ideas. We were repeatedly reinforced by the realization that Carlos' conceptualizations were invariably correct. So we joined with him in enjoying the exhilaration of confirming the daringly potent hypotheses that he formulated.

Another example was Carlos' recognition, based on his reading of the literature, that glycine has a dual, mutually opposing effect on nociception by stimulating two different receptor types—*N*-methyl-D-aspartate (NMDA), which increases nociception, and blocking of the strychnine receptor, which decreases nociception. He identified AP-5, which selectively blocks the NMDA receptor, and hypothesized that by blocking this pain-promoting receptor, the analgesic effect of vaginocervical stimulation would be intensified. With Mario Caba, visiting the IAB as a doctoral student, and Carlos' long-time colleague, Ana María López-Colomé, professor at UNAM, who was visiting the IAB via our exchange program, we tested the effect of AP-5 injected intrathecally, and observed a profound intensification and prolongation of the analgesia produced by vaginocervical stimulation (Beyer et al., 1992). Carlos then, again via the literature, found that glycine can be ingested in large amounts with no deleterious effects. So with Porfirio Gómora at Tlaxcala, they dissolved large amounts of glycine in the rats' drinking water (increasing their intake with added chocolate, which I suggested), and found that this treatment itself produced a potent analgesia. Porfirio has continued this line of research, as shown in his chapter in this volume.

17.1.4.1 Losing Yet Another Dear Friend ... Where is Hugo? He's Never Late

Carlos and I maintained phone contact every couple of months to catch up on events. He phoned me one day to tell me that our dear friend and colleague, Hugo Aréchiga, had died suddenly of a burst aortic aneurism. Carlos commented that Hugo was famous for always being impeccably dressed and precisely punctual, always arriving at scheduled meetings neither a minute early nor a minute late. But at the funeral for Hugo, Carlos described how, ironically, everyone felt that they were waiting...where is Hugo? Why isn't he here? Hugo is never late....

17.1.4.2 Mating Produces Analgesia; Stronger than Morphine

A fundamental guiding principle in our research, and one that we always kept returning to in our discussions, was the potential natural function and adaptive significance of our findings. We asked this of the analgesic effect of vaginocervical stimulation in rats and humans. We considered whether natural mating in rats would produce analgesia. With Gabriela González-Mariscal and Porfirio Gómora, we administered a very brief (100 ms) shock to the tail of female rats at the threshold intensity to elicit a vocalization squeak. Then we increased the intensity by 30% and applied a shock during each male's mount, intromission, and ejaculation. We counted the number of squeaks elicited by the shock under each of the behavioral conditions. We found a mild reduction in vocalizations during mounts, greater reduction during intromissions, and virtually no vocalizations during ejaculations. In order to calibrate the strength of this mating-induced analgesia, we used the same shock-induced vocalization procedure in a dose–response paradigm using morphine, rather than mating. It took more than twice the standard analgesic dose of morphine (4 mg/kg) to match the effect of intromissions, and at 15 mg/kg morphine, we were still obtaining vocalizations in response to the shock, by contrast with no vocalizations to the shock during ejaculations (Gómora et al., 1994). Thus, we concluded that the analgesia produced naturally during mating is approximately four times stronger than the standard rat/human analgesic dose of morphine! It is always a challenge to provide a rational basis for assuming an adaptive function for a behavioral or physiological process. In the case of mating-induced analgesia in rats, we speculate that the analgesia renders the female willing to accept the multiple intromissions prior to ejaculation that are necessary for pregnancy. Two findings by others are relevant to this conclusion. Norman Adler did a clever study for his doctoral dissertation. Normally, rats intromit an average of 8 sequential bouts prior to ejaculation. Norm waited for 6 intromissions, and then placed the males with new females. Those females thus received an average of only 2 intromissions before they received the ejaculate. Their pregnancy rate was significantly lower than the rats that received the typical 8 intromissions prior to ejaculation (Adler, 1969). Norm subsequently showed that the multiple intromissions increased progesterone secretion, which is necessary to stimulate uterine development to support implantation of the zygotes (Adler, Resko, and Goy, 1970). The second relevant finding was that excessive multiple intromissions induce female rats to reject the males (Hardy and DeBold, 1973). Consequently, we speculated that the natural analgesia of mating in rats makes the females willing to accept the multiple intromissions that are necessary for her pregnancy (Komisaruk, Beyer, and Whipple, 2006).

17.1.5 Enter Beverly

Subsequently, Beverly Whipple hypothesized that the birth canal stimulation that occurs during human childbirth would elevate pain thresholds. We found that indeed, pain thresholds, measured by calibrated force applied to the fingers, increased significantly during the first stage of labor, before the fetus' head starts to emerge through the cervix (Whipple, Josimovich, and Komisaruk, 1990). In the case of humans, we speculated that the pain of childbirth would be greater in the absence of the analgesic mechanism, possible adaptive functions being that the reduction in pain reduces the stress (stress can interfere with milk ejection) and facilitates bonding of the mother to the newborn (Komisaruk, Beyer, and Whipple, 2006).

17.1.5.1 We Interest Carlos in Human Sexuality Research; Writing Our Book

When, with Beverly, I started to do research on human sexual responses, this was a new challenge for Carlos. When he saw the results we were obtaining with functional magnetic resonance imaging (MRI), he became interested in exploring human sexuality in addition to the rat sexual behavior. In his characteristic pattern, he immersed himself in the human sexuality literature. It was at about that time that I received the request from Johns Hopkins University Press to write a book on orgasm. I had previously been asked to write the prologue to a book on orgasm by an author contracted by the Hopkins Press, but when that author withdrew from the project, the editor at the Hopkins Press asked me to write the book. It was obvious to me that the field is so extensive that it would be necessary to recruit Carlos and Beverly to contribute their expertise. So I asked Carlos if he would like to write about pharmacological and hormonal aspects of human sexuality and orgasm, asked Beverly if she would like to write about health and social aspects; I would write about neuroscience aspects. I was delighted that Carlos and Beverly agreed to join in the endeavor, which ended up taking about 5 years. Carlos suggested that we work on his segments together in Los Cabos in Baja California Sur, enjoying the environment. So we holed up in a quiet hotel in Cabo San Lucas in 2004 for over a week and focused on our writing.

17.1.5.2 I Get to Realize My Childhood Intrigue with Baja California Sur

Carlos' daughter, Gaby, had moved to Los Cabos several years earlier, with her husband, Juan, to get away from the congestion of Mexico City. When I was 8 years old, my aunt gave me a book on features and cultures of the world, which I read from cover to cover when I was recovering from pneumonia. I was particularly intrigued with the Baja California peninsula, a long, narrow desert in Mexico. For years after first reading about Baja, I had the fantasy that you have to be careful when you walked there because you could fall off either side. I fell in love with the desert when I was an undergraduate working the summer of 1959 as an assistant at the field research station in Portal, Arizona, that was run by the American Museum of Natural History. So when Carlos and I took a break from writing, and Gaby and Juan drove us around Los Cabos, she took us also to a sleepy little town, Todos Santos, about 50 miles north of Cabo San Lucas. The Pacific coast is about 1.5 miles from the town. Gaby asked Chuck Cimino, a realtor in town, to show us around near the coast. Chuck took Carlos, Gaby, Juan, and me to a hill overlooking the Pacific. The day was warm, sunny, and dry. A sea breeze wafted up the hill. There were cactus all around us and sandy soil at our feet. A jackrabbit sat nearby, watching us. I could see for miles all around, and hear the rumble of the crashing waves. There were palm trees nearby, Chuck explaining that an underground river created an oasis in this desert. A sierra mountain range was far to our back. As I looked around 360 degrees for miles in every direction, I counted 7 little houses. Totally enchanted, I asked Chuck, "Is this land for sale?" He said yes. I said, "I'll take it!" I never did anything so impulsive in my life. We went back to his office in town and I wrote a check as a deposit. I bought the land, a third of an acre. When I told my dear friend and colleague, María Cruz Rodríguez del Cerro, Profesora Catedrática of Psychobiology at UNED in Madrid, she said her sister, Rosa, was an architect. Rosa said she would design a house on the land. She did, and Juan introduced me to his friend, Ricardo Canedo, a house builder. I showed him the blueprints that Rosa prepared, told him how much I had in my home equity line, asked him if he could build it for that; he said yes, so we did it. Figure 18.13 shows Carlos contemplating as he looked out over the Pacific in my home in Todos Santos.

17.1.5.3 More Research Literature on Orgasm Than We Realized

Getting back to our book, I initially thought there wouldn't be much literature on orgasm. Actually, we ended up citing well over 600 relevant articles in our book! Our book *The Science of Orgasm* was finally published in 2006. We wrote it in Scientific American style and gave it a low-key humorous touch—a "plain brown wrapper" as the book jacket (Komisaruk, Beyer, and Whipple, 2006). It has now been translated into Spanish, Chinese, Japanese, Dutch, and Korean. It is being translated also into Turkish. After it was published, our editor, Vincent Burke, suggested that we write a less technical version for a more popular audience. Beverly suggested that we expand it to consider multicultural factors in human sexuality. She introduced us to Sara Nasserzadeh, who joined as an author, and we organized the book in a question-and-answer format. Vincent Burke suggested "The Orgasm Answer Guide" as the title, which we adopted (Komisaruk et al., 2009). It has now been translated into Spanish and German. Carlos continued his interest in human sexuality, and we were asked to write several popular magazine articles, including in Mexico, in Spanish, which Carlos organized (e.g., Beyer and Komisaruk, 2009).

17.1.6 Unfinished Work ... Carlos' Life Cut Short

We had planned to test the effects of ingesting glycine on pain thresholds in humans, but those plans were stymied by Carlos' hopeless, untimely, unexpected, sudden, excruciatingly sad death in October 2013.

This book is our expression of gratitude to the tremendous impact that this wonderful guy has had on us all.

17.2 MY REMEMBRANCES OF CARLOS, BY BEVERLY

17.2.1 At the Institute of Animal Behavior

I first met Carlos when I was working with Barry on my doctorate at the IAB, in psychobiology, with a major in neurophysiology. I was at the IAB, either full or part time, from the fall of 1983 until I defended my dissertation in February of 1986. I was the only student who was working with women at the IAB. I was conducting research building on my earlier published research concerning the rediscovery and naming of the Gräfenberg spot (or G spot) and women's sexual responses.

17.2.2 My Path to Research

A little history of how I came to the IAB: I did not plan to become a researcher. I was a nursing professor and a sexuality educator. At a meeting of the American Association of Sexuality Educators, Counselors and Therapists (AASECT) around 1977, I met Dr. John Perry, who had developed an electronic perineometer to measure pelvic muscle strength, which could also be used as a biofeedback device. I thought this device would be excellent to use to help teach Kegel exercises correctly to women who had stress urinary incontinence (SUI). This is when a woman loses urine when she coughs, jumps, or sneezes. So I began teaching Kegel exercises to treat SUI using the electronic perineometer to give the women feedback on how well they were doing the exercises. These exercises are named after Dr. Arnold Kegel, who found that the exercises helped prevent surgery in women with SUI (Kegel, 1949, 1952).

Some of the women who came to me to learn the Kegel exercises for their stress incontinence had very strong pelvic muscles; by contrast with stress incontinence, the pelvic muscles are very weak. The women with the strong pelvic muscles stated that the expulsion of fluid from their urethra occurred during sexual activity and was triggered by stimulation of a sensitive area felt through the anterior or front vaginal wall.

17.2.2.1 Rediscovery and Naming of the Gräfenberg Spot

A search of the literature led John Perry and me to Dr. Ernst Gräfenberg's article, "The role of the urethra in female orgasm" (Gräfenberg, 1950), in which he described a sensitive area felt through the anterior wall of the vagina, and an expulsion of fluid from the urethra that is different from urine, when this area is stimulated. In order to characterize this area, we invited physicians and nurse practitioners to perform a sexological examination of the vagina looking for areas of sensitivity. They found a sensitive area that swelled when it is stimulated in the 400 women examined. These women reported that this sensitive area was felt through the anterior vaginal wall at between 11 and 1 o'clock, which was confirmed by the sexological examination. This confirmed Dr. Gräfenberg's finding and led to our rediscovery of and naming this sensitive area the Gräfenberg spot or G spot (Addiego et al., 1981; Ladas, Whipple, and Perry, 1982, 2005).

More about Dr. Gräfenberg, his development of the first intrauterine device, and his work with this sensitive area we named after him, and his identification of female ejaculation can be found in a paper I published about his life (Whipple, 2000).

The G spot is typically located about halfway between the back of the pubic bone and the cervix, along the course of the urethra. It swells when it is stimulated, although it is not possible to palpate in an unstimulated state. That is why it is not found in a gynecological exam. Physicians do not sexually stimulate patients, and the region is also occluded by a bivalve speculum. We hypothesized that this area is composed of many tissues, organs, and nerve pathways (Ladas, Whipple, and Perry, 1982, 2005).

17.2.2.2 Female Ejaculation

We found that only some of the women to whom I was teaching Kegel exercises lost fluid (i.e., ejaculated) at orgasm or during sexual stimulation; they seemed to have very strong pubococcygeus muscles. So we designed a study to determine if there was a significant difference in the muscle strength of women who claimed to ejaculate and those who did not. We found a significant difference in the pelvic muscle strength of women who claimed to ejaculate compared to those who did not ejaculate; those who ejaculated had the stronger pubococcygeus muscles, a finding that we published in 1981 (Perry and Whipple, 1981).

The phenomenon of female ejaculation refers to expulsion of fluid from the urethra that is different from urine. Many women reported having surgery to correct this "problem" and others reported that they stopped having orgasms to prevent "wetting the bed." The fluid was described as looking like watered-down fat-free milk, tasting sweet, and usually about a teaspoon in volume.

A number of studies have been published in which the fluid expelled from the urethra was subjected to chemical analysis. In our first published studies, we found a significant difference between urine and female ejaculate in terms of the levels of prostatic acid phosphatase, urea, and creatinine (Addiego et al., 1981; Belzer, Whipple, and Moger, 1984). We have also found a significant elevation in glucose in the ejaculate (Addiego et al., 1981; Ladas et al., 1982; Belzer et al., 1984), and other researchers report a significant elevation in fructose (Cabello, 1997; Zaviacic, 1999). Dr. Francesco Cabello, from Spain, reported that he tested the hypothesis that all women ejaculate, although, because the amount is so small, and most women are lying on their backs during sexual activity, it may not be expelled and some may have retrograde ejaculation. He found a significant difference in prostatic-specific antigen (PSA) between preorgasmic and postorgasmic urine specimens (PSA is what they test for in blood when screening for prostate cancer in men and measuring disease progress and remission). Cabello also found PSA in the female ejaculate (Cabello, 1997). Dr. Milan Zaviacic from Slovakia has since reported on PSA in the female ejaculate, which is secreted by the female prostate (Zaviacic, 1999).

A study conducted in Guadalajara, Mexico by Dr. Alberto Rubio-Casillas from the Universidad de Guadalajara and Dr. Emmanuele Jannini from Italy demonstrated that female ejaculation and squirting/gushing are two different phenomena (Rubio-Casillas and Jannini, 2011). They stated that "the real female ejaculation is the release of a very scanty, thick, and whitish fluid from the female prostate, while the squirting is the expulsion of a diluted fluid from the urinary bladder" (Rubio-Casillas and Jannini, 2011, pp. 3502–3). They conducted biochemical studies of the two types of fluids as well as urine from the same subject, and there were significant differences between all three fluids, the squirting fluid being mainly diluted urine, but containing a small amount of PSA.

Zaviacic proposed that the paraurethral and Skene's glands are the female prostate gland and this is where the female ejaculation is coming from (Zaviacic, 1999). The name of these glands was officially changed to the female prostate gland by the Federative International Committee on Anatomical Terminology (FICAT) in 2001. We also believe that this tissue is part of the area that we have identified as the Gräfenberg spot or G spot.

We published our studies in 1981 (Addiego et al., 1981; Perry and Whipple, 1981) and in the first edition of *The G Spot and Other Discoveries about Human Sexuality* (Ladas et al., 1982). Dr. Vern Bullough, a sexuality researcher and nurse, spoke with me at a Society for the Scientific Study of Sexuality (SSSS) conference and told me, when we talked about my going for a PhD, that I must get my doctorate in a "hard science."

17.2.3 My Path to Rutgers University

In 1981 I wanted to go to the World Congress of Sexology in Israel but could not afford to go, so I wrote for copies of the published abstracts that were of interest to me. One person I communicated with was Dr. Barry Komisaruk. He had heard about our research at the conference from Dr. Julian Davidson and Dr. Benjamin Graber (John Perry and I wrote two of the chapters in Graber's book, *Circumvaginal Musculature and Sexual Function* (Perry and Whipple, 1982 a,b). Barry invited me to present a guest lecture about our research in a class in human sexuality that he was teaching at Rutgers.

Based on his prior behavioral and neurophysiological research in rats in which he found that vaginocervical stimulation in rats produces a strong analgesia, Barry was interested in determining if vaginocervical stimulation also blocks pain in women. And I wondered…is the G spot just for pleasure or does it have adaptive significance?

So this is where I went for my PhD in psychobiology with a major in neurophysiology (the "hard science" that Dr. Vern Bullough recommended). After obtaining approval from the Rutgers University IRB (i.e., Institutional Review Board for the protection of human subjects in research), we performed a series of studies in women, measuring pain thresholds by applying calibrated gradually increasing force to the fingers during vaginal self-stimulation.

17.2.3.1 Analgesia in Women with Stimulation of the Area of the Gräfenberg Spot

We found that the elevation in pain detection threshold increased by a mean of 47% when pressure was self-applied to the anterior vaginal wall (the area of the Gräfenberg spot). When stimulation was self-applied in a pleasurable manner, the pain threshold was greater (by 84%) than that in the resting control condition. The pain detection threshold increased by a mean of 107% when the women reported orgasm from stimulation of this area. There were no increases in tactile (or touch) thresholds. This demonstrated that the effect was an analgesic, rather than an anesthetic, effect and not a distracting effect. We published this study in 1985 and a follow-up study in 1988 (Whipple and Komisaruk, 1985, 1988). We then demonstrated that an analgesic effect also occurs naturally during labor (Whipple, Josimovich, and Komisaruk, 1990).

We believe that childbirth would be more painful without this natural pain-blocking effect, which is activated when the pelvic, the hypogastric, and possibly the sensory vagus nerves are stimulated as the cervix dilates and from pressure in the vagina produced by the emerging fetus.

17.2.4 Working with Carlos' Colleagues in Mexico

After we identified a natural analgesic effect in women during labor (Whipple et al., 1990), we then conducted a study in Mexico with Carlos' colleagues, Pablo Pacheco, Margarita Martínez-Gómez, and Laura Oliva-Zárate, on the inverse relationship between the level of chronic dietary intake of chile picante (a dietary source of capsaicin) and the intensity of analgesia produced by vaginal self-stimulation (Whipple et al., 1989). The rationale for that study was the following: (1) capsaicin was shown to permanently destroy c-fibers when injected to neonatal rats (Skofitsch et al., 1985); (2) the pelvic nerve consists mainly of c-fibers (Kawatani, Nagel, and DeGroat, 1986); (3) the pelvic nerve is the main contributor to the analgesic effect of vaginal stimulation (Cunningham et al., 1991), and (4) capsaicin injected to neonatal rats blocked the analgesic effect of vaginal stimulation (Rodriguez-Sierra et al., 1988). Pablo and Margarita said that there were many women in Mexico who had commenced eating hot chili peppers from the time that they were very young. I hypothesized that such women would have a weaker analgesic effect of vaginal self-stimulation than women who never ate chili peppers. Our findings supported the hypothesis (Whipple et al., 1989). That has led to the question of whether the pain of childbirth would be different in women who did or did not consume chilis during their upbringing. We are still interested in performing that study. In India, women are told not to eat hot spicy food for three months before they deliver. This was also confirmed to me verbally when I spoke to hundreds of OB/GYN physicians in Malaysia. They said women of Indian origin had a harder time during labor than women from Malaysia or China, the three main ethnic groups in Malaysia.

17.2.4.1 Time with Carlos and His Family

Carlos was a visiting professor while I was at the IAB. We became good friends and colleagues. His daughters, María Emilia and Gaby, visited and stayed at our home in Southern New Jersey many times. I still remember them walking on our lake when it was frozen, something that they had never experienced. We visited Carlos and his family in Mexico City and I also visited his labs in Tlaxcala and worked with some of his colleagues there. My husband Jim and I visited Carlos, his wife Josefina, and his daughters multiple times in Mexico City, and Jim and I stayed at Carlos' home at those times. Carlos' home was like a museum with his collection of masks on the wall, and his extensive collection of Mexican artifacts. He also took us to many museums in Mexico City because of our interest in Mexican culture.

Carlos, Barry, and I, along with other colleagues, conducted and published studies concerning vaginal stimulation in female rats. We published one of our joint studies on the somato-motor components of the pelvic and pudendal nerves of the female rat (Pacheco et al., 1989).

17.2.5 My New Human Physiology Laboratory at Rutgers University

After I completed my PhD, I obtained a second master's degree in nursing, because Rutgers College of Nursing wanted me on their faculty and offered me a large grant to build a human physiology laboratory, but I needed a master's degree in nursing. So I completed that degree the next year and started teaching and conducting further research at Rutgers University in 1987.

My research program has been devoted to validating the reports of pleasurable experiences from sensual and sexual stimulation in women. Another type of orgasmic response we measured in my new human physiology laboratory was orgasm from imagery alone. That is, no one, including the woman herself, touched her body, but she experienced orgasm. This study was based on Dr. Gina Ogden's doctoral research. We found that the physiological correlates of orgasm, that is, significant increases in blood pressure, heart rate, pupil diameter, and pain thresholds, were comparable during orgasm from genital self-stimulation and orgasm from imagery alone (Whipple, Ogden, and Komisaruk, 1992). We later conducted functional MRIs of the brain in women during self-stimulation and during imagery-induced orgasm (Komisaruk and Whipple, 2005a).

We continued our research program by validating the subjective reports of women with complete spinal cord injury (SCI) that do, indeed, experience orgasm. These women had been told by their health care professionals, based on the literature, that they could not experience orgasm or if they did, it was a "phantom orgasm." We have documented that women with complete SCI do indeed experience orgasm from self-stimulation of the anterior wall of the vagina, the cervix, and a hypersensitive area of their body above the level of their injury (Whipple, Gerdes, and Komisaruk, 1996; Whipple et al., 1996a,b; Whipple et al., 1998; Komisaruk and Whipple, 2005b).

The women in our study who had complete SCI above the level of entry into the spinal cord of all the known genital spinal nerves (pudendal, pelvic, and hypogastric) reported to us that they could feel the stimulator self-applied to their vagina and cervix, and the stimulation significantly elevated the pain thresholds measured at their fingers. To account for this unexpected and puzzling finding, we postulated the existence of a sensory pathway that bypasses the spinal cord, carrying sensory input from the vagina and cervix directly to the brain, which Barry postulated to be the vagus nerve. To test this hypothesis in laboratory rats, we worked with Carlos and his team, and in 1996 we published that vagotomy blocks responses to vaginocervical stimulation after genitospinal neurectomy in rats (Cueva-Rolon et al., 1996).

17.2.5.1 Brain Imaging Research

To test whether the vagus nerve provides a vaginal sensory pathway in women, we hypothesized that the brain regions to which the vagus nerve projects (the nucleus tractus solitarii [NTS] in the medulla oblongata) would be activated by vaginal or cervical self-stimulation in women with complete SCI above the level of the entry into the spinal cord of the genital sensory nerves. We tested the hypothesis that the vagus nerve can convey afferent activity from the cervix to the NTS by conducting preliminary positron emission tomography (PET) scans of the brain coupled with MRI to provide neuroanatomical localization (Whipple and Komisaruk, 2002).

This study provided preliminary evidence that vaginal or cervical self-stimulation activates the region of the NTS. We subsequently found that functional MRI provides a much higher resolution image of this region of the brain stem than PET, so we continued using functional MRI in women with complete SCI above the level of entry into the spinal cord of the genitospinal nerves, and found that some of the women experienced orgasm from the self-stimulation, enabling us to identify, for the first time, brain regions activated during orgasm (Komisaruk et al., 2004). Those findings were instrumental in getting Carlos interested in human sexuality, and probably most instrumental in stimulating Carlos' willingness to collaborate with us on our joint books, chapters, and popular articles on human sexuality.

17.2.6 Coauthoring Books with Carlos

Barry was contacted by Johns Hopkins University Press to write a book on orgasm, and he asked Carlos and me to collaborate, using our compatible different perspectives and expertise, to write the book jointly as coauthors. Barry took the lead and we collaborated and published *The Science of Orgasm* (Komisaruk, Beyer, and Whipple, 2006), which has been published now also in Spanish, Chinese, Japanese, Korean, and Dutch. We were then asked to write a follow-up book about orgasm for the general public. I recommended that we invite a coauthor who is a sexual and relationship therapist and is also from a different geographical area. Dr. Sara Nasserzadeh, a sexual and relationship therapist from Iran, joined our team. We intentionally oriented the book to be culturally sensitive. We coauthored *The Orgasm Answer Guide* (Komisaruk et al., 2009), which is now also published in Spanish and German. Carlos organized a conference at the Autonomous University of Tlaxcala in November, 2009, invited Barry, Sara, and me to present to his colleagues and students, and joined in, giving a presentation at his own conference. Thus it became the first ... but, sadly, the only ... time that all four of us presented our contributions together.

A few days before Carlos died, we were all in Puebla, Mexico for the inauguration of the "Centro de Investigación Dra. Beverly Whipple." This was sponsored by the Autonomous University of Puebla, three other universities in Puebla, and the government of Mexico. Many top researchers from Mexico and the United States as well as my family were present. Unfortunately, Carlos called to say he didn't feel well and could not be in Puebla. He was so missed, as he is now that he is no longer physically with us. Carlos, you are loved and missed.

REFERENCES

Addiego, F., E.G. Belzer, J. Comolli, W. Moger, J.D. Perry, and B. Whipple. 1981. Female ejaculation: A case study. *J Sex Res* 17:13–21.

Adler, N.T. 1969. Effects of the male's copulatory behavior on successful pregnancy of the female rat. *J Comp Physiol Psychol* 69:613–622.

Adler, N.T., J.A. Resko, and R.W. Goy. 1970. The effect of copulatory behavior on hormonal change in the female rat prior to implantation. *Physiol Behav* 5:1003–1007.

Ariëns Kappers, C.U., G.C. Huber, and E.C. Crosby. 1936. *The Comparative Anatomy of the Nervous System of Vertebrates, Including Man.* New York: Hafner Publ Co.

Belzer, E., B. Whipple, and Moger, W. 1984. On female ejaculation. *J Sex Res* 20:403–406.

Beyer, C., and B.R. Komisaruk. 2009. El Orgasmo y su fisiologia. *Ciencia* 60:14.

Beyer, C., B.R. Komisaruk, A.-M. Lopez-Colome, and M. Caba. 1992. Administration of AP5, a glutamate antagonist, unmasks glycine analgesic actions in the rat. *Pharm Biochem Behav* 41:229–232.

Beyer, C., L.A. Roberts, and B.R. Komisaruk, 1985. Hyperalgesia induced by altered glycinergic activity at the spinal cord. *Life Sci* 37:295–301.

Beyer, C., N. Vidal, and P.G. McDonald. 1969. Interaction of gonadal steroids and their effect on sexual behaviour in the rabbit. *J Endocrinol* 45:407–413.

Cabello, F. 1997. Female ejaculation: Myths and reality. In *Sexuality and Human Rights.* eds. J.J. Borras-Valls and M. Perez-Conchillo, 1–8. Valencia, Spain: Nau Libres.

Cueva-Rolon, R., G. Sansone, R. Bianca, L.E. Gomez, C. Beyer, B. Whipple, and B.R. Komisaruk. 1996. Vagotomy blocks responses to vaginocervical stimulation after genitospinal neurectomy in rats. *Physiol Behav* 60:19–24.

Cunningham, S.T., J.L. Steinman, B. Whipple, A.D. Mayer, and B.R. Komisaruk. 1991. Differential roles of hypogastric and pelvic nerves in analgesic and motoric effects of vaginocervical stimulation in rats. *Brain Res* 559:337–343.

Gómora, P., C. Beyer, G. González-Mariscal, and B.R. Komisaruk. 1994. Momentary analgesia produced by copulation in female rats. *Brain Res* 656:52–58.

Gonzalez-Diddi, M., B.R. Komisaruk, and C. Beyer. 1972. Differential effects of testosterone and dihydrotestosterone on the diverse uterine tissues of the ovariectomized rat. *Endocrinology* 91:1129–1132.

Gräfenberg, E. 1950. The role of the urethra in female orgasm. *Intl J Sexology* 3:145–148.

Hardy, D.F., and J.F. DeBold. 1973. Effects of repeated testing on sexual behavior of the female rat. *J Comp Physiol Psychol* 85:195–202.

Kawatami, M., J. Nagel, and W.C. DeGroat. 1986. Identification of neuropeptides in pelvic and pudendal nerve afferent pathways to the sacral spinal cord of the cat. *J Comp Neurol* 249:117–132.

Kegel, A.H. 1949. The physiologic treatment of poor tone and function of the genital muscles and of urinary stress incontinence. *West J Surg Obstet Gynecol* 57:527–535.

Kegel, A.H. 1952. Sexual functions of the pubococcygeus muscle. *West J Surg Obstet Gynecol* 60:521–524.

Komisaruk, B.R. 1967. Effects of local brain implants of progesterone on reproductive behavior in ring doves. *J Comp Physiol Psychol* 64:219–224.

Komisaruk, B.R. 1970. Synchrony between limbic system theta activity and rhythmical behavior in rats. *J Comp Physiol Psychol* 70:483–492.

Komisaruk, B.R., and B. Whipple. 2005a. Brain activity imaging during sexual response in women with spinal cord injury. In *Biological Substrates of Human Sexuality.* ed. J. Hyde. 109–145. Washington DC: American Psychological Association.

Komisaruk, B.R., and B. Whipple. 2005b. Functional MRI of the brain during orgasm in women. *Ann Rev Sex Research* 16:62–86.

Komisaruk, B.R., B. Whipple, S. Nasserzadeh, and C. Beyer-Flores. 2009. *The Orgasm Answer Guide*. Baltimore, MD: The Johns Hopkins University Press.

Komisaruk, B.R., B. Whipple, A. Crawford, S. Grimes, W.-C. Liu, A. Kalnin, and K. Mosier. 2004. Brain activation during vaginocervical self-stimulation and orgasm in women with complete spinal cord injury: fMRI evidence of mediation by the Vagus nerves. *Brain Res* 1024:77–88.

Komisaruk, B.R., and C. Beyer. 1972. Responses of diencephalic neurons to olfactory bulb stimulation, odor, and arousal. *Brain Res* 36:153–170.

Komisaruk, B.R., C. Beyer, C., and B. Whipple. 2006. *The Science of Orgasm*. Baltimore, MD: The Johns Hopkins University Press.

Komisaruk, B.R., and J. Olds. 1968. Neuronal correlates of behavior in freely moving rats. *Science* 161:810–813.

Komisaruk, B.R., and K. Larsson. 1971. Suppression of a spinal and a cranial nerve reflex by vaginal or rectal probing in rats. *Brain Res* 35:231–235.

Komisaruk, B.R., P.G. McDonald, D.I. Whitmoyer, and C.H. Sawyer. 1967. Effects of progesterone and sensory stimulation on EEG and neuronal activity in the rat. *Exper Neurol* 19:494–507.

Ladas, A.K, B. Whipple, and J.D. Perry. 1982. *The G Spot and Other Recent Discoveries about Human Sexuality*. New York: Holt, Rinehart and Winston (published in 22 languages).

Ladas, A.K, B. Whipple, and J.D. Perry. 2005. *The G Spot and Other Discoveries about Human Sexuality, Classic Edition*. New York: Owl Books.

Larsson, K., P. Södersten, and C. Beyer. 1973. Induction of male sexual behaviour by oestradiol benzoate in combination with dihydrotestosterone. *J Endocrinol* 57:563–564.

Lehrman, D.S., and R.P. Wortis. 1960. Previous breeding experience and hormone-induced incubation in the ring dove. *Science* 132:1667–1668.

Lisk, R.D. 1962. Diencephalic placement of estradiol and sexual receptivity in the female rat. *Am J Physiol* 203:493–496.

Macrides F, H.B. Eichenbaum, and W.B. Forbes. 1982. Temporal relationship between sniffing and the limbic theta rhythm during odor discrimination reversal learning. *J Neurosci* 2:1705–1717.

Mena, F., G. Anguiano, and C. Beyer. 1961. Release of oxytocin by stimulation of the caudal part of the hypothalamus. *Bol Inst Estud Med Biol Univ Nac Auton Mex* 19:119–124.

Morali, G., K. Larsson, and C. Beyer. 1977. Inhibition of testosterone-induced sexual behavior in the castrated male rat by aromatase blockers. *Horm Behav* 9:203–213.

Pacheco, P., M. Martinez-Gomez, B. Whipple, C. Beyer, and B.R. Komisaruk. 1989. Somatomotor components of the pelvic and pudendal nerves of the female rat. *Brain Res* 490:85–94.

Perry, J.D., and B. Whipple. 1981. Pelvic muscle strength of female ejaculators: Evidence in support of a new theory of orgasm. *J Sex Res* 17:22–39.

Perry, J.D., and B. Whipple. 1982a. Multiple components of the female orgasm. In *Circumvaginal Musculature and Sexual Function*. ed. B. Graber. 101–114. New York: S. Karger.

Perry, J.D., and B. Whipple. 1982b. Vaginal myography. In *Circumvaginal Musculature and Sexual Function*. ed. B. Graber. 61–73. New York: S. Karger.

Ramirez, V.D., B.R. Komisaruk, D.I. Whitmoyer, and C.H. Sawyer. 1967. Effects of hormones and vaginal stimulation on the EEG and hypothalamic units in rats. *Am J Physiol* 212:1376–1384.

Roberts, L.A., C. Beyer, and B.R. Komisaruk. 1985. Strychnine antagonizes vaginal stimulation-produced analgesia at the spinal cord. *Life Sci* 36:2017–2023.

Roberts, L.A., C. Beyer, and B.R. Komisaruk. 1986. Nociceptive response to altered GABAergic activity in the spinal cord. *Life Sci* 39:1667–1674.

Rodriguez-Sierra, J.F., G. Skofitsch, B.R. Komisaruk, and D.M. Jacobowitz. 1988. Abolition of vagino-cervical stimulation-induced analgesia by capsaicin administered to neonatal, but not adult rats. *Physiol Behav* 44:267–272.

Rubio-Casillas, A., and E.A. Jannini. 2011. New insights from one case of female ejaculation. *J Sex Med* 8:3500–3504.

Sawyer, C.H. 1947. Cholinergic drugs and the hypophysis in salamander larvae. *Anat Rec* 97:366.

Semba, K., and B.R. Komisaruk. 1978. Phase of the theta wave in relation to different limb movements in awake rats. *Electroenceph Clin Neurophysiol* 44:61–71.

Skofitsch, G., N. Zamir, C.J. Helke, J.M. Savitt, and D.M. Jacobowitz. 1985. Corticotropin releasing factor-like immunoreactivity in sensory ganglia and capsaicin sensitive neurons of the rat central nervous system: Colocalization with other neuropeptides. *Peptides* 6:307–318.

Whipple, B. 2000. Ernst Gräfenberg: From Berlin to New York. *Scand J Sexol* 3:43–49.

Whipple, B., and B.R. Komisaruk. 1985. Elevation of pain thresholds by vaginal stimulation in women. *Pain* 21:357–367.

Whipple, B., and B.R. Komisaruk. 1988. Analgesia produced in women by genital self-stimulation. *J Sex Res* 24:130–140.

Whipple, B., and B.R. Komisaruk. 2002. Brain (PET) responses to vaginal-cervical self-stimulation in women with complete spinal cord injury: Preliminary findings. *J Sex Marital Ther* 28:79–86.

Whipple, B., C.A. Gerdes, and B.R. Komisaruk. 1996. Sexual response to self-stimulation in women with complete spinal cord injury. *J Sex Res* 33:231–240.

Whipple, B., E. Richards, M. Tepper, and B.R. Komisaruk. 1996a. Sexual response in women with complete spinal cord injury. In *Women with Physical Disabilities: Achieving and Maintaining Health and Well-Being*. eds. D.M. Krotoski, M. Nosek, and M. Turk. 69–80. Baltimore, MD: Paul H. Brooks Publishing Co.

Whipple, B., E. Richards, M. Tepper, and B.R. Komisaruk. 1996b. Sexual response in women with complete spinal cord injury. *Sex Disabil* 14:191–201.

Whipple, B., E. Richards, M. Tepper, C. Gerdes, and B.R. Komisaruk. 1998. A quantitative and qualitative study concerning sexual response in women with complete spinal cord injury. In *Sexuality and Human Rights*. eds. J.J. Borras-Valls, and M. Perez-Conchillo. 267–271. Valencia, Spain: World Cong. Sexol.

Whipple, B., G. Ogden, and B.R. Komisaruk. 1992. Physiological correlates of imagery induced orgasm in women. *Arch Sex Behav* 21:121–133.

Whipple, B., J.B. Josimovich, and B.R. Komisaruk. 1990. Sensory thresholds during the antepartum, intrapartum, and postpartum periods. *Int'l J Nursing Stud* 27:213–221.

Whipple, B., M. Martinez-Gomez, L. Oliva-Zarate, P. Pacheco, and B.R. Komisaruk. 1989. Inverse relationship between intensity of vaginal self-stimulation-produced analgesia and level of chronic intake of a dietary source of capsaicin. *Physiol Behav* 46:247–252.

Zaviacic, M. 1999. *The Human Female Prostate: From Vestigial Skene's Paraurethral Glands and Ducts to Woman's Functional Prostate*. Bratislava, Slovakia: Slovak Academic Press.

18 Photo Gallery*

FIGURE 18.1 Ca. 1965. Carlos Beyer at Brain Research Institute, University of California, Los Angeles, California.

FIGURE 18.2 1980. At the wedding of *Carmen Clapp* and *Gonzalo Martínez de la Escalera.* From left to right: Flavio Mena and his wife Rosita, *Alonso Fernández-Guasti, Gabriela González-Mariscal, Carmen Clapp*, Carlos Beyer, and *Gonzalo Martínez de la Escalera.*

* **Authors of the present book are in italic.**

FIGURE 18.3 1980. With *Beverly Whipple* at the Conference on Reproductive Behavior at Rockefeller University, New York.

FIGURE 18.4 Ca. 1980. As Visiting Professor at the Institute of Animal Behavior, Rutgers University, Newark, New Jersey.

FIGURE 18.5 Ca. 1984. With *Barry Komisaruk at* CINVESTAV (Mexico City).

FIGURE 18.6 1986. With Jay Rosenblatt when Carlos received the gavel from the Conference on Reproductive Behavior, indicating he would host the next meeting.

FIGURE 18.7 1989. With Jay Rosenblatt at the ceremony when Rosenblatt received an *Honoris Causa Doctorate* from the National University for Long Distance Education (Madrid).

FIGURE 18.8 1999. With 5 of his former students, all members of the UAM-I class of 1978. From left to right: *Gonzalo Martínez de la Escalera, Carmen Clapp, Porfirio Gómora,* Carlos Beyer, Ricardo Mondragón, and *Gabriela González-Mariscal.*

FIGURE 18.9 Ca. 2000. With *Gabriela González-Mariscal* at the Society for Behavioral Neuroendocrinology Conference.

FIGURE 18.10 2004. With (from left to right): *Alonso Fernández-Guasti*, Flavio Mena, Javier Velázquez, *Gabriela González-Mariscal*, Carlos Beyer, *Gabriela Moralí,* José Ramón Eguíbar and *Barry Komisaruk*. (Veracruz, México, XLVII National Congress of Physiological Sciences)

FIGURE 18.11 2005. With *Maria Cruz Rodriguez del Cerro* and *Barry Komisaruk* in Todos Santos, Mexico.

FIGURE 18.12 2008. At Todos Santos, Mexico.

FIGURE 18.13 2005. At Todos Santos, Mexico.

Complete Bibliography of Carlos Beyer (-Flores)

PUBLICATIONS IN JOURNALS

1. Anguiano, G., C. Beyer, and M. Alcaraz. 1956. Mecanismo fisiológico de la hiperglicemia causada por el veneno de alacrán. *Bol. Inst. Estud. Méd. Biol.* (Méx.) 14:93–101.
2. Anguiano, G., and C. Beyer. 1959. Tratamiento de la intoxicación alacránica. *Rev. Fac. Med.* (Méx.) 1:663–675.
3. Aguado, S.E., and C. Beyer 1960. EEG changes caused by scorpion venom in the cat. *Bol. Inst. Estud. Méd. Biol.* (Méx.) 18:81–86.
4. Chavira, R.A., C. Beyer, G. Anguiano, and F. Mena. 1960. Constriction of the pupil in cats elicited by stimulation of the anterior perforated substance. *Bol. Inst. Estud. Méd. Biol.* (Méx.) 18:87–89.
5. Beyer, C. 1960. Cambios en el nivel de la glucosa sanguínea producidos por estimulación del lóbulo frontal. *Bol. Inst. Estud. Méd. Biol.* (Méx.) 18:63–71.
6. Beyer, C., G. Anguiano, and F. Mena. 1961. Oxytocin release in response to stimulation of cingulate gyrus. *Am. J. Physiol.* 200:625–627.
7. Beyer, C., G. Anguiano, and F. Mena. 1965. Changes in uterine motility in the cat induced by hypertonic solutions. *Bol. Inst. Estud. Méd. Biol.* (Méx.) 23:19–31
8. Chavira, R.A., G. Anguiano, C. Beyer, and F. Mena. 1961. Cerebral centers of pupillary activation. *J. Internat. Coll. Surgeons.* 36:385–390.
9. Mena, F., G. Anguiano, and C. Beyer. 1961. Liberacion de oxitocina por estimulación del hipotálamo caudal. *Bol. Inst. Estud. Méd. Biol.* (Méx.) 19:119–124.
10. Anguiano, G., R.A. Chavira, C. Beyer, and F. Mena. 1962. Pupilodilatación producida por soluciones hipertónicas. *Bol. Inst. Estud. Méd. Biol.* (Méx.) 20:105–111.
11. Beyer, D., F. Mena, P. Pacheco, and M. Alcaraz. 1962. Blockage of lactation by brain-stem lesions in the cat. *Am. J. Physiol.* 202:465–468.
12. Beyer, C., J.S. Tindal, and C.H. Sawyer. 1962. Electrophysiological study of projections from mesencephalic central gray matter to forebrain in the rabbit. *Exp. Neurol.* 6:435–450.
13. Tindal, J.S., C. Beyer, and C.H. Sawyer. 1963. Milk ejection reflex and maintenance of lactation in the rabbit. *Endocrinology.* 72:720–724.
14. Chavira, R.A., G. Anguiano, C. Beyer, and S. De Buen. 1963. Catarata experimental por soluciones salinas hipertónicas. *Bol. Inst. Estud. Méd. Biol.* (Méx.) 21:59–63.
15. Mena, F., and C. Beyer. 1963. Effect of high spinal section on established lactation in the rabbit. *Am. J.Physiol.* 205:313–316.
16. Beyer, C., T. Yaschine, and F. Mena. 1964. Alterations in sexual behaviour induced by temporal lobe lesions in female rabbits. *Bol. Inst. Estud. Méd. Biol.* (Méx.) 22:379–386.
17. Beyer, C., and C.H. Sawyer. 1964. Effects of vigilance and other factors on nonspecific acoustic responses in the rabbit. *Exp. Neurol.* 10:156–169.
18. Beyer, C. 1964. Neural control of oxytocin and prolactin release. World Health Organization, Switzerland. Paper Mch/Rres, 49:1–9.
19. Beyer, C., and F. Mena. 1965. Induction of milk secretion in the rabbit by removal of the telencephalon. *Am. J. Physiol.* 208:289–292.
20. Beyer, C., and F. Mena. 1965. Blockage of milk removal in the cat by periventricular diencephalic lesions. *Am. J. Physiol.* 208:585–588.

21. Beyer, C., and F. Mena. 1965. Effect of ovariectomy and barbiturate administration on lactation in the cat and the rabbit. *Bol. Inst. Estud. Méd. Biol.* (Méx.) 23:89–99.
22. Barry, J., C. Beyer, B. Flerko, J.C. Daidlaw, L. Martini, R.P. Michael, A. Queriw, A.E. Rakoff, and K. Shizume. 1965. Informes técnicos en neuroendocrinologia de la reproduccion humana. *World Health Org. Techn. Rep.* 304:1–21.
23. Beyer, C., V.D. Ramírez, D.I. Whitmoyer, and C.H. Sawyer. 1967. Effects of hormones on the electrical activity of the brain in the rat and rabbit. *Exp. Neurol.* 18:313–326.
24. Yaschine, T., F. Mena, and C. Beyer. 1967. Gonadal hormones and mounting behavior in the female rabbit. *Am. J. Physiol.* 213:867–872.
25. Mena, F., and C. Beyer. 1968. Effect of spinal cord lesions on milk ejection in the rabbit. *Endocrinology.* 83:615–617.
26. Mena, F., and C. Beyer. 1968. Induction of milk secretion in the rabbit by lesions in the temporal lobe. *Endocrinology.* 83:618–620.
27. Beyer, C., M.L. Cruz, and N. Rivaud. 1969. Persistence of sexual behavior in ovariectomized-adrenalectomized rabbits treated with cortisol. *Endocrinology.* 85:790–793.
28. Alcaraz, M., C. Guzmán-Flores, M. Salas, and C. Beyer. 1969. Effect of estrogen on the responsivity of hypothalamic and mesencephalic neurons in the female cat. *Brain Res.* 15:439–446.
29. Beyer, C., and N. Rivaud. 1969. Sexual behavior in pregnant and lactating domestic rabbits. *Physiol. Behav.* 4:753–757.
30. Beyer, C., N. Vidal, and P. Mcdonald. 1969. Interaction of gonadal steroids and their effect on sexual behaviour in the rabbit. *J. Endocrinol.* 45:407–413.
31. Mena, F., and C. Beyer. 1969. Effect of large doses of oxytocin on milk secretion in intact and spinal cord transected rats. *The Physiologist* 12:300.
32. Beyer, C., N. Rivaud, and M.L. Cruz. 1970. Initiation of sexual behavior in prepuberally ovariectomized rabbits. *Endocrinology.* 86:171–174.
33. McDonald, P., N. Vidal, and C. Beyer. 1970. Sexual behavior in the ovariectomized rabbit after treatment with different amounts of gonadal hormones. *Horm. Behav.* 1:161–172.
34. Beyer, C., P. McDonald, and N. Vidal. 1970. Failure of 5-alpha dihydrotestosterone to elicit estrous behavior in the ovariectomized rabbit. *Endocrinology.* 86:939–941.
35. Beyer, C., M.L. Cruz, and J. Martínez-Manautou. 1970. Effect of chlormadinone acetate on mammary development and lactation in the rabbit. *Endocrinology.* 86:1172–1174.
36. Beyer, C., and F. Mena. 1970. Parturition and lactogenesis in rabbits with high spinal cord transection. *Endocrinology.* 7:195–197.
37. McDonald, P., C. Beyer, F. Newton, B. Brein, R. Baker, H.S. Tan, C. Sampson, P. Kitching, R. Greenhill, and D. Pritchard. 1970. Failure of 5-alpha-dihydrotestosterone to initiate sexual behaviour in the castrated male rat. *Nature.* 227:964–965.
38. Beyer, C., N. Vidal, and A. Mijares. 1970. Probable role of aromatization in the induction of estrous behavior by androgens in the ovariectomized rabbit. *Endocrinology.* 87:1386–1389.
39. Anguiano, G., F. Mena, and C. Beyer. 1970. Effect of electrical stimulation and sections of the neuraxis on uterine motility in the cat. *Bol. Inst. Estud. Méd. Biol.* (Méx.) 26:311–314.
40. Beyer, C., J. Almanza, L. De la Torre, and C. Guzmán-Flores. 1971. Brain stem multiunit activity during relaxation behavior in the female cat. *Brain Res.* 29:213–222.
41. Beyer, C., J. Almanza, L. De la Torre, and C. Guzmán-Flores. 1971. Effect of genital stimulation on the brain stem multi-unit activity of anaestrous and estrous cats. *Brain Res.* 32:143–150.
42. Beyer, C., and B.R. Komisaruk. 1971. Effects of diverse androgens on estrous behavior, lordosis reflex, and genital tract morphology in the rat. *Horm. Behav.* 2:217–225.
43. Beyer, C., G. Moralí, and R. Vargas. 1971. Effect of diverse estrogens on estrous behavior and genital tract development in ovariectomized rats. *Horm. Behav.* 2:273–277.

44. Beyer, C., G. Moralí, and M.L. Cruz. 1971. Effect of 5-α dihydrotestosterone on gonadotropin secretion and estrous behavior in the female Wistar rat. *Endocrinology*. 89:1158–1161.
45. Beyer, C., and N. Vidal. 1971. Inhibitory action of mer 25 on androgen-induced oestrous behaviour in the ovariectomized rabbit. *J. Endocrinol.* 51:401–402.
46. Koranyi, L., C. Beyer, and C. Guzmán-Flores. 1971. Effect of ACTH and hydrocortisone on multiple unitactivity in the forebrain and thalamus in response to reticular stimulation. *Physiol. Behav.* 7:331–335.
47. Koranyi, L., C. Beyer, and C. Guzmán-Flores. 1971. Multiple unit activity during habituation sleep-wakefulness cycle and the effect of ACTH and corticosteroid treatment. *Physiol. Behav.* 7:321–329.
48. Beyer, C., and B.R. Komisaruk. 1971. Reply to androgen-induced receptivity. *Horm. Behav.* 2:357–358.
49. Komisaruk, B.R., and C. Beyer. 1972. Responses of diencephalic neurons to olfactory bulb stimulation, odor and arousal. *Brain. Res.* 36:153–170.
50. Komisaruk, B.R., and C. Beyer. 1972. Differential antagonism by MER-25 of behavioral and morphological effects of estradiol benzoate in rats. *Horm. Behav.* 3:63–70.
51. González-Diddi, M., B.R. Komisaruk, and C. Beyer. 1972. Differential effects of testosterone and dihydrotestosterone on the diverse uterine tissues of the ovariectomized rat. *Endocrinology*. 91:1129–1132.
52. Beyer, C., R.B. Jaffey, and V.L. Gay. 1972. Testosterone metabolism in target tissues effects of testosterone and dihydrotestosterone injection and hypothalamic implantation on serum lh in ovariectomized rats. *Endocrinology*. 91:1372–1375.
53. Cruz, M.L., and C. Beyer. 1972. Effect of septal lesions on maternal behavior and lactation in the rabbit. *Physiol. Behav.* 9:361–365.
54. Pérez-Palacios, G., A.E. Pérez, M.L. Cruz, and C. Beyer. 1973. Comparative uptake of 3h-androgens by the brain and the pituitary of castrated male rats. *Biol. Reprod.* 8:395–399.
55. Beyer, C., K. Larsson, G. Pérez-Palacios, and G. Moralí. 1973. Androgen structure and male sexual behavior in the castrated rat. *Horm. Behav.* 4:99–108.
56. Larsson, K., P. Södersten, and C. Beyer. 1973. Induction of male sexual behaviour by oestradiol benzoate in combination with dihydrotestosterone. *J. Endocrinol.* 57:563–564.
57. Beyer, C., and N. Rivaud. 1973. Differential effect of testosterone and dihydrotestosterone on the sexual behavior of prepuberally castrated male rabbits. *Horm. Behav.* 4:175–180.
58. Larsson, K., P. Södersten, and C. Beyer. 1973. Sexual behavior in male rats treated with estrogen in combination with dihydrotestosterone. *Horm. Behav.* 4:289–299.
59. Pérez-Palacios, G., K. Larsson, and C. Beyer. 1973. Physiological implication of the metabolism of androgen in brain. *Acta Physiol. Latinoam.* 23:460–462.
60. Beyer, C. 1974. Regulacion neuroendócrina de la conducta sexual en el humano. *Rev. Méd* (Méx.) 13:187–192.
61. Moralí G., K. Larsson, G. Pérez-Palacios, and C. Beyer. 1974. Testosterone, androstenedione, and androstenediol: effects on the initiation of mating behavior of inexperienced castrated male rats. *Horm. Behav.* 5:103–110.
62. Mena, F., D. Aguayo, G. Reyes, and C. Beyer. 1974. Effect of suckling and associated factors on ovarian compensatory hypertrophy during lactation in the rat. *J. Endocrinol.* 1:317–330.
63. Beyer, C., M.L. Cruz, V. Gay, and R. Jaffe. 1974. Effects of testosterone and dihydrotestosterone on fsh serum concentration and follicular growth in female rats. *Endocrinology*. 95:722–727.
64. Mena, F., C. Beyer, and C.E. Grosvenor. 1974. On the mechanism by which oxytocin depresses milk ejection and milk secretion in rats. *Am. J. Physiol.* 227:1249–1254.

65. Pérez, A.E., A. Ortiz, M. Cabeza, C. Beyer, and G. Pérez-Palacios. 1975. In vitro metabolism of 3h-androstenedione by the male rat pituitary, hypothalamus, and hippocampus. *Steroids*. 25:53–62.
66. Larsson, K., G. Pérez-Palacios, G. Moralí, and C. Beyer. 1975. Effects of dihydrotestosterone and estradiol benzoate pretreatment on testosterone-induced sexual behavior in the castrated male rat. *Horm. Behav.* 6:1–8.
67. Cervantes, M., L. De la Torre, and C. Beyer. 1975. Analysis of various factors involved in EEG synchronization during milk drinking in the cat. *Brain Res.* 91:89–98.
68. Pérez-Palacios, G., K. Larsson, and C. Beyer. 1975. Biological significance of the metabolism of androgens in the central nervous system. *J. Steroid. Biochem.* 6:999–1006.
69. Beyer, C., De la Ttorre, L., K. Larsson, and G. Pérez-Palacios. 1975. Synergistic actions of estrogen and androgen on the sexual behavior of the castrated male rabbit. *Horm. Behav.* 6:301–306.
70. Kubli, C., M. Cervantes, and C. Beyer. 1976. Changes in multiunit activity and eeg induced by the administration of natural progestins to flaxedil immobilized cats. *Brain Res.* 114:71–81.
71. Beyer, C., G. Moralí, F. Naftolin, K. Larsson, and G. Pérez-Palacios. 1976. Effect of some antiestrogens and aromatase inhibitors on androgen induced sexual behavior in castrated male rats. *Horm. Behav.* 7:353–363.
72. Beyer, C. 1976. Factores neuroendócrinos y conducta. *Bol. Inst. Estud. Méd. Biol.* (Méx.) 29:181–185.
73. Larsson, K., P. Södersten, C. Beyer, G. Moralí, and G. Pérez-Palacios. 1976. Effects of estrone, estradiol and estriol combined with dihydrotestosterone on mounting and lordosis behavior in castrated male rats. *Horm. Behav.* 7:379–390.
74. Pacheco, P., C. Beyer, G. Mexicano, and K. Larsson. 1976. Effects of genital stimulation upon spinal reflex activity of female cats under various hormonal conditions. *Physiol. Behav.* 17:699–703.
75. Beyer, C., G. Moralí, K. Larsson, and P. Södersten. 1976. Steroid regulation of sexual behavior. *J. Steroid Biochem.* 7:1171–1176.
76. Pérez, A.E., C. Beyer, K. Larsson, and G. Pérez-Palacios. 1977. In vitro conversion of 5-alpha-androstenediol to testosterone by the central nervous system and pituitary of the male rat. *Steroids.* 29:627–633.
77. Moralí, G., K. Larsson, and C. Beyer. 1977. Inhibition of testosterone-induced sexual behavior in the castrated male rat by aromatase blockers. *Horm. Behav.* 9:203–213.
78. Contreras, J.L. and C. Beyer. 1979. A polygraphic analysis of mounting and ejaculation in the New Zealand white rabbit. *Physiol. Behav.* 23:939–943.
79. Cervantes, M., R. Ruelas, and C. Beyer. 1979. Progesterone facilitation of EEG synchronization in response to milk drinking in female cats. *Psychoneuroendocrinology* 4:245–251.
80. Kubli-Garfias, C., L. Medrano-Conde, C. Beyer, and A. Bondani. 1979. In vitro inhibition of rat uterine contractility induced by 5-alpha and 5-beta progestins. *Steroids.* 34:609–617.
81. Beyer, C., J. Velázquez, K. Larsson, and J.L. Contreras. 1980. Androgen regulation of the motor copulatory pattern in the male New Zealand white rabbit. *Horm. Behav.* 14:179–190.
82. Beyer, C., E. Canchola, and K. Larsson. 1981. Facilitation of lordosis behavior in the ovariectomized estrogen primed rat by dibutyryl cyclic AMP. *Physiol. Behav.* 26:249–251.
83. Beyer, C., J.L. Contreras, G. Morali, and K. Larsson. 1981. Effects of castration and sex steroid treatment on the motor copulatory pattern of the rat. *Physiol. Behav.* 27:727–730.
84. Beyer, C., and E. Canchola. 1981. Facilitation of progesterone induced lordosis behavior by phosphodiesterase inhibitors in estrogen primed rats. *Physiol. Behav.* 27:731–733.
85. Beyer, C., P. Gómora, E. Canchola, and Y. Sandoval. 1982. Pharmacological evidence that LHRH action on lordosis behavior is mediated through a rise in cyclic AMP. *Horm. Behav.* 16:107–112.
86. González-Mariscal, G., A. Fernández-Guasti, and C. Beyer. 1982. Anesthetic pregnanes counteract androgen-induced defeminization. *Neuroendocrinology.* 34:357–362.

87. Beyer, C., A. Fernández-Guasti, and G. Rodríguez-Manzo. 1982. Induction of female sexual behavior by GTP in ovariectomized estrogen primed rats. *Physiol. Behav*. 28:1073–1076.
88. Beyer, C., J.L. Contreras, K. Larsson, M. Olmedo, and G. Moralí. 1982. Patterns of motor and seminal vesicle activities during copulation in the male rat. *Physiol. Behav*. 29:495–500.
89. Cervantes, M., R. Ruelas, and C. Beyer. 1983. Serotonergic influences on EEG synchronization induced by milk drinking in the cat. *Pharmacol. Biochem. Behav.* 18:851–855.
90. Fernández-Guasti, A., G. Rodríguez-Manzo, and C. Beyer. 1983. Effect of guanine derivatives on lordosis behavior in estrogen primed rats. *Physiol. Behav.* 31:589–592.
91. Velázquez-Moctezuma, J., J.E. Monroy, C. Beyer, and E. Canchola. 1984. Effects of REM deprivation on the lordosis response induced by gonadal steroids in ovariectomized rats. *Physiol. Behav.* 32:91–94.
92. Soto, M.P., M. Reynoso, and C. Beyer. 1984. Sexual dimorphism in the motor mounting pattern of the New Zealand white rabbit: steroid regulation of vigor and rhythmicity of pelvic thrusting. *Horm. Behav.* 18:225–234.
93. Moralí, G., L. Carrillo, and C. Beyer. 1985. Neonatal androgen influences sexual motivation but not the masculine copulatory motor pattern in the rat. *Physiol. Behav.* 34:267–275.
94. Fernández-Guasti, A., K. Larsson, and C. Beyer. 1985. Potentiative action of alpha- and beta-adrenergic receptor stimulation in inducing lordosis behavior. *Pharmacol. Biochem. Behav.* 22:613–617.
95. Fernández-Guasti, A., K. Larsson, and C. Beyer. 1985. Prevention of progesterone induced lordosis behavior by alpha or beta adrenergic antagonists in ovariectomized estrogen-primed rat. *Pharmacol. Biochem. Behav.* 22:279–282.
96. Roberts, L.A., C. Beyer, and B.R. Komisaruk. 1985. Strychnine antagonizes vaginal stimulation-produced analgesia at the spinal cord. *Life Sci.* 36:2017–2023.
97. Beyer, C., L. Roberts, and B.R. Komisaruk. 1985. Hyperalgesia induced by altered glycinergic activity at the spinal cord. *Life Sci.* 37:875–882.
98. Fernández-Guasti, A., K. Larsson, and C. Beyer. 1985. Comparison of the effects of different isomers of bicuculline infused in the preoptic area on male rat sexual behavior. *Experientia.* 41:1414–1416.
99. Moralí, G., G. Hernández, and C. Beyer. 1986. Restoration of the copulatory pelvic thrusting pattern in castrated male rats by the intracerebral implantation of androgen. *Physiol. Behav.* 36:495–499.
100. Fernández-Guasti, A., K. Larsson, and C. Beyer. 1986. Lordosis behavior and GABAergic neurotransmission. *Pharmacol. Biochem. Behav.* 24:673–676.
101. Fernández-Guasti, A., K. Larsson, and C. Beyer. 1986. Effect of bicuculline on sexual activity in castrated male rats. *Physiol. Behav.* 36:235–237.
102. Rodríguez-Manzo, G., M.L. Cruz, and C. Beyer. 1986. Facilitation of lordosis behavior in ovariectomized estrogen-primed rats by medial preoptic implantation of 5 beta, 3 beta, pregnanolone: a ring a reduced progesterone metabolite. *Physiol. Behav.* 36:277–281.
103. Roberts, L.A., C. Beyer, and B.R. Komisaruk. 1986. Nociceptive responses to altered GABAergic activity at the spinal cord. *Life Sci.* 36:1667–1674.
104. Fernández-Guasti, A., K. Larsson, and C. Beyer. 1986. GABAergic control of masculine sexual behavior. *Pharmacol. Biochem. Behav.* 24:1065–1070.
105. Beyer, C., and G. González-Mariscal. 1986. Elevation in hypothalamic cyclic AMP as a common factor in the facilitation of lordosis in rodents: a working hypothesis. *Ann. N.Y. Acad. Sci.* 474:270–281.
106. Beyer, C., C. Banas, P. Gómora, and B.R. Komisaruk. 1988. Prevention of the convulsant and hyperalgesic action of strychnine by intrathecal glycine and related amino acids. *Pharmacol. Biochem. Behav.* 29:73–78.
107. Sandoval, Y., B.R. Komisaruk, and C. Beyer. 1988. Possible role of inhibitory glycinergic neurons in the regulation of lordosis behavior in the rat. *Pharmacol. Biochem. Behav.* 29:303–307.

108. Beyer, C., G. González-Mariscal, J.R. Eguíbar, and P. Gómora. 1989. Lordosis facilitation in estrogen primed rats by intrabrain injection of pregnanes. *Pharmacol. Biochem. Behav.* 31:919–926.
109. González-Mariscal, G., and C. Beyer. 1989. Blockage of LHRH-induced lordosis by alpha- and beta-adrenergic antagonists in ovariectomized estrogen primed rats. *Pharmacol. Biochem. Behav.* 31:573–577.
110. Moralí, G., C. Beyer, and B.R. Komisaruk. 1989. Copulatory pelvic thrusting in the male rat is insensitive to the perispinal administration of glycine and GABA antagonists. *Pharmacol. Biochem. Behav.* 32:169–173.
111. Pacheco, P., M. Martínez-Gómez, B. Whipple, C. Beyer, and B.R. Komisaruk. 1989. Somato-motor components of the pelvic and pudendal nerves of the female rat. *Brain Res.* 490:85–94.
112. McCarthy, M.M., C. Beyer, and B.R. Komisaruk. 1989. Barbiturate-induced analgesia: permissive role of a GABA-A agonist. *Pharmacol. Biochem. Behav.* 32:897–900.
113. Beyer, C., C. Banas, O. González-Flores, and B.R. Komisaruk. 1989. Blockage of substance P-induced scratching behavior in rats by the intrathecal administration of inhibitory aminoacid agonists. *Pharmacol. Biochem. Behav.* 34:491–495.
114. González-Mariscal, G., O. González-Flores, and C. Beyer. 1989. Intrahypothalamic injection of RU486 antagonizes the lordosis induced by ring A-reduced progestins. *Physiol. Behav.* 46:435–438.
115. Hudson, R., González-Mariscal, G., and C. Beyer. 1990. Chin marking behavior, sexual receptivity and pheromone emission in steroid-treated ovariectomized rabbits. *Horm. Behav.* 24:1–13.
116. López-colomé, A.M., M.M. McCarthy, and C. Beyer. 1990. Enhancement of ^{3}H-muscimol binding to brain synaptic membranes by progesterone and related pregnanes. *Eur. J. Pharmacol.* 176:279–303.
117. McCarthy, M.M., M. Caba, B.R. Komisaruk, and C. Beyer. 1990. Modulation by estrogen and progesterone of the effect of muscimol on nociception in the spinal cord. *Pharmacol. Biochem. Behav.* 37:123–128.
118. González-Mariscal, G., A.I. Melo, and C. Beyer. 1990. Variations in chin marking behavior of New Zealand female rabbits throughout the whole reproductive cycle. *Physiol. Behav.* 48:361–365.
119. McCarthy, M.M., D.B. Masters, J.M. Fiber, A.M. López-Colomé, C. Beyer, B.R. Komisaruk, and H.H. Feder. 1991. GABAergic control of receptivity in the female rat. *Neuroendocrinology.* 53:473–479.
120. Porter, R.H., F. Lévy, P. Poindron, M. Litterio, B. Schaal, and C. Beyer. 1991. Individual olfactory signatures as major determinants of early maternal discrimination in sheep. *Dev. Psychobiol.* 24:151–158.
121. Beyer, C., M. Caba, C. Banas, and B.R. Komisaruk. 1991. Vasoactive intestinal polypeptide (VIP) potentiates the behavioral effect of substance P intrathecal administration. *Pharmacol. Biochem. Behav.* 39:695–698.
122. Beyer, C., B.R. Komisaruk, A.M. López-Colomé, and M. Caba. 1992. Administration of AP5, a glutamate antagonist, unmasks glycine analgesic actions in the rat. *Pharmacol. Biochem. Behav.* 42:229–232.
123. González-Mariscal, G., P. Gómora, M. Caba, and C. Beyer. 1992. Copulatory analgesia in male rats ensues from arousal, motor activity, and genital stimulation: blockage by manipulation and restraint. *Physiol. Behav.* 51:775–781.
124. Martínez-Gómez, M., R. Chirino, C. Beyer, B.R. Komisaruk, and P. Pacheco. 1992. Visceral and postural reflexes evoked by genital stimulation in urethane-anesthetized female rats. *Brain Res.* 575:279–284.

125. González-Mariscal, G., A.I. Melo, A. Zavala, and C. Beyer. 1992. Chin-marking behavior in male and female new zealand rabbits: onset, development and activation by steroids. *Physiol. Behav.* 52:889–893.
126. González-Mariscal, G., A.I. Melo, and C. Beyer. 1993. Progesterone, but not LHRH or prostaglandin E2, induces sequential inhibition of lordosis to various lordogenic agents. *Neuroendocrinology*. 57:940–945.
127. González-Mariscal, G., A.I. Melo, A. Zavala, R. Chirino, and C. Beyer. 1993. Sex steroid regulation of chin-marking behavior in male new zealand rabbits. *Physiol. Behav.* 52:889–893.
128. Caba, M., G. González-Mariscal, and C. Beyer. 1994. Perispinal progestins enhance the antinociceptive effects of muscimol in the rat. *Pharmacol. Biochem. Behav.* 47:177–182.
129. Masters, D., F. Jordan, C. Beyer, and B.R. Komisaruk. 1993. Release of aminoacids into regional superfusates of the spinal cord by mechanostimulation of the reproductive tract. *Brain Res.* 621:279–290.
130. Poindron, P., M. Caba, P. Gómora, D. Krehbiel, and C. Beyer. 1994. Responses of maternal and non-maternal ewes to social and mother-young separation. *Behav. Process.* 31:97–110.
131. González-Mariscal, G., V. Díaz-Sánchez, A.I. Melo, C. Beyer, and J.S. Rosenblatt. 1994. Maternal behavior in New Zealand white rabbits: quantification of somatic events, motor patterns and steroid plasma levels. *Physiol. Behav.* 55:1081–1089.
132. Beyer, C., and G. González-Mariscal. 1994. Effects of sex steroids on sensory and motor spinal mechanisms. *Psychoneuroendocrinology*. 19:517–527.
133. González-Mariscal, G., P. Gómora, and C. Beyer. 1994. Participation of opiatergic, GABAergic, and serotonergic systems in the expression of copulatory analgesia in male rats. *Pharmacol. Biochem. Behav.* 49:303–307.
134. Gómora, P., C. Beyer, G. González-Mariscal, and B.R. Komisaruk. 1994. Momentary analgesia produced by copulation in female rats. *Brain Res.* 656:52–58.
135. Beyer, C., O. González-Flores, and G. González-Mariscal. 1995. Ring A-reduced progestins potently stimulate estrous behavior in rats: paradoxical effect through the progesterone receptor. *Physiol. Behav.* 58:985–993.
136. Komisaruk, B.R., C. Bianca, G. Sansone, L.E. Gómez, R. Cueva-Rolón, C. Beyer, and B. Whipple. 1996. Brain-mediated responses to vaginocervical stimulation in spinal cord-transected rats: role of the vagus nerves. *Brain Res.* 708:128–134.
137. Cueva-Rolón, R., G. Sansone, C. Bianca, L.E. Gómez, C. Beyer, B. Whipple, and B.R. Komisaruk. 1996. Vagotomy blocks responses to vaginocervical stimulation after genitospinal neurectomy in rats. *Physiol. Behav.* 60:19–24.
138. Caba, M., R. Silver, G. González-Mariscal, A. Jiménez, and C. Beyer. 1996. Oxytocin and vasopressin immunoreactivity in rabbit hypothalamus during estrus, late pregnancy and postpartum. *Brain Res.* 720:7–16.
139. González-Mariscal, G., A.I. Melo, P. Jiménez, C. Beyer, and J.S. Rosenblatt. 1996. Estradiol, progesterone, and prolactin regulate maternal nest-building in rabbits. *J. Neuroendocrinol.* 8:901–907.
140. González-Mariscal, G., M.E. Albonetti, E. Cuamatzi, and C. Beyer. 1997. Transitory inhibition of scent-marking by copulation in male and female rabbits. *Anim. Behav.* 53:323–333.
141. Beyer, C., O. González-Flores, and G. González-Mariscal. 1997. Progesterone receptor participates in the stimulatory effect of LHRH, prostaglandin E2, and cyclic AMP on lordosis and proceptive behaviors in rats. *J. Neuroendocrinol.* 9:609–614.
142. González-Mariscal, G., A.I. Melo, R. Chirino, P. Jiménez, C. Beyer, and J.S. Rosenblatt. 1998. Importance of mother/young contact at parturition and across lactation for the expression of maternal behavior in rabbits. *Dev. Psychobiol.* 32:101–111.

143. González-Flores, O., N. Sánchez, G. González-Mariscal, and C. Beyer. 1998. Ring A reduction of progestins is not essential for estrous behavior facilitation in estrogen-primed rats. *Pharmacol. Biochem. Behav.* 60:223–227.
144. Beyer, C., O. González-Flores, J.M. Ramírez-Orduña, and G. González-Mariscal. 1999. Indomethacin inhibits lordosis induced by ring A-reduced progestins: possible role of 3α-oxoreduction in progestin-facilitated lordosis. *Horm. Behav.* 35:1–8.
145. Caba, M., B.R. Komisaruk, and C. Beyer. 1998. Analgesic synergism between AP5 (an NMDA receptor antagonist) and vaginocervical stimulation in the rat. *Pharmacol. Biochem. Behav.* 61:45–48.
146. Beyer, C. 1998. Regulación neuroendócrina de la conducta estereotipada. *Rev. Acad. Méd. Cat.* (Spain). 14:246–262.
147. Segovia, S., A. Guillamón, M.C.R. Del cerro, E. Ortega, C. Pérez-Laso, M. Rodríguez-Zafra, and C. Beyer. 1999. The development of brain sex differences: a multisignalling process. *Behav. Brain Res.* 105:69–80.
148. González-Mariscal, G., A.I. Melo, A.F. Parlow, C. Beyer, and J.S. Rosenblatt. 2000. Pharmacological evidence that prolactin promotes rabbit maternal behavior by acting since late gestation. *J. Neuroendocrinol.* 12:983–992.
149. Caba, M., K.Y.F. Pau, C. Beyer, A. González, R. Silver, and H.G. Spies. 2000. Coitus-induced activation of c-fos and gonadotropin-releasing hormone in hypothalamic neurons in female rabbits. *Mol. Brain Res.* 78:69–79.
150. González-Mariscal, G., P. Jiménez, C. Beyer, and J.S. Rosenblatt. 2003. Androgens stimulate specific aspects of maternal nest-building and reduce food intake in rabbits. *Horm. Behav.* 43:312–317.
151. Beyer, C., O. González-Flores, M. García-Juárez, and G. González-Mariscal. 2003. Non-ligand activation of estrous behavior in rodents: cross talk at the progesterone receptor. *Scand. J. Psychol.* 44:219–227.
152. Moralí, G., M.P. Soto, J.L. Contreras, M. Arteaga, M.D. González-Vidal, and C. Beyer. 2003. Detailed analysis of the male copulatory motor pattern in mammals: biological basis. *Scand. J. Psychol.* 44:277–286.
153. Caba, M., C. Beyer, G. González-Mariscal, and J.I. Morrell. 2003. Immunocytochemical detection of estrogen receptor alpha in the female rabbit forebrain: topography and regulation by estradiol. *Neuroendocrinology.* 77:208–222.
154. Beyer, C., and A. Macías. 2003. Analgesia genital: un sistema de modulación no opiáceo del dolor. *Gac. Méd. Méx.* 139:s21–s25.
155. Caba, M., M.J., Rovirosa, C. Beyer, and G. González-Mariscal. 2003. Immunocytochemical detection of progesterone receptors in the female rabbit forebrain: topography and regulation by estradiol and progesterone. *J. Neuroendocrinol.* 15:1–10.
156. González-Mariscal, G., R. Chirino, C. Beyer, and J.S. Rosenblatt. 2004. Removal of the accessory olfactory bulbs promotes maternal behavior in virgin rabbits. *Behav. Brain Res.* 152:89–95.
157. González-Flores, O., N. Sánchez, M. García-Juárez, F.J. Lima-Hernández, G. González-Mariscal, and C. Beyer. 2004. Estradiol and testosterone modulate the anesthetic action of the GABA-A agonist THIP but not of the neurosteroid 3-alpha, 5-alpha-pregnanolone in the rat. *Psychopharmacology.* 172:283–290.
158. González-Flores, O., I. Camacho, E. Domínguez, J.M. Ramírez-Orduña, C. Beyer, and R. Paredes. 2004. Progestins and place preference conditioning after paced mating. *Horm. Behav.* 46:151–157.
159. González-Mariscal, G., J.C. Flores-Alonso, R. Chirino, J.S. Rosenblatt, and C. Beyer. 2004. Intracerebroventricular injections of prolactin conteract the antagonistic effect of bromocryptine on rabbit maternal behaviour. *J. Neuroendocrinol.* 16:949–955.

160. González-Mariscal, G., R. Chirino, J.S. Rosenblatt, and C. Beyer. 2005. Forebrain implants of estradiol stimulate maternal nest-building in ovariectomized rabbits. *Horm. Behav.* 47:272–279.
161. González-Flores, O., J.M. Ramírez-Orduña, F.J. Lima-Hernández, M. García-Juárez, and C. Beyer. 2006. Differential effect of kinase A and C blockers on lordosis facilitation by progesterone and its metabolites in ovariectomized estrogen-primed rats. *Horm. Behav.* 49:398–404.
162. Gómora-Arrati, P., C. Carmona, G. Domínguez, C. Beyer, and J.S. Rosenblatt. 2006. GABA receptor agonists in the medial preoptic area and maternal behavior in lactating rats. *Physiol. Behav.* 87:51–65.
163. Segovia, S., A. García-Flagueras, B. Carrillo, P. Collado, H. Pinos, C. Pérez-Laso, C. Vinader-Caerols, C. Beyer, and A. Guillamón. 2006. Sexual dimorphism in the vomeronasal system of the rabbit. *Brain Res.* 1102:52–61.
164. González-Flores, O., C. Beyer, F.J. Lima-Hernández, P. Gómora-Aarrati, M. Gómez-Camarillo, K.L. Hoffman, and A.M. Etgen. 2007. Facilitation of estrous behavior by vaginal cervical stimulation in female rats involves alpha-1-adrenergic receptor activation of the nitric oxide pathway. *Behav. Brain Res.* 176:237–243.
165. Chirino, R., C. Beyer, and G. González-Mariscal, 2007. Lesion to the main olfactory epithelium facilitates maternal behavior in virgin rabbits. *Behav. Brain Res.* 180:127–132.
166. Beyer, C., K.L. Hoffman, and O. González-Flores. 2007. Neuroendocrinology of estrous behavior in the rabbit: some comparisons with the rat. *Horm. Behav.* 52:2–11.
167. González-Flores, O., J.R. Ramírez-Orduña, F.J. Lima-Hernández, M. García-Juárez, and C. Beyer. 2007. Lordosis facilitation by LHRH, PGE2, or db cAMP require activation of the kinase a signaling pathway in estrogen primed rats. *Pharmacol. Biochem. Behav.* 86:179–175.
168. Komisaruk, B.R., C. Beyer, and B. Whipple. 2008. Orgasm: an integration of body, nervous system and mind. *The Psychologist.* 21:100–103.
169. Gómora-Arrati, P., C. Beyer, F.J. Lima-Hernández, M.E. Gracia, A.M. Etgen, and O. González-Flores. 2008. GNRH mediates estrous behavior induced by ring A-reduced progestins and vaginocervical stimulation. *Behav. Brain Res.* 187:1–8.
170. González-Flores, O., A.M. Etgen, B.R. Komisaruk, P. Gómora-Arrati, A. Macías, F.J. Lima-Hernández, M. García-Juárez, and C. Beyer. 2008. Antagonists of the protein kinase A and mitogen-activated protein kinase systems and of the progestin receptor block the ability of vaginocervical/flank-perineal stimulation to induce female rat sexual behaviour. *J. Neuroendocrinol.* 20:1361–1367.
171. Melo, A.I., R. Chirino, A. Jiménez, E. Cuamatzi, C. Beyer, and G. González-Mariscal. 2008. Effect of forebrain implants of testosterone or estradiol on scent-marking and sexual behavior in male and female rabbits. *Horm. Behav.* 54:676–683.
172. González-Flores, O., P. Gómora-Arrati, M. García-Juárez, M. Gómez-camarillo, F.J. Lima-Hernández, C. Beyer, and A.M. Etgen. 2009. Nitric oxide and ERK/MAPK mediation of estrous behavior induced by GNRH, PGE2 and db-cAMP in rats. *Physiol. Behav.* 96:606–612.
173. González-Flores, O., A.M. Etgen, B.R. Komisaruk, P. Gómora-Arrati, A. Macías, F.J. Lima-Hernández, M. García-Juárez, and C. Beyer. 2009. Antagonists of the protein kinase A and mitogen-activated protein kinase system and the progestin receptor block the ability of vaginocervical/flank perineal stimulation to induce female rat sex behavior. *J. Neuroendocrinol.* 20:1361–1367.
174. González-Mariscal, G., A. Jiménez, R. Chirino, and C. Beyer. 2009. Motherhood and nursing stimulate c-fos expression in the rabbit forebrain. *Behav. Neurosci.* 123:731–739.

175. González-Flores, O., C. Beyer, P. Gómora-Arrati, M. García-Juárez, F.J. Lima-Hernández, A. Soto-Sánchez, and A.M. Etgen. 2010. A role for src kinase in progestin facilitation of estrous behavior in estradiol-primed female rats. *Horm. Behav.* 58:223–229.
176. García-Juárez, M., C. Beyer, A. Soto-Sánchez, R. Domínguez-Ordoñez, P. Gómora-Arrati, F.J. Lima-Hernández, J.R. Eguíbar, A.M. Etgen, and O. González-Flores. 2011. Leptin facilitates lordosis behavior through GNRH-1 and progestin receptors in estrogen-primed rats. *Neuropeptides.* 45:63–67.
177. Gómez-Camarillo, M.A., C. Beyer, R.A. Lucio, M. García-Juárez, A. González-Arenas, I. Camacho-Arroyo, B.R. Komisaruk, and O. González-Flores. 2011. Differential effects of progesterone and genital stimulation on sequential inhibition of estrous behavior and progesterone receptor expression in the rat brain. *Brain Res. Bull.* 85:201–206.
178. Lima-Hernández, F.J., C. Beyer, P. Gómora-Arrati, M. García-Juárez, J.L. Encarnación-Sánchez, A.M. Etgen, and O. González-Flores. 2012. SRC kinase signaling mediates estrous behavior induced by 5β-reduced progestins, GNRH, PGE2 and vaginocervical stimulation in estrogen-primed rats. *Horm Behav.* 62:579–584.
179. Beyer, C., B.R. Komisaruk, O. González-Flores, and P. Gómora-Arrati. 2013. Glycinamide, a glycine pro-drug, induces antinociception by intraperitoneal or oral ingestion in ovariectomized rats. *Life Sci.* 92:576–581.
180. García-Juárez, M., C. Beyer, P. Gómora-Arrati, R. Domínguez-Ordoñez, F.J. Lima-Hernández, J.R. Eguíbar, Y.L. Galicia-Aguas, A.M. Etgen, and O. González-Flores. 2013. Lordosis facilitation by leptin in ovariectomized, estrogen-primed rats requires simultaneous or sequential activation of several protein kinase pathways. *Pharmacol. Biochem. Behav.* 110:13–18.

CHAPTERS IN BOOKS

1. Beyer, C. 1965. Variaciones en la excitabilidad del sistema nervioso central mediante hormonas. In *III Simposio panamericano de farmacología y terapéutica*, Excerpta Medica Foundation International Congress Series 127:353–361.
2. Beyer, C. 1967. Variaciones periódicas en la actividad del mecanismo cerebral relacionado con la conducta sexual en ausencia de hormonas gonadales. In: *IV Simposio panamericano de farmacología y terapéutica*, Excerpta Medica Foundation International Congress Series 185:61–66.
3. Beyer, C., and F. Mena. 1968. Neural substrate of oxytocin and prolactin secretion during lactation. In *Progress in Endocrinology*, 952–958. Amsterdam: Excerpta Medica Foundation.
4. Beyer, C., and F. Mena. 1969. Neural factors in lactation. In *Physiology and Pathology of Adaptation Mechanisms*, 310–344. New York: Pergamon Press.
5. Beyer, C., and C.H. Sawyer. 1969. Hypothalamic unit activity related to control of the pituitary gland. In *Frontiers in Neuroendocrinology*, 255–287. eds. W.F. Ganong, and L. Martini. Oxford: Oxford University Press.
6. Beyer, C. 1970. Control de la secrecion de gonadotropinas hipofisiarias por el sistema nervioso central. In *Endocrinología de la Reproducción*, ed. J. Martínez-Manautou, 1–12. México D.F.: La Prensa Médica Mexicana.
7. Beyer, C. 1971. Regulación neuroendócrina de gonadotropinas en los mamíferos. In *Cursos del VI congreso mexicano de ginecología y obstetricia,* 203–207. México, D.F.
8. Beyer, C. 1972. Effect of estrogen on brain stem neuronal responsivity in the cat. In *Steroid Hormones and Brain Function*, eds. C.H. Sawyer, and R. Gorski, 121–126. Los Angeles, CA: University of California Press.
9. Beyer, C., and P. McDonald. 1973. Hormonal control of sexual behaviour in the female rabbit. In *Advances in Reproductive Physiology, vol. 6*, ed. J. Bishop, 185–219. London: Logos Press.

10. Beyer, C., and B.R. Komisaruk. 1974. Effects of diverse androgens on estrous behavior, lordosis reflex, and genital tract morphology in the rat. In *Mating Reflexes*, ed. L.M. Kow, 15–23. New York: MSS Information Corporation.
11. Beyer, C., G. Moralí, and M.L. Cruz. 1974. Effect of 5-alpha dihydrotestosterone on gonadotropin secretion and estrous behavior in the female Wistar rat. In *Mating Reflexes*, ed. L.M. Kow, 24–31. New York: MSS Information Corporation.
12. Beyer, C., G. Moralí, and M.L. Cruz. 1974. In *Gonadotropins, Current Research*, ed. J.P. Hearn, 84–92. New York: MSS Information Corporation.
13. Beyer, C., N. Vidal, and A. Mijares. 1974. Probable role of aromatization in the induction of estrous behavior by androgens in the ovariectomized rabbit. In *Mating Reflexes*, ed. L.M. Kow, 33–40. New York: MSS Information Corporation.
14. Beyer, C., and M.L. Cruz. 1974. Mecanismos de acción de andrógenos sobre el sistema hipotálamo hipofisiario en mamíferos. In *Problemas actuales de ciencias fisiológicas*, 245–259. México D.F.: Monografías de la Sociedad Mexicana de Ciencias Fisiologicas.
15. Beyer, C., and M. Cervantes. 1975. Electrofisiología del sistema neuroendócrino. In *Neuroendocrinología*, eds. O. Schiaffini, A. Oriol Bosch, L. Martini, and M. Motta, 35–73. Barcelona: Toray.
16. Beyer, C. 1976. Neuroendocrine mechanisms in sexual behavior. In *Subcellular Mechanisms in Reproductive Neuroendocrinology*, eds. F. Naftolin, K. Ryan, and J. Davis, 471–484. Amsterdam: Elsevier.
17. Beyer, C. 1979. Factores biológicos en la sexualidad humana. In *Fundamentos de endocrinología clínica*, eds. C. Malacara, M. García Viveros, and C. Valverde, 234–238. México D.F.: La Prensa Médica Mexicana.
18. Moralí, G., and C. Beyer. 1979. Neuroendocrine control of mammalian estrous behavior. In *Endocrine Control of Sexual Behavior*, ed. C. Beyer, 33–75. New York: Raven Press.
19. Beyer, C., K. Larsson, and M.L. Cruz. 1979. Neuronal mechanisms probably related to the effect of sex steroids on sexual behavior. In *Endocrine Control of Sexual Behavior*, ed. C. Beyer, 365–387. New York: Raven Press.
20. Beyer, C., E. Canchola, M.L. Cruz, and K. Larsson. 1980. A model for explaining estrogen progesterone interactions in the induction of lordosis behavior. In *Endocrinology 1980*, eds. I.A. Cumming, J.W. Funder, and F.A.O. Medelsohn, 615–618. Canberra: Australian Academy of Sciences.
21. Larsson, K., and C. Beyer. 1981. Some aspects of the neuroendocrine regulation of mammalian sexual behaviour. In *Neuroendocrine Regulation and Altered Behavior*, eds. P.D. Hrdide, and R.L. Singhal, 97–118. London: Croom Helm Ltd.
22. Larsson, K., and C. Beyer. 1981. Sex steroid hormones and sexual behavior. In *Steroid Hormone Regulation of the Brain*, ed. K. Fuxe, 279–291. New York: Pergamon Press.
23. Beyer, C., and H.H. Feder. 1987. Sex steroids and afferent input: their roles in brain sexual differentiation. *Annual Review of Physiology* 49:349–364. New York: Academic Press.
24. Ramírez, V.D., and C. Beyer. 1988. The ovarian cycle of the rabbit: its neuroendocrine control. In *The Physiology of Reproduction*, eds. E. Knobil, and J.D. Neill, 1873–1892. New York: Raven Press.
25. González-Mariscal, G., and C. Beyer. 1991. Posible participación de mecanismos extragenómicos en la facilitación de la conducta de lordosis en roedores. In *Tópicos selectos en biología de la reproducción*, ed. R. Domínguez-Casalá, 327–346. México D.F.: Miguel Angel Porrúa.
26. Beyer, C., and G. González-Mariscal. 1991. Functional implications of progesterone metabolism: effects on psychosexual development, brain sexual differentiation

and perception. In *Behavioral Biology, Neuroendocrine Axis*, eds. T. Archer, and S. Hansen, 151–165. New Jersey: Lawrence Erlbaum Associates.

27. Beyer, C., and G. González-Mariscal. 1991. Effects of progesterone and natural progestins in brain. In *Reproduction, Growth and Development*, eds. A. Negro-Vilar, and G. Pérez Palacios, 199–208. New York: Raven Press.
28. Beyer, C., and G. Moralí. 1992. Motor aspects of masculine sexual behavior in rats and rabbits. In *Advances in the Study of Behavior* 21:201–238. New York: Academic Press.
29. Beyer, C., and G. González-Mariscal. 1993. Steroidogenic regulation of male motor copulatory patterns. In *Progress in Endocrinology*, eds. R. Mornex, C. Jaffiol, and J. Leclere, 178–180. New York: Parthenon Publishing.
30. Poindron, P., M. Caba, P. Gómora, D. Krehbiel, and C. Beyer. 1994. Réactions a la separation sociale et à la privation du jeune chez la brebis et consequences pour l'entretien des ongules domestiques à la parturition. In *Comportement et adaptation des animaux domestiques aux contraintes de l'elevage*, 183–194. France: Institut National de la Recherche Agronomique.
31. Beyer, C., and O. González-Flores. 1999. Mecanismos moleculares de la acción de las hormonas esteroides en mamíferos. In *Biomedicina y genética molecular*, eds. P. Gariglio, and M.E. Orozco. México D.F.: Editorial Limusa.
32. Aréchiga, H., and C. Beyer. 1999. Perspectivas de desarrollo de la investigación en ciencias biológicas en méxico. In *Las ciencias naturales en México*, eds. H. Aréchiga, and C. Beyer. 15–33. México D.F.: Fondo de Cultura Económica.
33. Beyer, C., O. González-Flores, M. García-Juárez, and G. González-Mariscal. 2003. Facilitación hormonal de la conducta de estro en roedores: un modelo de integración de señales membranales e intracelulares en el receptor de progesterona. In *Fisiología, ecología y comportamiento: una apuesta multidisciplinaria*, eds. M. Martínez-Gómez, Y. Cruz, R. Hudson, and R.A. Lucio, 107–117. México D.F.: CONACYT-UNAM.
34. Komisaruk, B.R., C. Beyer, and B. Whipple. 2009. Sexual pleasure. In *Pleasure and the Brain*, eds. K.C. Berridge, and M. Kringelbach. 169–177. Oxford: Oxford University Press.
35. Beyer, C. 2010. Descentralización de la ciencia en méxico: problemas y algunas posibles soluciones. In *El debate de la ciencia en México*, 55–61. México D.F.: Foro Consultivo Científico y Tecnológico.
36. Mateos, J.L., and C. Beyer. 2012. Los inicios de la investigación en el Centro Médico Nacional del Instituto Mexicano del Seguro Social. In *La reforma de la investigación en el IMSS 2006–2012*, 31–44.
37. González-Flores, O., K.L. Hoffman, A.I. Melo, and C. Beyer. 2014. Effects of estrogen fluctuations during the menstrual cycle on cognitive functions: cellular systems and brain areas involved. In *Estrogens and Cognition, Psychobiological and Clinical Aspects*, ed. I. González-Burgos, 1–25. Kerala, India: Research Signpost.

CO-AUTHORSHIP OF BOOKS

1. Komisaruk, B.R., C. Beyer, and B. Whipple. 2006. *The Science of Orgasm*. Baltimore, MD: Johns Hopkins University Press.
2. Komisaruk, B.R., B. Whipple, S. Nasserzadeh, and C. Beyer. 2009. *The Orgasm Answer Guide*. Baltimore, MD: Johns Hopkins University Press.

EDITED BOOKS

1. Beyer, C. 1979. *Endocrine Control of Sexual Behavior*. New York: Raven Press.

Index

Note: Page numbers followed by f and t refer to figures and tables, respectively.

1H-[1,2,4]oxadiazolo[4,3-a]quinoxalin-1-one (ODQ), 120
5-Alpha-Dihydrotestosterone effects on brain sexual differentiation of female rat, 15–17
5-Alpha-reduction role in rats, 13–17
  5-Alpha-Dihydrotestosterone effects on sexual differentiation, 15–17
  testosterone and 5-Alpha-Dihydrotestosterone effect on gonadotropin secretion and lordosis behavior, 13–15
5′-AMP-activated protein kinase (AMPK), 121
5α-reductase (5α–R) enzymes, 266
5-HT neurotransmission, 70
16 kDa peptide hormone, 120
16 kDa prolactin-immunoreactive protein, 255
26S proteasome pathway, 123

A

α_1 Antagonists, 39–40
A15 ventral (A15v), 152
Ablation effects, 38
Accelerometric–polygraphic technique, 67
Accelerometric recording, 50, 52–53, 55
Accessory olfactory bulb (AOB), 203–204
  environmental prenatal stress alters sexual dimorphism, 216–218
Accessory sex gland, 264
Accessory sexual glands, 85
Activation functions (AFs), 115
Adenylyl cyclase, 119
Adipo Q receptors, 116
Ad lib paradigm, 88
Adrenalectomy, 8
Adrenocorticotropin hormone (ACTH), 165
AF-1, 115
AF-2, 115
Allodynia, 283
Almanza, Javier, 47
American Association of Sexuality Educators, Counselors and Therapists (AASECT), 304
Aminita muscaria, 233
Aminoterminal 16 kDa fragment, 254
Analgesia
  glycinamide, precursor of glycine, counteracts allodynia, 280–283, 282f
  glycine analogs, 280
  glycine by vaginal stimulation produces, 279–280
  sexual behavior in female rats 7, 279
Androgen, 263, 266
  replacement effect of rabbits and rats
    on estrous behavior, 9–13, 10f, 11f, 12t, 13f
    male sexual behavior, 17
Angiogenesis, 254–255
Angiogenesis-dependent pathologies, 256
Angiogenesis/neovascularization, 254
Animal personality, 176
ANS. *See* Autonomic nervous system (ANS)
Anterior cingulate cortex (ACC), 211
Antiangiogenic prolactin fragments, 255
Antinociceptive effect of glycinamide, potential therapeutic implication, 283–284
Antiprolactin antibody, 255
Antisense oligonucleotides, 118
AOB. *See* Accessory olfactory bulb (AOB)
Aortic aneurism, 301
Apocrine secretion, 265
Apoptosis, 269
Arginine vasopressin (AVP), 233
Aromatizable androgens, 167
Aromatizable testosterone, 296
Aromatization research, 296
Arousal, 89
Atlanto-occipital space, 280
Autoimmune diseases, 256
Autonomic nervous system (ANS), 267, 270–272, 271f

B

Beach, Frank, 33–34
Bed nuclei of the stria terminalis (BNST), 156
Bed nucleus of the accessory olfactory tract (BAOT), 204
  environmental prenatal stress alters sexual dimorphism, 216–218

Bed nucleus of the stria terminalis (BST), 204
Behavioral effects, maternal care as preventive factor in EPS, 218–220
Behavioral events, copulation, 84
Behavioral neuroendocrinology, 4, 166–178
  physiology to personality, 172–178
    neuropsychiatric disorders, rabbit behavior, 174–176
    ontogeny of individual differences in phenotype, 176–178
    reflex activity of pelvic region, female rabbit, 172–174
  rabbit, 166–172
    hormonal modulation, 167–170
    lactation, 166–167
    maternal nest building behavior, 170–172
Benign prostatic hyperplasia (BPH), 264
Beyer, Carlos, 34, 113–114, 187, 251–252, 317f
  Brain Research Institute, 313f
  with Barry Komisaruk at CINVESTAV, 315f
  with Beverly Whipple, 314f
  Carmen Clapp and Gonzalo Martínez de la Escalera wedding, 313f
  with *Gabriela González-Mariscal*, 317f
  with his former students, 316f
  at Institute of Animal Behavior, Rutgers University, 314f
  with Jay Rosenblatt, 315f, 316f
  with Maria Cruz Rodriguez del Cerro and Barry Komisaruk, 318f
  remembrances by Barry, 290–304
    Beverly, 302–303
    intellect against pain, 300–302
    research credentials, 290–292
    research field creation, 296–300
    testosterone as estrogen, 292–296
    at Todos Santos, Mexico, 318f, 319f
    unfinished work, 303–304
  remembrances by Beverly, 304–309
    coauthoring books with Carlos, 309
    human physiology laboratory at rutgers university, 308–309
    at IAB, 304
    path to research, 304–306
    path to Rutgers University, 306–307
    working with Carlos' colleagues in Mexico, 307–308
Bilateral implantation of DHT, 14
Bilevel chamber, 135
Birth canal stimulation, 302
Bisphenol A (BPA), 215
Blindness, 255
Brain
  imaging research, 309
  microelectrodes, 295
  PR expression, 116
  rhythms, 294
Bulbospongiosus muscles, 84, 95
B upstream segment (BUS), 115

C

Calcitonin gene-related peptide (CGRP), 268
CAMP. *See* Cyclic adenosine monophosphate (cAMP)
Camp response element-binding protein (CREB), 238
Cannabinoid 1 (CB1) receptors, 108, 109f
Capillary blood vessels, 254
Capsaicin, 307
CAPTUMAR program, 68
Castration, 109, 168
  replacement effects, 56–62
    females, 57–58, 58f
    golden hamsters, 60–61, 60f
    guinea pigs, 61–62
    males, 56–57, 57f
    rabbit, 56
    rats, 58–60, 59f
Catheter, 280–281
Cell proliferation, 267–269
Cellular mechanisms, 114
  to estrous behavior regulation, 118–125
    leptin and female sexual behavior, 120–122
    NO pathway, 119–120
    PKA, 119
    src tyrosine kinase, MAPK pathway, and lordosis behavior, 122–125
    vaginocervical stimulation, 122
Cellular proliferation, 267
Cellular tyrosine kinase molecule, 122
Central nervous system (CNS), 233, 280
Central Pattern Generator (CPG), 66, 70–71
Centro de Investigación y Estudios Avanzados (CINVESTAV), 294
Cerebellar body, 104
Cerebellar cortex, 104
Cerebellar neurons, 105
Cerebellar neurosteroids, 109
Cerebellar vermis, 106
Cerebellum, 103–105, 104f
Cervantes, Miguel, 47
Cervical stimulation, 296
cGMP (cyclic guanosine monophosphate), 119–120, 123
CHECAMAR program, 68–69
Chemosensory system, 203
Chinning, 169
Chloride–cation–cotransporters (CCC), 238
Chronobiology, 188
Chronostasis, 188, 192
  and suckling-entrained oscillator, 192–193

Cingulate cortex, 5
CINVESTAV (Centro de Investigación y Estudios Avanzados), 294
Circadian oscillations, 150
Circadian system, 192–193
Circumvaginal Musculature and Sexual Function (book), 306
Clock genes, 151
Clonidine, 71
Cloth harness, 67–68
CNS (central nervous system), 233, 280
Cognition, 191
Complex system (CS), 188–189
  characteristic, 194
Concentric layers, cytoarchitecture, 203
Conditioned place preference (CPP), 134–135, 139
Constitutive clock, 192–193
Contractile uterine responses, 5
Contreras, José Luis, 47
Coolidge Effect, 89–91, 137
  adaptive/evolutionary significance, 87t, 92–93
  behavioral and seminal parameters, 87t
  sexual behavior in males and females under, 136–138
Copulation, 84
  motor patterns and genital responses, 84
  satiety include seminal expulsion, 87t, 91–92
  sexual incentive and, 89–90
Copulatory behavior, 35, 37–38, 46, 52, 54
Copulatory motor patterns, 84
  hormonal regulation in mammals, 45–62
    castration and hormone replacement effects, 56–62
    male copulation, motor and genital components, 46–47
    masculine copulatory motor pattern characteristics, 47–56
  neuronal and neurochemical correlates, 65–76
    pelvic thrusting, computerized capture and analysis, 67–69, 68f, 69f
    spinal neurochemical modulation, 70–71
    VTA and pedunculopontine nucleus, 72–76, 73f, 75f
Copulatory testing, 88
Corpus luteum, 114
Cortex, 293
Corticotropin releasing hormone (CRH), 165
CPG (Central Pattern Generator), 66, 70–71
Cross-fostering procedure, EPS, 219f, 220t
Cross-talk, 114
c-Src tyrosine kinase molecule, 122
Cutaneous stimulation, 281
Cyclic adenosine monophosphate (cAMP), 116, 119, 123, 253
Cyclic guanosine monophosphate (cGMP), 119–120, 123
Cytoplasm, 265
Cytoplasmic signaling pathways, 118

D

Darting, 114
Deeper intravaginal thrusts, 84
Dehydroepiandrosterone (DEHA), 270
DHT propionate (DHTp), 168
Diazepam (DZ), 212
Digging behavior, 171, 190
Dihydrotestosterone (DHT), 167, 263, 266
Direct spermatobioscopy, 86
Dopamine (DA), 133–134, 137, 266
Dopaminergic regions, 152
Dopaminergic system, 134
Dose–response paradigm, 301
Drive, 89
Drosophila melanogaster, 136
Dry grass, 174
Dynamic systems, 188
Dysregulation, 239

E

Earthquake, 298–299
EEG. *See* Electroencephalographic (EEG)
Ejaculation, 84–87, 95
  kinds of ejaculate, 86–87, 87t
  restoration of, 94t, 95–97
  seminal plug and transcervical sperm transport, 85–86
Ejaculatory fluids, 85
Ejaculatory responses, 50
Electric circuit, copulation, 46–47
Electroejaculation, 86
Electroencephalographic (EEG)
  activity, 74
  signals, 67
Electromyographic recordings, 47
Electronic perineometer, 304
Electrophysiological expressions of "relaxation behavior", 23–26, 24f, 25f, 26f
"Electrophysiological study of projections from mesencephalic central gray matter to forebrain in the rabbit" (journal article), 164
Emission phase, 84
Endocannabinoids, 108
Endocrine-disrupting chemicals, 214–215
Endocrine system, 291, 293
Endoplasmic reticulum, 265
Endothelial NOS (eNOS), 119
  cells, 255

Environmental perinatal stress, 215–220
environmental prenatal stress alters sexual dimorphism, 216–218
maternal care as preventive factor in behavioral effects, 218–220
Environmental prenatal stress (EPS), 202, 215
Estradiol, 114–115, 123, 132, 169, 190–191
Estradiol benzoate (EB), 9–10
Estrogen, 115, 238
effect on responsiveness to sensory/sexually relevant stimuli, 21–23, 22f, 23t
receptor, 293
regulation of kinases, 240f
Estrogen-inducible PRs, 115
Estrogen–progesterone interactions, estrous behavior expression, 9
Estrous behavior, 8
Ethamoxytriphetol antiestrogen (MER-25), 12, 18–19, 293
effects on estrous behavior, 12t
Excitatory effects, GABA, 239f
Excitatory transmitter, 233
Extravaginal pelvic thrust, 52–54, 61

F

Federative International Committee on Anatomical Terminology (FICAT), 306
Female ejaculation, 305–306
Female rabbit, 189–193
nest-building, 189–191
behaviors expression of, 190f
stage, 189–190
nursing, 191–193
chronostasis and suckling-entrained oscillator, 192–193
Female sexual behavior, hormonal modulation, 167–168
Female sexual behavior of rabbits and rats, hormonal regulation, 7–17
5-alpha-reduction role, 13–17
5-Alpha-Dihydrotestosterone effects on sexual differentiation of brain, 15–17
testosterone (T) and 5-Alpha-Dihydrotestosterone (DHT) effect on gonadotropin secretion and lordosis behavior, 13–15
estrous behavior
androgen replacement effect, aromatization role, 9–13
estrogen–progesterone interactions, 9
pseudomale behavior display, 7–8
removing endocrine glands effects on, 8–9
during various reproductive conditions, 8
lordosis and pseudomale behavior in rabbits, 7–13
Fertilization, 131, 265
Fibrous zone, 264
Fixed period method, 88
Follicle stimulating hormone (FSH), 165
Food-entrained oscillator (FEO), 192
Fos immuno reactivity (Fos-IR), 106–107

G

GABA. *See* Gamma-aminobutyric acid (GABA)
GABA-A receptor, 232–233
antagonist, 165
GABA-B receptor, 233
GABA transporters (GATs), 234
GABA transporter 1 (GAT1), 234
Galactolysis, 252
Galactopoiesis, 252
Gametic cells, 85
γ-acetylene GABA (GAG), 71
Gamma-aminobutyric acid (GABA), 71–72, 231–243, 300
action modulation at multiple levels, 234–236
brain development, 239–241
central to
reproduction, hormonal modulation, 241–243
SCN and circadian rhythms, 235f
effects of paradoxical, 233
excitatory neurotransmitter, 237–239
and GABA-A receptors structure, 232
neural circuitry of lordosis, 242f
neurosteroids and, 237
overview, 231–233
and suprachiasmatic nucleus, 233–234
transaminase inhibitors, 71
γ-aminobutyric acid-A (GABA-A) receptors, 117
Ganglion cells, 268, 271
Gating process, 293
Gene expression, regulation, 117
Genital stimulation, 22
Genitospinal nerves, 309
Glutamic acid decarboxylase (GAD), 235
Glycinamide, 280–281
Glycine, 279–280, 300
GnRH. *See* Gonadotropin-releasing hormone (GnRH)
Golgi apparatus, 265
Gonadotropin-releasing hormone (GnRH), 39–40, 118–119, 122, 165, 254
neurons, 16–17, 241
Gräfenberg spot (G spot), 304–305
analgesia in women with stimulation, 306–307
rediscovery and naming of, 304–305

H

Hair pulling, 171
Hedonically positive events/stimuli, 89

Heimer, Lennart, 35
Heterotypical behaviour, 10
Heterotypical (pseudomale) sexual behavior, 49–52
rabbit, 49
rat, 52
Histone trimethylation, 269
Homotypical (male) sexual behavior, 47–52
rabbit, 47–49, 48f, 49f
rat, 50–52, 51f
Hormonal influence, 132
Hormonal modulation, rabbit, 167–170
female sexual behavior, 167–168
male sexual behavior, 168–169
sexually dimorphic behavior patterns, 169–170
Hormonal regulation, 266–267
copulatory motor pattern in mammals, 45–62
castration and hormone replacement effects, 56–62
male copulation, motor and genital components, 46–47
masculine copulatory motor pattern characteristics, 47–56
female sexual behavior of rabbits and rats, 7–17
5-alpha-reduction role, 13–17
lordosis and pseudomale behavior in rabbits, 7–13
male sexual behavior of rabbits and rats, 17–21
androgen replacement effects and 5-alpha reduction role, 17
aromatization role in testosterone effect, 18–21, 18t, 19f, 20f
Hormone, 120
replacement effects, 56–62
females, 57–58, 58f
golden hamsters, 60–61, 60f
guinea pigs, 61–62
males, 56–57, 57f
rabbit, 56
rats, 58–60, 59f
Hormone response elements (HREs), 115
Human endothelial cells, 255
Human MB, stress effects on, 220–221
Human mothering, vomeronasal input, 209–211
Hyperalgesia, 283
HyperPRL, 268, 271f
Hypogastric Nerve (HgN), 267–268
Hypothalamic nuclei, 151
Hypothalamic regulatory factor, 253
Hypothalamic supraoptic and paraventricular nuclei, 5
Hypothalamo–hypophyseal portal system, 266
Hypothalamo–neurohypophyseal system, 254
Hypothalamus, 115–116, 118, 121, 253, 293

I

Immune system, 292
Immunocytochemistry, 116, 234
Impregnation, 93, 96
Incentive motivation, 89
Incertohypothalamic dopaminergic (IHDA), 152, 155
Indirect spermatobioscopy, 86
Inducible NOS (iNOS), 119
Induction, 206
In-fostering procedure for EPS, 219f
Inhibitory effects, GABA, 239f
Inner transition zone, 264
Insulin-like growth factor-I (IGF-I), 123
Intercalation, 237
Intermediate BST (BSTI), 204
Intracerebroventricular (icv), 118, 120–121
Intrathecal (i.t.) catheter, 280–281
Intravaginal thrusting, 50–54
Ischiocavernosus muscles, 84
Isoproterenol, 71

J

Jackrabbit, 303
Janus kinases (JAK), 121
Jet lag, 233

K

Ketamine, 175
Ketohydroxyoestrin, 166
Komisaruk, Barry
Carlos remembrances by, 290–304
Beverly, 302–303
intellect against pain, 300–302
research credentials, 290–292
research field creation, 296–300
testosterone as estrogen, 292–296
unfinished work, 303–304
at Todos Santos, Mexico, 318f, 319f
Kruskal–Wallis ANOVA, 281

L

Lactation
behavior in female cat and rabbit, neural regulation, 4–7
neural and hormonal factors, 5–7
overview, 4–5
rabbit, 166–167
suckling stimulation by kits during, 151
synchronization of neuroendocrine brain areas during, 152
Lactogenesis, 252
Lagomorphs, 34, 39, 158
Larsson, Knut, 35, 47, 294

Larval salamanders, 291
Lateral BST (BSTL), 204
Lateral olfactory tract (LOT), 204
Lateral septum (LS), 156
Leptin and female sexual behavior, 120–122
Ligand-binding domain (LBD), 115
Ligand-dependent transcription factors, 115
Liking sexual behavior, 134–135
Limbic system, 295
Lipophilic molecule, 117
Littermates, 177
Lobule VII, 105
Locus coeruleus (LC), GABA-A effects on, 212–213
Long-Term Depression (LTD), 108
Lordosis, 114
  and pseudomale behavior in rabbits, 7–13
    androgen replacement effect, 9–13, 10f, 11f, 12t, 13f
    estrogen–progesterone interactions, 9
    estrous behavior during reproductive conditions, 8
    pseudomale behavior display, 7–8
    removing endocrine glands effects, 8–9
Lordosis neural control, 38–39
Lordosis quotient (LQ), 9
Luteinizing hormone (LH), 165, 241
Luteinizing hormone releasing hormone (LHRH), 165

M

Macula pellucida, 164
Magnetic resonance imaging (MRI), 107, 302, 308
Main olfactory bulb (MOB), 203
Main olfactory system (MOS), 202–203
Male
  copulation, motor and genital components, 46–47
  hormonal modulation of rabbit, 168–169
  sexual behavior, neural control, 35–38, 36f, 37f
  sexual behavior of rabbits and rats, hormonal regulation, 17–21
    androgen replacement effects and 5-alpha reduction role, 17
    aromatization role in testosterone effect, 18–21
Mammalian behavioral elements, 209
Mammary gland, 252
Manautou, Martínez, 6
Mann–Whitney U-test, 281
MAPK. *See* Mitogen-activated protein kinase (MAPK)
Masculine copulatory motor pattern characteristics, 47–56
  golden hamster, 52–54, 53f, 54f
  guinea pig, 54–55, 55f
  mouse, 55–56
  rabbit, 47–49, 48f, 49f
  rat, 50–52, 51f
Masculine copulatory patterns, 90
Master clock, 150–151, 150f
Maternal behavior (MB), 203
  mammals, 205–206
  patterns, 207f
  rhythmic pattern in structures, 156–159, 157f, 158f, 159f
Maternal brain, 188
  permanent *vs.* transitory changes, 193–194
Maternal nest building
  behavioral components, 171
  behavior of rabbit, 170–172
MB. *See* Maternal behavior (MB)
McDonald, Peter, 9
Medial anterior BST (BSTMA), 204
Medial BST (BSTM), 204
Medial central BST (BSTMC), 204
Medial posterior BST (BSTMP), 204
Medial prefrontal region (mPFC), 73–74
Medial preoptic area (MPOA), 56, 72–73, 139, 188, 191, 205
  neuroendocrine and behavioral role in rabbits, 33–41
    endocrine role of female, 39–40
    lordosis, neural control, 38–39
    neural control of male sexual behavior, 35–38, 36f, 37f
Membrane-associated PR (mPR), 116
Mena, Flavio, 4, 34, 252
MER-25. *See* Ethamoxytriphetol antiestrogen (MER-25)
Merocrine secretion, 265
Mesencephalic locomotor region (MLR), 72–73
Mesencephalic reticular formation, 21–24, 26–27
Mesocorticolimbic dopaminergic pathway, 72
Mesolimbic system, 133, 139
Metabolically active steroids, 213
Metaplasia, 269
MicroRNAs, 269
Milacemide, 280
Mild hyperPRL, 268–269
Mitogen-activated protein kinase (MAPK), 116, 121–123
  pathway, 267, 272
Molecular-signaling network, 122
Morphine, 301
MOS (main olfactory system), 202–203
Motor learning, 107–109, 108f, 109f
Mounting behavior, 8–11, 15
MPOA. *See* Medial preoptic area (MPOA)
μ opioid receptors, 139
MUA. *See* Multiunit activity (MUA)

Multiunit activity (MUA), 22–24, 26, 72–73, 75, 108
Myostrichomorpha, 33

N

Neoplasia, 269
Nervous system, 103
Nest building of female rabbit, 189–191
  behaviors expression of, 190f
  stage, 189–190
Neural and hormonal factors in lactation and maternal behaviour, cat and rabbit, 5–7
  ovarian steroid hormones role, 6
  oxytocin and prolactin secretion, 5–6
  septal lesions effect, 6–7, 7f
  steroid hormones effect on adult rabbits mammary glands, 6
Neural network, 103
  training, 268
Neural processes in reproduction-related behavioral phenomena, 21–28
  electrophysiological expressions of relaxation behavior, 23–26, 24f, 25f, 26f
    characterization, 23–24, 24f
    electrographic signs in lactating mothers, 25–26, 26f
    progesterone facilitatory effect on EEG synchronization, 24–25, 25f
  estrogen effect on responsiveness, 21–23, 22f, 23t
  progesterone effect, 26–28, 27f
Neural regulation of lactation and maternal behavior in female cat and rabbit, 4–7, 7f
  neural and hormonal factors, 5–7
    ovarian steroid hormones role, 6
    oxytocin and prolactin secretion, 5–6
    septal lesions effect, 6–7, 7f
    steroid hormones effect on adult rabbits mammary glands, 6
Neural structures, 151
Neural systems, 139
Neuroactive steroids, 213–214, 214f
Neuroendocrine synchronization model, rabbit doe as, 149–159
  lactation
    brain areas during, 152
    suckling stimulation by kits during, 151
  master clock, 150–151, 150f
  OT and dopaminergic cells, 152–156, 153f, 154f, 155f
  rhythmic pattern in maternal behavior structures, 156–159, 157f, 158f, 159f
Neuroendocrinology, 109–110, 253, 293
Neuromodulator, 122, 133, 268, 272
Neuronal gating, 293
Neuronal NOS (nNOS), 119
Neuropeptide Y (NPY), 40, 268, 272
Neurophysiology, 304, 306
Neuropsychiatric disorders, rabbit behavior in, 174–176
Neurosteroids, 117, 213
  and GABA, 237
Neurotransmitter receptors, modulation, 117
Neurotransmitters, 70–71, 76, 120, 279, 300
Nipple-search pheromone, 169
Nitric Oxide (NO) pathway, 118–120
N-methyl-D-aspartate (NMDA), 300
  receptor, 117, 284
    antagonists, 175–176
NO-cGMP pathway, 120–121
Non-contact stimulation, 105–106, 106f
Non-photic zeitgeber, 193
Noradrenaline, 71
NO synthase (NOS), 119–120
Nucleus tractus solitarii (NTS), 309
Nursing of female rabbit, 191–193
  chronostasis and suckling-entrained oscillator, 192–193
  components, 191–192

O

OB-Rs receptor, 121
Obsessive–compulsive disorder (OCD), 174–175
Obsessive–compulsive spectrum disorder, 174
Ocular angiogenesis, 255
Ocular effects, 255
Odd-skipped related transcription factor 1 (OSR1), 238
Olfaction, 105
Olfactory, 206
  epithelium, 202
  system, 202–203
Olfactory bulbs, 15–16, 293
Ontogeny of individual differences in phenotype, 176–178
Opiates, 138
Orbitofrontal cortex (OFC), 211
Oryctolagus cuniculus, 164
OT. *See* Oxytocin (OT)
Ovarian hypertrophy, 14
Ovarian steroid hormones role in lactation, 6
Ovariectomized (ovx), 114, 116, 118, 190
  rats, 280–281
Ovariectomy, 6, 8–9, 11, 14, 16, 280
Ovulation, 164–165
ovx. *See* Ovariectomized (ovx)
Oxytocic, 165–166
Oxytocin, 165–166
Oxytocin (OT), 152–156, 153f, 154f, 155

P

Pacemaker, 295
Pacheco, Pablo, 251–252
P action, mechanisms, 117–118
  gene expression, regulation, 117
  neurotransmitter receptors, modulation, 117
  signaling cascades activation, 117–118
Pain thresholds, 302
Paradoxical GABA effects, 233
Paraventricular (PVN) neural structure, 151–152, 153f, 155–156
Parental behavior (PB), 202
Pascal programming language, 67
Peculiar nursing behavior, 166
Pedunculopontine nucleus (PPN), 66, 72
Pelvic ganglion, 267–268
Pelvic nerve (PvN), 267–268, 307
Pelvic thrusting, computerized capture and analysis, 67–69, 68f, 69f
Penile events, 90
Penis–cerebellum neural pathway, 106
Perineal tapping, 22–23
Peripheral nervous system, regulation, 267–268
Peripheral zone, 264
Periventricular hypophyseal dopaminergic (PHDA), 152, 155
PFC. *See* Prefrontal cortex (PFC)
Pharmacological antagonism of DHT, 266
Pheromone, 210
Physiological events, copulation, 84
Physiology to personality, behavioral neuroendocrinology, 172–178
  neuropsychiatric disorders, rabbit behavior to, 174–176
  ontogeny of individual differences in phenotype, 176–178
  reflex activity of pelvic region, female rabbit as, 172–174
Picrotoxin (PTX), 165, 212
Pitocin, 165
Pitressin, 165
Pituitrin, 165
PKA (protein kinase A), 118–119, 121
Placental lactogen, 255
Polygraphic analysis, 52, 54
Polygraphic and accelerometric technique, 47
Positron emission tomography (PET), 107, 309
Postejaculatory interval, 86
Postsynaptic GABA-B receptors, 233
PR-A isoform, 115–116
PR-B isoform, 115–116, 118
Prefrontal cortex (PFC), 66, 73–74, 76
Pregnant rats groups, 218
Preoptic area functions, 41
  endocrine role, 39–40
Preoptic testosterone implants, 37–38
Presynaptic GABA-B receptors, 233
Primiparous mothers, 193
PRL. *See* Prolactin (PRL)
Progesterone, 114–115, 117, 190–191
  effect on brain electrical activity, 26–28, 27f
Progestin receptors (PRs)
  estrous behavior by different agents, cellular mechanisms, 118–125
    leptin and female sexual behavior, 120–122
    NO pathway, 119–120
    PKA, 119
    SRC tyrosine kinase, MAPK pathway, and lordosis behavior, 122–125, 124f
    vaginocervical stimulation and female sexual behavior, 122
  expression in brain, 116
  isoforms and female sexual behavior display in rodents, 118
  P action, mechanisms, 117–118
    gene expression, regulation, 117
    neurotransmitter receptors, modulation, 117
    signaling cascades activation, 117–118
Prolactin
  mRNA and 16 kDa protein, 255
  secretion, regulation of, 253
  structure–function relationship, 254
Prolactin (PRL), 152, 263–264, 266, 268–270, 270f
Prolactin Inhibitory Factor (PIF), 266
Proliferating basal cells, 265
Prostaglandin (PG), 119
Prostate
  anatomy, 264–265, 264f
  cancer, 263–264
  gland function, 266–268
    hormonal regulation, 266–267
    peripheral nervous system, regulation, 267–268
  histology, 265–266
Prostatic diseases, 268–272
  ANS, 270–272, 271f
  PRL, 268–270, 270f
Prostatic ducts, 265
Prostatic dysplasia, 269
Prostatic-specific antigen (PSA), 265–266, 305
Protein kinase A (PKA), 118–119, 121
Protein kinase G (PKG), 118, 120
Protein phosphorylation, 119
PRs. *See* Progestin receptors (PRs)
Pseudomale behavior, 46, 49, 167
  display in rabbits, 7–8
Purina rat pellets, 280
Purkinje dendrites, 108
Purkinje neurons, 104, 107–108, 110
PVN. *See* Paraventricular (PVN) neural structure

R

Rabbit
behavioral neuroendocrinology, 166–172
hormonal modulation, 167–170
lactation, 166–167
maternal nest building behavior, 170–172
preoptic area endocrine role, 39–40
in sexual behavior, 33–35, 34f
Ramstergig species, 33
Reflex activity of pelvic region, female rabbit, 172–174
Reinforcement intensity, 89
Reproductive neuroendocrine axis, 253–254
Reproductive Neuroendocrinology, rabbit as model animal in, 163–166
ovulation, 164–165
oxytocin and gonadal steroid hormones, 165–166
Rhinencephalon, 293, 295
Rhythmic pattern in maternal behavior structure, 156–159
Rosenblatt, Jay, 187
Rostrocaudal diffusion, 281

S

Sawyer, Charles, 5, 291–292
Schizophrenia, 175
The Science of Orgasm (book), 303, 309
SCN. *See* Suprachiasmatic nuclei/nucleus (SCN)
Sea trout oocytes, 116
Secreting cells, 265
Seminal coagulation process, 85
Seminal expulsion, 85
Seminal plasma, 85
Seminal vesicle pressure (SVP), 47
Sensitization, 206
Septal lesion effect on maternal behavior and lactation in rabbits, 6–7, 7f
Sex Hormone Binding Globulin (SHBG), 267
Sexual arousal
habituation, 138
mechanism, 89
Sexual behavior, 106–107, 107f, 131
Sexual cerebellum, 103–110
cerebellum, 103–105, 104f
motor learning, 107–109, 108f, 109f
neuroendocrinology, 109–110
non-contact stimulation, 105–106, 106f
sexual behavior, 106–107, 107f
Sexual dimorphism, VNS, 203–205
AOB, 203–204
BAOT, 204
hypothalamic nuclei receiving vomeronasal input, 205
medial and posteromedial cortical amygdaloid nuclei, 204
patterns, 207
components of BST, 204
posteromedial division, BST, 204
sexual dimorphism of VNS and neurofunctional hypothesis, 207–209, 208f
vomeronasal organ, 203
Sexual exhaustion, 136
Sexually dimorphic behavior patterns, rabbit, 169–170
Sexually dimorphic nucleus of the preoptic area (SDN-POA), 212
Sexual motivation, 89
Sexual reproduction, 103
Sexual reward, biological basis, 138–139
Sexual satiety, 87t, 88–93
Coolidge Effect, 90–91
adaptive/evolutionary significance, 92–93
copulating, 88–89
recovery from, 93–97, 94t
copulatory behavior, reestablishment of, 87t, 93–95, 94t
ejaculation, restoration of, 95–97
satiety include seminal expulsion, copulation, 91–92
sexual behavior under, 136–138
sexual incentive and copulation, 89–90
Shunting inhibition, 233
Skeletal muscle movements, 84
Skene's paraurethral gland, 264–265
Sleep–wake cycles, 233
Social behavior network, 133, 139
SON. *See* Supraoptic neurons (SON)
Soulairac, André, 35
SPECTRO program, 69
Spinal cord injury (SCI), 308–309
Spinal neurochemical modulation, 70–71
Sprague-Dawley female rats, 280
Src tyrosine kinase, MAPK pathway, and lordosis behavior, 122–125, 124f
Ste20p-related Proline Alanine-rich Kinase (SPAK), 238
Steinach, Eugen, 34
Stereotypic motor patterns, 84
Stereotypic sequence behavior patterns, MB, 206
Steroid, 238
hormones effect on adult rabbits mammary glands, 6
receptors, 115
states, 236
Steroid-acute-regulatory-protein (StAR), 215
Steroidal agents, 114

Steroid hormone metabolism effects on brain and behavior, 3–28
hormonal regulation of
female sexual behavior. *See* Female sexual behavior of rabbits and rats, hormonal regulation
male sexual behavior of rabbits and rats, 17–21
neural processes in reproduction-related behavioral phenomena, 21–28
electrophysiological expressions of relaxation behavior, 23–26, 24f, 25f, 26f
estrogen effect on responsiveness, 21–23, 22f, 23t
progesterone effect, 26–28, 27f
neural regulation of lactation and maternal behavior, 4–7, 7f
background, 4–5
neural and hormonal factors, 5–7
Stressor device, 216f
Stress urinary incontinence (SUI), 304
Strychnine, 280–282
injection, 283
Strychnine receptor, 300
Suckling, 177
stimulation, 6
Suckling-entrained oscillator (SEO), 192
Suprachiasmatic nuclei/nucleus (SCN), 149–151, 158, 233
Supraoptic neurons (SON), 151–152, 154f, 155–156, 207, 255

T

T aromatization, 12, 17, 20
T-butyl bicyclophosphorothionate (TBPS), 117
Telencephalon, 5, 8
Temporal lobe lesions, 7–8
Testosterone, 168, 263–264, 292–293
absence, 265
aromatization role in effect of, 18–21, 18t, 19f, 20f
effect on gonadotropin secretion and lordosis behavior of rats, 13–15
levels, 132
in blood, 267
PRL, 266–267
Testosterone propionate (TP), 10, 268
Theta
frequency, 74
hippocampal generator, 74
rhythm, 293–295, 300
Thyroid-stimulating hormone (TSH), 165
Tonic glycine-induced inhibition, 280
Trichotillomania, 174
Tuber cinereum, 165
Tuberoinfundibular dopaminergic (TIDA), 152, 155
Tubular coiled structure, 85
Tyrosine hydroxylase (TH), 152

U

Unitary whole, 191
Universidad Autónoma Metropolitana (UAM-I), 251
Urinary behavior patterns, 173
Urinary incontinence, 173
Uterine horn, 86
Uterus, 252

V

Vaginal-cervical stimulation, 296, 299–300
Vaginal plethysmography, 138
Vaginal stimulation, 295
Vaginal stimulation-induced pseudopregnancy, 291
Vaginocervical stimulation (VCS), 118, 120, 123, 138, 300–301
female sexual behavior, 122
Vasoactive intestinal peptide (VIP), 233, 253, 268
Vasoinhibins, 255–256
Vasopressin, 165–166
VCS. *See* Vaginocervical stimulation (VCS)
Ventral BST (BSTV), 204
Ventral tegmental area (VTA), 72–76, 73f, 75f
Ventromedial hypothalamus (VMH), 14, 23, 26–27
Ventromedial nucleus (VMN), 116
Ventromedial nucleus of the hypothalamus/ ventromedial hypothalamus (VMH), 139
VNS. *See* Vomeronasal system (VNS)
Vocalization threshold (VT), 281, 283
Vomeronasal organ (VNO), 202, 203
in adult humans, 210
Vomeronasal system (VNS)
human mothering, vomeronasal input, 209–211
MB in mammals, 205–206
overview, 202–203
and PB, neurotransmitter role in sex difference, 211–215
endocrine-disrupting chemicals, 214–215
locus coeruleus and AOB, GABA-a effects, 212–213
neuroactive steroids, 213–214, 214f
sexual dimorphism, 203–205
AOB, 203–204
BAOT, 204
functional implication, 207–209

hypothalamic nuclei receiving vomeronasal input, 205
medial and posteromedial cortical amygdaloid nuclei, 204
posteromedial division, BST, 204
vomeronasal organ, 203
Vomeropherins, 210
von Frey fiber stimulation, 282, 282f
VTA. *See* Ventral tegmental area (VTA)

W

Wanting sexual activity, 133–134
Water ad libitum, 280
West Coast Sex Conference, 299
Whipple, Beverly
Carlos remembrances by, 304–309
coauthoring books with Carlos, 309
human physiology laboratory at Rutgers University, 308–309
at IAB, 304
path to research, 304–306
path to Rutgers University, 306–307
working with Carlos' colleagues in Mexico, 307–308

Z

Zigzag locomotion, 114

Milton Keynes UK
Ingram Content Group UK Ltd.
UKHW021633071024
449327UK00020BA/1288